The Arctic Basin

Results from the Russian Drifting Stations

Ivan E. Frolov, Zalman M. Gudkovich,
Vladimir F. Radionov, Alexander V. Shirochkov
and Leonid A. Timokhov

The Arctic Basin

Results from the Russian Drifting Stations

Springer

Published in association with
Praxis Publishing
Chichester, UK

PRAXIS

Professor Ivan E. Frolov
Professor Zalman M. Gudkovich
Dr Vladimir F. Radionov
Dr Alexander V. Shirochkov
Dr Leonid A. Timokhov
Arctic and Antarctic Research Institute (AARI)
St Petersburg
Russia

SPRINGER–PRAXIS BOOKS IN GEOPHYSICAL SCIENCES
SUBJECT *ADVISORY EDITOR*: Dr Philippe Blondel, C.Geol., F.G.S., Ph.D., M.Sc., Senior Scientist, Department of Physics, University of Bath, Bath, UK

ISBN 978-3-642-06328-2 e-ISBN 978-3-540-37665-1

Springer is part of Springer-Science + Business Media (springeronline.com)

Bibliographic information published by Die Deutsche Bibliothek

Die Deutsche Bibliothek lists this publication in the Deutsche Nationalbibliografie; detailed bibliographic data are available from the Internet at http://dnb.ddb.de

Printed in Germany

Cover design: Jim Wilkie
Project management: Originator Publishing Services, Gt Yarmouth, Norfolk, UK

Printed on acid-free paper

Contents

Preface

The Arctic has always attracted investigators – courageous people devoted to science. Before the 19th century expeditions were dispatched to search for new lands and to study the Arctic nature were of episodic character. The First International Polar Year (IPY) 1882/1883 was held as a result of the joint efforts of scientists from 12 countries including Russia – the basis for the observation network system in the Arctic was created. Geophysical, meteorological and some biological observations began at 13 points of the northern Polar area on a systematic basis.

The next 50 years were the time of the great expeditions and significant geographical discoveries: F. Nansen on board the ship *Fram* (1893–1896); F. Vilkitsky at the mouths of the Ob′ and Yenisey Rivers (1894–1897); E. Toll on board the schooner *Zarya* (1900–1902); B. Vilkitsky on board both the ships *Taimyr* and *Vaigach* (1913–1915); R. Amundsen and O. Sverdrup on board the ship *Maud* (1918, 1921, 1922, 1923); O. Schmidt and V. Viese on board the ice-breaking ship *Georgy Sedov* (1929–1931); and the ice-breaking ship *A. Sibiryakov* (1932); and O. Schmidt on board the ship *Cheluskin* (1933).

In 1932/1933, the second International Polar Year was held. By that time the epoch of the geographical discoveries in the Arctic was mainly completed: extensive knowledge on the nature of the Arctic and its hydrometeorological and ice regime was gained; the grounds of the hydrometeorological support system for navigation along the Northern Sea Route were created including airborne ice reconnaissance and methods of ice forecasting.

During the period of the Second IPY, the hydrometeorological network in the polar regions was significantly expanded: 115 basic stations participated from the USSR in the program with 50 of them being opened during the Second IPY. The measurements made provided new information on the characteristics and dynamics of the polar atmosphere and the ice and oceanic regime.

The area northward of 50°N.

Source: AMAP (2003).

Establishment in May, 1937, of the first drifting 'North Pole-1' (NP-1) station in the Arctic Ocean and organization of the first 'Sever' airborne high-latidudinal expedition (AHE) continued the Second IPY in the study of the Arctic. The experience of these expeditions convincingly proved the possibility of organizing and conducting (directly from the ice) long-term complex meteorological, oceanographic and ice observations.

World War II interrupted for a long time the large-scale studies in the Arctic. The 'Sever' AHEs were resumed in 1948, and organization of drifting stations in 1950.

To date, whilst the Russian research 'NP-33' station drifts in the vicinity of the North Pole, it can be stated that our knowledge of the Arctic Ocean for the last 70 years has achieved a level differing little from the level of knowledge of the other oceans and even exceeding it for some parameters. In the 20th century, detailed information on the seabed relief of the Arctic Ocean and its structure was obtained.

Based on these data, numerous atlases of the state of the natural environment and climate of the Arctic were prepared and published, the systems of hydrometeorological and ice support for the economic activities in the northern areas were developed and systems for remote sensing determination and monitoring of hydrometeorological characteristics in general were created including Earth satellites.

Studies showed that exactly in the polar regions global climatic changes have the largest manifestations. Significant temperature fluctuations were observed at different stages of modern evolution of the Earth's climate: cooling at the end of the 19th century, warming in the 1920s–1940s (most evident in the Arctic) and cooling in the 1950s–1970s. Significant warming has been observed during the last two decades of the 20th century, which in the opinion of a number of scientists will also continue in the 21st century.

What takes place in the Arctic? Will the Arctic ice melt? How and to what extent do the global climate changes depend on natural and anthropogenic factors?

There are so far no answers to these and other questions. That is why the International Oceanographic Commission (IOC) and the International Council of Scientific Unions (ICSU) under the aegis of the World Meteorological Organization join their efforts in holding the International Polar Year 2007/2008.

The authors hope that during this period an international drifting station will be organized in the Arctic Ocean using experience gained from Russian polar explorers.

The readers of the book will be acquainted with the brief history of establishing the drifting polar stations and the main scientific results of the analysis of meteorological, ice, oceanographic and geophysical data collected in the high-latitudinal expeditions.

Contributors

Aleksandrov, Evgeniy I.
Arctic and Antarctic Research Institute, St Petersburg, Russia
alexmet@aari.nw.ru

Baskakov Georgiy A.
Arctic and Antarctic Research Institute, St Petersburg, Russia
aaricoop@aari.nw.ru

Bayborodova, Victoria R.
Arctic and Antarctic Research Institute, St Petersburg, Russia
alexmet@aari.nw.ru

Blagoveshchenskaya, Nataly F.
Arctic and Antarctic Research Institute, St Petersburg, Russia
nataly@aari.nw.ru

Bryazgin, Nikolay N.
Arctic and Antarctic Research Institute, St Petersburg, Russia
alexmet@aari.nw.ru

Dementyev, Alexander A.
Arctic and Antarctic Research Institute, St Petersburg, Russia
alexmet@aari.nw.ru

Ivanov, Vladimir V.
Arctic and Antarctic Research Institute, St Petersburg, Russia
v_ivanov@aari.nw.ru

Karklin, Valeriy P.
Arctic and Antarctic Research Institute, St Petersburg, Russia
karklin@aari.nw.ru

Klyachkin, Sergey V.
Arctic and Antarctic Research Institute, St Petersburg, Russia
svkl@aari.nw.ru

Korovin, Vladimir P.
Russian State Hydrometeorological University, St Petersburg, Russia
smilena@rol.ru

Lebedev, German A.
Arctic and Antarctic Research Institute, St Petersburg, Russia
lebedev@aari.nw.ru

Melnikov, Igor A.
Shirshov Institute of Oceanograph, RAS, Moscow, Russia
migor@online.ru

Nikiforov, Evgeniy G.
Arctic and Antarctic Research Institute, St Petersburg, Russia
aaricoop@aari.nw.ru

Petryashov, Victor V.
Zoological Institute, RAS, St Petersburg, Russia
marine@zin.ru

Pivovarov, Sergey V.
Arctic and Antarctic Research Institute, St Petersburg, Russia
pivovarov@mail.lanck.net

Sirenko, Boris I.
Zoological Institute, RAS, St Petersburg, Russia
marine@zin.ru

Smagin, Vitaliy M.
Arctic and Antarctic Research Institute, St Petersburg, Russia
smagin@aari.nw.ru

Smirnov, Alexey V.
Zoological Institute, RAS, St Petersburg, Russia
synapta@zin.ru

Ugryumov, Alexander I.
Russian State Hydrometeorological University, St Petersburg, Russia
smilena@rol.ru

Acknowledgements

The Preface and Chapter 1 of the book were prepared by I. Ye. Frolov, A. I. Ugryumov, and V. P. Korovin, and briefly describe the history of environmental studies in the Arctic for the last two and a half centuries. Main attention is devoted to the history of research at the North Pole drifting stations in 1937–1938, 1950–1951, 1954–1991 and 2003–2004.

Chapter 2 'Meteorological investigations' presents the quantitative estimates and describes the main regularities of the distribution of meteorological parameters above the central Arctic basin. The Chapter is written by V. F. Radionov and V. V. Ivanov (Section 2.3) with participation of Ye. I. Aleksandrov (Sections 2.1.2 and 2.1.3), N. N. Bryazgin (Sections 2.1.4 and 2.1.6) and A. A. Dementyev (Section 2.1.5). The figures were prepared by Ye. I. Aleksandrov and V. P. Baiborodova.

Chapter 3 'Sea ice cover' was written by Z. M. Gudkovich and I. Ye. Frolov and generalizes the results of investigations of the ice cover parameters and sets forth the peculiarities of thermal and dynamic processes occurring in it. V. P. Karklin and S. V. Klyachkin also participated in the preparation of this chapter.

Chapter 4 'Oceanographic studies' was submitted by L. A. Timokhov. Individual sections were written by: G. A. Baskakov (Sections 4.1, 4.2), Ye. G. Nikiforov (Section 4.3), G. A. Lebedev (Sections 4.4.4, 4.4.6, 4.4.7), V. M. Smagin and S. V. Pivovarov (Section 4.5), and I. A. Mel′nikov (Section 4.6) jointly with B. I. Sirenko (Section 4.6.4), V. V. Petryashev (Section 4.6.4), and A. V. Smirnov (Section 4.6.4).

Chapter 5 'Geophysical observations' was submitted by A. V. Shirochkov with Section 5.3.3 being written jointly with N. F. Blagoveshchenskaya.

The conclusion was written by I. Ye. Frolov and V. F. Radionov.

The authors would like to thank all who participated in writing and preparing the book.

The book is based on the results of research at the North Pole drifting stations and their generalization published in numerous journals, volumes of articles and monographs – many of which are included in the references of this book.

The authors are also grateful to several generations of national polar investigators, staff of the Arctic and Antarctic Research Institute (AARI), and other institutes and organizations. It is their selfless labour and high scientific erudition that have made a substantial contribution to the development of science on the nature of the Earth's polar regions. The materials of their observations, analysis and generalizations served as a basis for obtaining the results published in this book.

The authors thank Professor Ola M. Johannessen (Founding Director, Nansen Environmental and Remote Sensing Centre and Geophysical Institute, Bergen, Norway) who initiated the preparation of this book.

Figures

Tables

Abbreviations

AA	auroral absorption
AARI	Arctic and Antarctic Research Institute
AHE	airborne high-latitude expedition
AIDJEX	Arctic Ice Dynamics Joint Experiment (USA–Canada–Japan)
AO	Arctic Ocean
AWI	Alfred Wegener Institute
DARMS	drifting automatic radio-meteorological station
FMO	Flying Meteorology Observatory
GARP	Global Atmospheric Research Program
GCA	general circulation of the atmosphere
HF	high frequency
ICSU	International Council of Scientific Unions
INSROP	Northern Sea Route International Programme
IOC	International Oceanographic Commission
IPY	International Polar Year
LUF	lower usable frequency
MLT	Moscow Local Time
MUF	maximum usable frequency
NSR	Northern Sea Route
PCA	polar cap absorption
POLEX	Polar Experiment
RGMAA	Rossiyskiy Gosudarstvenniy Muzey Arktiki i Antarktiki (Russian State Museum of the Arctic and Antarctic)
RMSD	root-mean-square deviation
SID	sudden ionospheric disturbance
TOC	total ozone content
USSR	Union of Soviet Socialist Republics
UT	Universal Time (Greenwich)
VLF	very low frequency

1

A brief history of Arctic exploration

1.1 FIRST ARCTIC EXPEDITIONS

First evidence of the white Arctic desert was obtained in the 9th–13th centuries by Vikings during their voyages to the shores of Iceland, Greenland, America and the White Sea, and a little later in the 11th–16th centuries by Russian pomors during the voyages to Spitsbergen–Grumant and Novaya Zemlya.

The geography of the northern expeditions gradually expanded – an understanding about the boundaries of the northern lands and the Arctic Seas was formed and knowledge on the hydrometeorological regime and the hydrographic peculiarities of the Arctic and the Arctic Ocean was accumulated (see map in the Preface).

In 1764 and 1766, Mikhail Lomonosov in *Arkhangelsk* initiated organization of an expedition aiming to reach the Pacific Ocean across the North Pole. The expedition was headed by Captain–Commodore Vasily Chichagov. The expedition programme included a wide complex of scientific studies. Chichagov's expedition did not fulfill its aim as the sailing route was developed on the basis of an erroneous assumption of the presence of open ice in the central Arctic Ocean. However, in a scientific respect the expedition contributed much to gaining knowledge of the nature of the high latitudes of the Arctic. This was the first Arctic expedition based on scientific calculation and foresight. In 1766, the expedition reached 80°34′N – the Spitsbergen area – a World record of that time for freely floating ships.

The life and studies of Fritjof Nansen (Figure 1.1) – a famous Norwegian traveller, scientist and public man – have determined for many decades the strategy and tactics of polar research in the central Arctic. Nansen was the first in history to propose and implement the idea of drifting a scientific station in ice (Nansen, 1902). This became the theoretical basis for the organization of all Soviet and US drifting stations in the 20th century.

(a)

(b)

Figure 1.1. (a) Photo portrait of Fritjof Nansen. (b) The schooner *Fram* drifting in the ice of the Arctic Ocean during the Nansen expedition in 1893–1896.
Photo from the RGMAA archives.

After the desperate lunge of Robert Peary to the North Pole, whose main aim was to establish the discoverer's fame, traverses on foot were rarely undertaken except at the very end of the century – exclusively for sporting purposes.

The scientific studies of the central Arctic for understandable reason were carried out using the 'Nansen way': on drifting ice or on board ships frozen into the ice. In this way studies could accommodate incomparably more equipment than a traverse on foot would allow.

This was only one, material aspect of Nansen's idea. Another geographical aspect was that if the initial station drift point was successfully chosen, it could cross, together with the ice floes, the entire central Arctic, and if you were fortunate, would pass across the North Pole.

This great scientific foresight was based on the assumption of the existing general drift of Arctic ice from east to west – from the seas of East Siberia to the Pole and further on to the Greenland Sea.

The Norwegian meteorologist Henrik Mohn suggested the existence of a constant current from the shores of East Siberia to the passage between Spitsbergen and Greenland.

Mohn's arguments in favour of the existence of the transpolar current and the transport of drifting Arctic ice convinced far from everyone, but there was one man – the young Norwegian scientist Fritjof Nansen, who recognized in Mohn's hypothetical concept the only practical possibility to reach the North Pole and simultaneously carry out extensive studies of the near-Pole area of the Arctic Ocean. According to Nansen's project, the expedition ship had to be intentionally frozen into ice in the vicinity of the New Siberian Islands and put to the disposal of the polar current that would bring it to the Pole or to a nearby area.

Nansen, with energy peculiar to him, began arranging the expedition. Using money allocated by the Norwegian government and partly collected by subscription, a ship was built with a displacement of 420 tons, a length of 39 m, a breadth of 11 m and with 220 hp – specially adapted for a long stay in ice. Its egg-shaped hull of and strengthened frames prevented the ice from crushing it – in fact ice under pressure only pushed the ship upwards. Nansen called his ship the *Fram* ('Forward' in Norwegian).

On 24 June, 1893, Captain Otto Sverdrup took the *Fram* out of Kristiania (Oslo) to sea with 12 crew and a 5-year supply of food and gear. On 4 August, the *Fram* entered the Kara Sea. After rounding the Yamal Peninsula, it headed for Cape Cheluskin. Passing along the western shore of the Taimyr Peninsula, the seafarers discovered a number of islands (Sverdrup, Scott–Hansen, Ringnes, Mohn, Nordenskjold, Firnley and Geiborg). On 21 September, 1893, the ship was frozen into the ice at 78°50′N and 133°37′E to the north-west of the New Siberian Islands. The historical drift of the *Fram* began across the central Arctic, during which regular hydrometeorological and magnetic observations were carried out, the flora and fauna of the Arctic Ocean were studied and the ice drift speed and direction were investigated. The use of the ship as an observation platform provided quite comfortable living conditions for polar explorers.

In January 1894, the *Fram* reached 83°24′N, however, during the next days it became clear that the north drift direction had ceased. Understanding that the *Fram* would not drift across the Pole, Nansen made the decision to continue to the Pole on foot. On 14 March, 1895, together with lieutenant Fredrik Johansen, Nansen left the *Fram*, which was at that time at 84°N and 101°55′E. Three Siberian dog sledges began moving to the Pole. Severe ice conditions and a constant southward ice drift only allowed them to reach 86°13′06″N (a record for that time) at 95°E by 7 April. A strong south ice drift, which was practically impossible to overcome on foot, made them reconsider their plan and the travelers turned to the south in the direction of the Franz Josef Land. Conducting scientific observations at all times during the traverse, Nansen and Johansen, with great difficulty, reached the northern group of the islands of Franz Josef Land in August, where they had to overwinter on Jackson Island after first preparing supplies of bear, seal and walrus meat and building a winter cabin. The travellers abandoned their winter cabin on 19 May, 1896, leaving behind a brief message about the expedition at the end of which Nansen wrote: 'Today we are starting to the south-west along the land aiming to reach Spitsbergen by the shortest route. We believe that we are at Jilles Land'. Having great difficulty advancing southward they met, by chance, on 18 June 1896 on Nordbruk Island with the expedition of Frederic Jackson that was exploring Franz Josef Land.

On board his ship *Windward*, Jackson took the travelers to the motherland and in the middle of August Nansen and Johansen arrived at Vardo. Soon after, Otto Sverdrup brought the *Fram* there without losing a single person during a three-year unprecedented voyage. After Nansen and Johansen had left the *Fram*, the ship had continued its drift and on 15 November, 1895, had reached 85°56′N, 66°31′E. At that point the drift direction changed to the south-west. 1896 was marked by strong frosts and ice shears. Ridging was observed almost daily. On 21 February an enormous ice billow approached the ship and covered part of the stern. The *Fram* again weathered the test.

The people on board the ship began to prepare for becoming free from ice captivity: the engine was examined and by exploding pyroxylin, the stern and bow of the *Fram* became free of the ice. In July, polynyas began to open increasingly more frequently. The engine was started and the ship moved whenever possible in open water southward. Finally, on 13 August, on the day when Nansen and Johansen arrived to Vardo, the *Fram* broke away from the ice and exited to open water. On 23 August, 1896 (at 4 p.m.), Nansen's expedition was completed.

The expedition on board the *Fram* opened a new epoch in the study of the high latitudes of the Arctic Ocean. Nansen and his comrades not only justified this practical way of studying the central Arctic, but also made a number of fundamental scientific discoveries shedding light for the first time on the central Arctic's real nature. It turned out that the near-Pole area of the Arctic is an ocean with no land. Nansen also denied the theory of the 'free sea' of August Peterman, proving that the Arctic basin is ice-covered all year round and moves under the action of winds and currents. Nansen and his staff were able to determine the presence of depths greater than 3,000 m in the Arctic basin. In the 200–800-m depths, an intermediate layer of relatively warm Atlantic water was revealed. Above this layer there was colder freshened water. Evidence of animal organisms inhabiting the ocean strata was collected.

The drift of the *Fram* convincingly confirmed the validity of Mohn's hypothesis, transforming it to scientific fact. In Nansen's opinion, enormous masses of drifting ice are exported by currents and winds – from the seas washing the north-east coast of Siberia, crossing the polar basin in the North Pole area and then moving on to the Greenland Sea. Based on the analysis of the expedition observations on board the *Jeannette* and the *Fram*, Nansen considered the duration of the Transarctic ice drift to be five–six years. During this time the ice cover undergoes significant changes. The ice floes, under the influence of the wind and currents, constantly collide, break and ridge – piling up to form unusually high ice billows.

Intense scientific exploration, both in the coastal and seaward areas of the Arctic Seas, was continued at the beginning of the 20th century. Here one should remember: the expeditions on board the ship *Zarya* headed by Edward Toll (1900–1902); and the explorer Alexander Kolchak with whose name the monograph devoted to Arctic sea ice characteristics is connected, and which is still of scientific value (Kolchak, 1909).

The first-quarter of the 20th century was marked by a number of known expeditions that make a deep impact in the history of the studies of the Arctic Seas and their ice regimes. Such expeditions include the voyage of Georgy Sedov (1912–1914) on board the *Foka* to Franz Josef Land, which was planned to be used in the future as a base for reaching the North Pole. The expedition of Georgy Brusilov on board the ship *Sv. Anna* was beset by ice in October 1912 near the shores of the Yamal Peninsula and was brought by the drift, at the beginning of 1914, to the Arctic basin. Part of the team, headed by navigator Valerian Al'banov, abandoned the ship in the hope of reaching Franz Josef Land on sledges and kayaks. Unfortunately, of the 24 people who participated in Brusilov's expedition only 2 returned. The ship's log, saved by Al'banov, contained valuable evidence of the north-western Kara Sea and the ice drift in this area.

A great contribution to Arctic science was made by the Russian expedition of Boris Vilkitsky on board two icebreaking transports *Taimyr* and *Vaigach*, built at the Nevsky ship-building plant. As a result of 3 years' efforts (1913–1915), the expedition sailed, for the first time, from the Pacific Ocean to the Atlantic Ocean and made a number of outstanding geographical discoveries including the discovery of the Northern Land. In the same years, the Canadian expedition of Vilhjalmur Stefansson explored the eastern Arctic. The expedition ship *Karluk*, under the command of Captain Robert Bartlett, made a forced drift from the Beaufort Sea to the Chukchi Sea (1913–1914) which ended in tragedy.

Especially fruitful was the Norwegian expedition, organized by Roald Amundsen on board the ship *Maud*, which after making a through voyage by the North-East Passage from west to east (1918–1920) then repeated the drift of the *Jeannette* from the Chukchi Sea to the New Siberian Islands (1922–1924). The results of observations of this expedition, analyzed by Sverdrup, were a new stage in the study of the hydrometeorological and ice regime of the Arctic shelf (Sverdrup, 1928).

By the early 1930s, the economic development of the territories adjoining the Arctic Seas, expansion of the network of polar stations and development of the airborne ice reconnaissance methods set new objectives for national Arctic science. Organization in the early 1920s of special scientific centres, in particular the

Northern Research Commercial Expedition (further Institute for Studies of the North, now the Arctic and Antarctic Research Institute) contributed to their fulfillment.

So the *Fram* expedition cleared the way for the establishment of scientific stations on the drifting ice of the central Arctic. The first of them was organized in 1937, when a group of Soviet polar explorers landed directly at the Pole. Why did one have to wait for 40 years after the legendary voyage of Nansen?

The *Fram* was primarily a completely unique ship designed specially for drifting in ice. After Nansen's expedition it sailed to the ice of the Canadian Arctic archipelago and Antarctica proving its reliability many times. As for conventional vessels, even of the icebreaking type, it was risky to use them for staying in heavy pack ice.

There are several examples in the history of Arctic exploration, when ships met such an end. The most impressive of them was the voyage of the ship *Cheluskin* in 1933–1934 (Figure 1.2). On 10 August, 1933, the *Cheluskin*, under the command of a known Arctic captain Vladimir Voronin, headed by a prominent scientist, organizer of the studies of the Arctic Otto Schmidt on board, departed Murmansk to overcome all Arctic Seas of Siberia and exit to the Pacific Ocean during one navigation season. The *Cheluskin* was close to success: in November 1933 the ship was in front of the Bering Strait, but could not cross it – the ship was beset by ice and entrained to the central Chukchi Sea where on 13 February, 1934 ice crushed it and the *Cheluskin* was sunk.

The expedition members and the crew had time to abandon the sinking vessel (one man perished) and remained on the ice floe organizing an ice camp on it. The expedition was saved by the USSR polar pilots who evacuated 104 people from the ice camp.

For courage and resolve all expedition members and the crew of the *Cheluskin* were awarded the order of the Red Star and 7 polar pilots became heroes of the Soviet Union.

Consequently, the people who devoted their lives to the studies of the Arctic and who wanted to penetrate its central area, did not trust ships and refused to use them as a tool for the observations of natural phenomena in the central Arctic. However, one alternative remained – drop polar explorers directly on the ice floe and use it as a platform for observations. This, however, required facilities for delivering the people and equipment which could only be fulfilled by aircraft. By the mid-1930s, aviation achieved the necessary level of development for such flights – a fact proven by the rescue of the *Cheluskin* expedition crew.

Another cause for the 40-year 'delay' in the organization of research drifting stations was an uncertainty about the Transarctic ice transport. The *Fram* voyage in spite of its scientific achievements was a single example of the Transarctic drift. Does polar ice always behave itself exactly so? The drift of the first Soviet station 'North Pole-1' in 1937–1938 showed that ice in the western Arctic drifted in the same direction as 40 years ago – from the North Pole southward to the Greenland Sea – what about the eastern Arctic?

The answer to this question was obtained in the same years. Man had already explored the North-East Passage along the Arctic coast of Siberia – now called the

(a)

(b)

Figure 1.2. (a) The ship *Cheluskin* is beset in the ice of the Chukchi Sea in 1933; (b) ice camp and crew of the *Cheluskin* in March 1934.
(a) Photo by Ya. Gakkel. (b) Photo by P. Novitsky.

Northern Sea Route – which served for the delivery of different cargos to the Siberian ports. It happened that during the 1937 navigation season, more than 20 ships were frozen into the ice along the Northern Sea Route, including ice-breakers and ice-breaking ships. The most remarkable were the drifts of the ships '*Malygin*, *Sadko* and *Georgy Sedov* that began on 23 October in the area where the *Fram* drifted 40 years earlier.

Based on the experience of Nansen, everyone assumed that the drift would be along the same westward route. The Arctic, however, offered a surprise: sea currents entrained their victims to the east. There were 217 people on board 3 vessels, which allowed research studies on a large-scale. The *Sadko* that was equipped with a winch for measuring large depths and many other scientific instruments was at their center.

By 1938, the ships were at 78°20′N, 141°43′E. It was decided to evacuate most of the polar explorers and escort the ships from the ice. On 3 April, a group of aircraft landed on a prepared ice airfield in the area of the drifting ships, and for two weeks the pilots transported 184 people to the mainland and delivered about 7 tons of food supplies and gear for those remaining. After the evacuation of people, the ice drift changed to the west.

During the 1938 navigation season, the ice-breakers *Yermak* and *I. Stalin* escorted the ships *Malygin* and *Sadko* to open water. The *Sedov*, due to significant damage of the steering device, had to remain in ice. The food supplies and gear were reloaded onto it and the ship became a drifting research station. On board the *Sedov* 15 volunteers remained. Konstantin Badigin was the captain during the drift and hydrographer Viktor Buinitsky supervised all research studies.

During the course of the drift that was slightly more to the north than that of the *Fram*'s drift, the polar explorers collected unique data on nature and life at the high latitudes of the Arctic. From the very beginning of the drift, astronomic and hydrometeorological observations were carried out on board the ship. The largest depth of the Arctic Ocean was measured at 4,975 m at 86°24′N, 38°35′E. On 29 August, 1939 the ship reached the northernmost point of its drift – 86°40′N, 47°55′E – 79 km north of the extreme drift point of the *Fram*. Having made several zigzags, the *Sedov* together with the ice sailed southward to a wide strait between Greenland and Spitsbergen.

The drift of the *Sedov* that continued for 812 days and that ended on 13 January, 1940 in the Greenland Sea at 80°30′N, 1°50′E, convincingly showed that the general ice drift direction in the Arctic basin determined by Nansen was still preserved some 40 years later. One could hope with great probability that the drifting research stations landed on the ice in the East Siberian or the Laptev Seas would cross the entire central Arctic, having at their disposal not less than 2 years for comprehensive studies of its nature.

It is obvious that Nansen himself in spite of the successful drift of the *Fram* did not consider a ship the most suitable means for the study of the central Arctic. Immediately after his famous expedition he expressed the idea of organizing research stations on the ice floes. Vilhjalmur Stefansson was the first to attempt to realize Nansen's idea by organizing in 1917–1918 observations on the drifting ice of the American sector of the Arctic in the Beaufort Sea.

Stefansson himself could not take part in the expedition – in early 1918 he fell ill with typhus and pneumonia and was delivered to Fort Yukon. Stefansson requested his closest assistant Sturker Storkerson be the leader of the first research station on a drifting ice floe.

According to the plan, the drift was to continue for a year. On 15 March, 1918, Storkerson with four volunteers and a support team, which he sent back within a month, headed north. Finding a suitable ice floe more than 400 km^2 and several hundreds of kilometres from Alaska, the winterers made camp. During the drift they carried out hydrometeorological observations and hunted seals which represented fuel and food for themselves and their 16 dogs. The ice floe drifted predominantly westward between 72°30′N and 74°N; travelling about 740 km. In autumn Storkerson had attacks of asthma and he decided to return to the mainland. On 8 November, 1918, after staying on the ice for 238 days, the polar explorers left for the Colville River mouth – some 7° more westward than their place of disembarkation, never fulfilling the expedition programme to the end.

Later Nansen himself being the President of the International Society of Arctic Study by means of Aircraft Communication – 'Aeroarctic' proposed a project which would establish a research station on the drifting ice of the central Arctic. The opening of such a station was planned at the time of the Second International Polar Year (1932–1933). The main obstacle in the way of this idea was the difficulty with the delivery of people and equipment to the polar area. Three variants for resolving this issue were proposed: using an airship, landing paratroopers and landing of aircraft. Each of the variants had its supporters and opponents, but it was impossible to implement them during the planned period.

After the death of Nansen in 1930, his ideas were met with wide response in the USSR. The growing interest of the Soviet Union in the central Arctic basin was related to the systematically performed work on the investigation and exploration of the Arctic. Specialists were particularly attracted by the possibility of landing polar explorers on ice directly in the vicinity of the North Pole, which would contribute not only to the study of nature in the central Arctic, but give priority to the USSR in establishing the first research station at the Pole.

1.2 THE 'NORTH POLE-1' DRIFTING STATION

Academician Otto Schmidt and Professor Vladimir Viese were passionate supporters of establishing a station in the North Pole area. The real project for setting up a research station at the Pole was first considered as early as in 1929 – proposed by Viese. With time, the initial projects, after a detailed analysis, were gradually improved but no practical steps were undertaken until 1935.

On 27 October, 1935, an evening session was held at the State Geographical Society in Leningrad devoted to the outcome of the high-latitudinal expedition on board the ice-breaking ship *Sadko*. On 13 September, 1935 the ship reached 82°41′N, which was several minutes north of that the record set by *Malygin* in 1932 (82°27′N) and the US ship *Roosevelt* in 1908 (82°30′N). There were many speakers at the session, who discussed the results of the voyage and presented scientific reports, but in addition there was an air of expectation. Finally in the end of the session, O. Yu. Schmidt asked for the floor. 'Today dear Comrades, – he said – I want to tell you important news. In the nearest time, a group of winterers will be landed on the drifting ice in the Arctic from the airplane to the north of the lands known to us for establishing a research station with the purpose of complete and diverse investigation of the behavior of ice, water and wind at the very heart of the Arctic in the expanses of the central area of the Arctic Ocean. Outstanding polar explorers, including Weiprecht and Nansen dreamt about such an expedition in their time'. The speech of Schmidt was a true sensation, and after the session it was discussed emotionally at the Geographical Society.

In 1936, the Main Administration of the Northern Sea Route submitted a detailed project regarding the organization of the scientific station on drifting ice in the vicinity of the North Pole. By that time the USSR had already organized a large network of polar stations on the shore of the Arctic Ocean and its numerous islands.

Ivan Papanin who had previously been in charge of polar stations for several years was appointed the leader of the station; the research team was comprised of oceanographer Petr Shirshov, participant of the voyages on board the *Sibiryakov* and the *Cheluskin*; astronomer and magnetologist Yevgeny Fedorov, who wintered over several times with Papanin; and radio/wireless operator Ernst Krenkel, who wintered over many times at polar stations and participated in Arctic voyages.

It was decided to deliver the expedition to the Pole by four-engined transport aircraft (ANT-6) designed by Andrey Nikolaevich Tupolev and especially adapted for flights in Arctic conditions. Each of the airplanes flying into the Arctic were given a call sign title prefixed with a capital letter N (Nord) followed by a number. A renowned polar pilot Mikhail Vodopianov was given the task of preparing such powerful machines. In March 1936, pilots Mikhail Vodopianov and Vasily Makhotkin, on board two heavy aircraft *N-127* and *N-128*, made a reconnaissance flight along the Moscow–Novaya Zemlya–Franz Josef Land route and farther towards the Pole. During this flight, Vodopianov on board airplane *N-127* reached 83°45′N and found that it was possible to land aircraft on the ice of the central Arctic. In July 1936, pilots Valery Chkalov, Georgy Baidukov and Aleksandr Belyakov on board the aeroplane '*ANT-25* made a flight along the Moscow–Franz Josef Land–Kamchatka–Nikolayevsk-on-Amur route of 9,374 km, breaking the World range record.

The Russian heavy machines were tested and prepared for the flight. Scientists developed a detailed work programme for the drifting station. Rudolf Island on Franz Josef Land was chosen as the expedition base. During the 1936 navigation season, Papanin, on board the ice-breaking ship *V. Rusanov*, delivered equipment and food supplies to the polar station and fuel for the airborne expedition. Twenty-four people who remained there for wintering headed by Yakov Libin prepared the airfield for receiving heavy airplanes and built workshops, warehouses and living quarters. A 300-W radio station and a radio beacon were installed for aircraft direction finding. Papanin and his staff were occupied with the provision of the expedition with instruments, clothing and special food products in Moscow and in the Main Administration of the Northern Sea Route (Glavsevmorput). At their insistence not long before the flight to the Pole, the radio equipment, the tents and scientific instruments were tested near Moscow.

On the morning of 22 March, 1937, the airborne expedition for landing four polar explorers on the ice floe, with Otto Schmidt appointed its leader, departed Moscow on board transport aeroplanes *TB-3* and a light two-engine airplane for remote reconnaissance *R-6*.

Within several hours, the aeroplanes landed at Kholmogory near the White Sea where their wheels were replaced by 8-m runners with a weight of 400 kg each. Due to poor weather conditions the continuing flight to the Pechora mouth had to be postponed until 30 March. The next landing in Naryan-Mar was made directly on the ice of the frozen Pechora River. Here, the expedition had to wait for almost two weeks for improved meteorological conditions. The expedition had no luck with the weather: the snow storm and storm winds on Novaya Zemlya were so strong that they threatened to carry away and destroy the aircraft. The flight to Rudolf Island – the expedition base – took place on 18 April. Almost 900 km separated them from the North Pole. A total of 42 people participated in the airborne expedition.

Further advance of the expedition was delayed by a complicated hydrometeorological situation. Finally on 5 May, Pavel Golovin's aeroplane *N-166* (*R-6*) was able to make a reconnaissance flight toward the Pole. Five hours after the takeoff a message was received by radio that the airplane was above the Pole and that there were suitable conditions for landing heavy aircraft. This aeroplane was the first Soviet aircraft to fly above the Pole. After staying in the air for 11 hours, Golovin successfully landed on Rudolf Island where he was met with great enthusiasm. Based on this reconnaissance flight and after a joint discussion, the flight to the Pole was scheduled on 21 May.

On 21 May at 4 hours 52 minutes the flagman airplane *N-170*, with the head of the expedition Otto Schmidt, the entire team of the drifting station (it received the name of the 'North Pole' station) and some equipment, took off from Rudolf Island and at 11 hours 35 minutes landed on the ice floe at 89°25′N, 78°40′W. Immediately after landing, the work on the organization of a permanent polar station on the ice floe, which was drifting toward the Greenland Sea, began (Figures 1.3 and 1.4).

On the morning of 22 May, Yevegeny Fedorov broadcast the first meteorological radiogram: 'The North Pole 22 May 06 hours Moscow time. Pressure 761. Temperature minus 12. Wind 8 m, west (by Greenwich meridian) gusty. The sun can be seen. Visibility of 1 km. Weak snow' – the first meteorological message from the North Pole in history! It was received on Rudolf Island. After that meteo-wires were transmitted without interruption to the mainland and included in the international weather report of the northern hemisphere.

On 25 May, news was received on the ice floe that the remaining three aircraft transporting prefabricated huts, food supplies and scientific equipment had taken off from Rudolf Island. On that day only Vasily Molokov arrived at the Pole in *N-171*. The other two airplanes were forced to land several tens of kilometres from the target. Until radio communication had been established with them, which took

(a)

(b)

Figure 1.3. (a) Head of the 'North Pole-1' drifting station Ivan Papanin (on the left) and second pilot of the aeroplane *N-169* Mikhail Kozlov before loading the aircraft at the air field in Kholmogory (22 March 1937). (b) Ice camp of 'North-Pole 1'.
Photos from the RGMAA archives.

(a)

(b)

Figure 1.4. (a) Geophysicist Yevgeny Fedorov began meteorological observations at the North Pole (May 1937); (b) Otto Schmidt (on the left) and captain of the ice-breaker *Yermak* Vladimir Voronin (Kronstadt, February 1938).
(a) Photo of M. A. Troyanovsky from the RGMAA archives. (b) Photo from Loskutov.

almost 20 hours, their comrades were very uneasy – the deployment of the polar station was pointless without the cargo that was carried by these aircraft. Luckily, pilot Anatoly Alekseyev on aeroplane *N-172* arrived at the Pole on 28 May – the last to arrive on 5 June was pilot Ilya Mazuruk.

Life was in full swing at the Pole. Tents of different colors appeared in the white desert as if by magic. Houses were built from bricks of compact snow, intended for food warehouses and a galley. The problem of meals for the significant number of expedition participants (42 people) was solved very simply – the crew of each airplane cooked food for themselves.

On 6 June at two in the morning in the bright light of the polar day, the opening of the World's first research station at the North Pole took place. Simultaneously this was a farewell to 38 participants of the expedition that delivered the four winterers to the Pole. Now they remained alone as the personnel of the 'North Pole' station. At 3 hours 37 minutes on 6 June, all four aeroplanes took off and headed for Rudolf Island. A severe and stubborn struggle for survival against nature and daily continuous observations began at the drifting ice floe.

The scientific work programme of the station included observations in meteorology, oceanography, the Earth's magnetism, gravimetry and astronomic determinations, and observations of the propagation of short-radiowaves in high Arctic latitudes.

All studies planned in the extensive programme were of great interest because they were being undertaken for the first time. However, ice and sea depths were given predominant attention. Petr Shirshov was the only hydrologist – the rest of course helped him in his difficult work.

As a result of making the first oceanographic station, a core of dark greenish-grey silt was recovered from the seabed and a positive temperature with a maximum at 400 m (0.99°C) was discovered beneath a layer of cold Arctic water. Beginning at 750 m, the temperature gradually decreased achieving a minimum at a depth of 2,930 m (−0.70°C). In the near-bottom layer, thermometers could not withstand a hydrostatic pressure and were crushed.

The chemical analysis of the water samples confirmed the opinion that a warmer water layer at a depth of 250–600 m was a result of inflow of Atlantic water. A constant positive water temperature in these depths gives an opportunity to plankton, which is a basis for life at sea, to develop and exist. This discovery refuted Nansen's idea, which had prevailed since the time of the *Fram* expedition – that the central Arctic Ocean is lifeless, presenting a dead desert.

During the drift period, 38 hydrological series – with temperature measurements and water sampling for hydrochemistry analysis – were made in standard horizons to the bottom. Twenty-two hydrobiological stations contributed to the determination of the composition of sea organisms at different ocean depths. Six-hundred measurements of ice drift speed and direction and current speed were carried out within the water column. From July to September such observations were carried out on a daily basis – 5–6 times in 24 hours. Throughout the entire drift, gravimetry observations were made at 22 points, each during a period of 2–3 days, and they were comprised of several 8-hour series. Fifty-five series of magnetic declination determinations and 14 daily series of measurements of the Earth's magnetic field oscillations were made.

In July and August, several series of observations of atmospheric electricity were conducted. With the onset of polar night, hourly observations of polar lights were carried out. During the time of the drift it was possible, using astronomical observations, to obtain about 150 accurate determinations of the geographical coordinates of the drifting ice floe.

The life of four inhabitants of the Pole – all of which were tried and experienced polar explorers – was in constant and intense labor. Labor-consuming types of scientific observations were performed by all expedition staff. The Head of the expedition, Papanin, was in charge of all practical activities. He often cooked food for the entire 'Pole population'.

As it was anticipated, the current carried the ice floe southward along the zero meridian during the first six months, with the sea depth decreased from 4,395 to 2,380 m. In November the ice floe changed drift direction, being carried on the current to the east shores of Greenland. Approaching the island, the ocean depth first increased and then began to drop sharply. The ice floe slowly melted and cracked. On 1 December, the ice floe was at 82°46′N – not far from the shores of Greenland.

To help the polar explorers, the ice-breaking ships *Taimyr* and *Murman* hurried on from Murmansk and a powerful ice-breaker *Yermak* went from Kronstadt. On the morning of 12 February, the winterers noticed to the east reflections of ship floodlights. On 18 February, the floodlights were even more visible and by midday on 19 February, it was possible to see the rescuers without binoculars. At 2 p.m., seamen from the *Taimyr* and *Murman* disembarked to the ice.

After several hours, all property of the drifting station and its heroic winterers were on board the *Taimyr* and *Murman* at 70°54′N, 19°48′W. On the way back the *Taimyr* and *Murman* met with the *Yermak* and personnel of the 'North Pole' station was transferred to it. On 15 March, the ice-breaker delivered four courageous polar explorers to Leningrad. Thus, entrained by winds and currents to the south-west, the ice floe of the Papanin team drifted 2,500 km for 274 days. The 'North Pole' station had achieved its objective.

Over these nine months of fruitful work, the station team collected unique scientific data. In the Pole area, the measured water column depth was 4,395 m; at 84°N the winch cable suddenly stopped at a depth of 2,384 m, being the true depth at this place. Later during the additional measurements in this area, three underwater sea mounts were detected. A chart of sea currents was drawn with numerous data obtained by oceanographic stations from the ice floe. Observations of the ice floe drift and further calculations confirmed the suggestion that the main mass of Arctic ice is annually exported to the Atlantic Ocean by the East Greenland Current: each year the Arctic relinquishes 10,000 km^3 of ice to the Atlantic Ocean. Biological studies revealed that the central Arctic was far from being lifeless as earlier believed. Geographical observations allowed supplementing the magnetic map of the Arctic. Meteorological observations taken several times a day were plotted on the synoptic charts prepared in Moscow. Data from the central Arctic provided, for the first time, practical assistance to aviation, marine fleets and all weather users.

An extensive complex of studies carried out for 9 months of intense heroic labor shed quite a new light on the nature of the Arctic Ocean and the character of different processes occurring in the Arctic basin. One main achievement was that the experience of operation of the 'North Pole' station proved that in spite of climate severity and the risk with which prolonged life of man on sea ice is connected, the Arctic basin can be investigated using drifting ice fields and aviation capabilities as the main means of transport.

1.3 RUSSIAN DRIFTING STATIONS FROM NP-2 TO NP-33

A natural wish of the Soviet polar explorers after the triumphal end of the 'North Pole' drift station was to organize a new drifting station that could pass along the *Fram* route. The expedition had to land on the ice in the East Siberian or the Chukchi Seas.

Preparation for opening the next NP station (as the 'North Pole' drifting stations were called now in the abbreviated form) actually started, but the war interfered with the peaceful plans, postponing continuation of polar studies for a long 12 years.

On 2 April 1950, a team of the NP-2 station, the second in the history of the Soviet drifting stations, was landed on the floating Arctic ice. The head of the station was a known polar explorer and scientist Mikhail Somov (Figure 1.5). Fifteen more people were on the ice floe with him: meteorologists, oceanographers, aerologists, specialists on sea ice studies, a radio operator, a mechanic and even a cameraman.

The technical equipment of NP-2 was much more complete and significantly improved compared with that used at the NP-1 station. In addition to advance research equipment, the polar explorers had at their disposal three radio stations, an electric power generator with a gasoline engine, a wind electric power generator and even a truck for transportation of heavy cargoes.

Supplies to the station were also well-organized: for 376 days of the drift, 71 aircraft from the mainland landed on the ice airfield; a total of 96 tons of cargo was delivered to the ice floe.

The drift of NP-2 was in general in the supposed direction – poleward in spite of numerous zigzags and loops that the ice floe made under the influence of variable currents and winds. For the year from April 1950 to April 1951, the polar explorers 'sailed' 2,600 km on the ice floe with a mean speed of 6.9 km a day; the general poleward drift (without taking into account zigzags and loops) comprised only 635 km and had a speed of 1.7 km a day.

At the very beginning of April 1951, there was a sudden and large ice shear. As a result, a large bulky and cumbersome ice floe carrying the NP-2 broke up in almost no time into numerous small fragments with no one being suitable for sustaining NP life. First, the station was evacuated to the neighbouring floe and then it had to be closed. The team of the NP-2 station was transported to the mainland by aeroplanes. The station was closed on 11 April, 1951.

The NP-2 station did not reach the North Pole, but the vast material it collected was the first contribution to the post-war study of the central Arctic that did not cease until 1991.

(a)

(b)

Figure 1.5. (a) Near a fracture in the ice in the vicinity of the 'North Pole-2' drifting station (Summer 1950); (b) head of the 'North Pole-2' drifting station Mikhail Somov in the tent near a laboratory table performing a hydrochemical analysis of the samples of the 'next' oceanographic stations (1950).
Photos from the RGMAA archives.

However, the history of NP-2 did not end here. After the evacuation of people, tents and some equipment impossible to transport remained in the camp. For several years these remains of the expedition drifted in the Arctic Ocean.

After evacuation of NP-2, ice in this area sharply changed drift direction – first to the east and then to the south. The camp remains were carried to Alaska where ice again moved westward. The former ice floe of NP-2 was soon found in the area from where the station drift originally began. Thus, the abandoned camp of NP-2, moving clockwise made a full circle in the western sector of the Arctic.

This was a real scientific discovery. It meant that there were two sufficiently independent systems of general ice drift in the central Arctic. The first one being the Transarctic drift system from east to west, from the Chukchi Sea to the Greenland Sea, discovered due to the finding of cargo from the *Jeannette* that perished in 1881 and further investigated by Nansen. The second being the system of drift through the anticyclonic ring in the western Canadian sector of the Arctic discovered by the movement of the remains of the NP-2 camp. The time of full turnover in this anticyclonic drift is estimated at four years.

It should be noted that the hypothesis about the existence of the anticyclonic ice drift in the western sector of the Arctic was suggested as early as the beginning of the 20th century by the Russian polar investigator A. Kolchak – the same Kolchak who during the Russian revolution of 1917 and the civil war became one of the main leaders of the White Army resisting the advance of Bolsheviks in Russia.

After the discovery of the second ice drift system in the Arctic Ocean, the organizers of drifting stations faced a complicated question: where to land the NP expeditions on ice so that they could assuredly move to the pole or be assuredly entrained in the anticyclonic drift? The formulation of the scientific programme of the station, and the degree of its fulfillment, depended directly on this. Figuratively speaking, the main problem of organizing the NP stations was 'how to make a good shot' – the drift either via the North Pole or western gyre.

A long history of operation of NP stations did not produce a precise answer to this question. The point is that the ice drift systems, as any other large natural objects, have a tendency to experience long-term fluctuations in time and space, which are not always clear and known in advance. Therefore, landing of expeditions on the ice floe at one and the same point, say in the Chukchi Sea, but in different years, does not yet guarantee their motion in ice along one and the same trajectory.

Opening a new NP station, polar explorers had to achieve five main objectives without which life and efficient scientific working practices at the station would have been simply impossible. These objectives were: the choice of a reliable ice platform; construction and arrangement of a working space; organization of full-value nutrition; choice of comfortable and warm clothing; and finally establishment of uninterrupted communication with the mainland, from where different supplies were delivered and help could be sent if necessary.

Before landing on the ice of each NP station the team was of course preceded by substantial preparatory work. First of all when preparing a new expedition, polar explorers took into account the experience gained and documented of the previous stations – it is one thing to imagine theoretically what should be taken to survive under the severe conditions of the Arctic drift and another to confront these conditions face to face.

Therefore, there were many improvisations to the first NP expeditions, in spite of thought-out supply, arising in response to unexpected questions, which the Arctic unceasingly posed. Polar explorers diligently learned how to conquer the Pole and although dramatic situations occurred practically at all NP stations, people learned with time how to respond to them with increasing effectiveness. The development of vehicles also contributed to this: the only means of transport at the NP-1 station was a sledge and manpower; tractors and trucks appeared at the NP-3 station (Figure 1.6). A tractor justified itself and became an integral part of all other NP stations, while the truck proved unsuitable for operation in the central Arctic. Instead, a device for quickly making the ice even (an ice milling machine), caterpillar snowmobiles 'Buran' looking like motorcycles and snowplough machines appeared later on. Helicopters also assisted polar explorers in their work. At the NP-29 station in 1987, a computer complex consisting of one base and two auxiliary computers was used.

In general, living and working conditions became more comfortable with time. Realization of scientific work, measurements and observations also began from elementary simple equipment which improved with time.

The most laborious type of measurements were those of the ocean state parameters from the surface to the bottom. Oceanographic studies and ocean depth measurements at the first drifting stations were made using manual hydraulic winches (Figure 1.7). Water sampling for hydrochemical analysis in the prescribed horizons was performed using Nansen bottles, to which deepwater reversing thermometers were attached for water temperature measurements. This was done by lowering a rope to depth, with a series of bottles attached which were used to take water samples and temperature measurements at prescribed depths. The rope was then raised and the bottles were taken from it in the reverse order. Performing work using such methodology required enormous efforts. Sometimes, to complete one set of measurements an oceanographic station took several hours. During the recovery of the bottles, people had to work in pairs to operate the winch levers and would change pairs every ten minutes.

Producing electrical power was achieved by diesel generators allowing application of electrical winches for oceanographic studies – facilitating to a greater extent the implementation of research. The Nansen bottles were gradually replaced by automatic sounding systems measuring *in situ* water salinity, temperature and pressure, allowing immediate access to data on the vertical distribution of these elements from the surface to the bottom. At the same time systems began to measure other hydrophysical and hydrochemical characteristics in addition. For measurements of current characteristics, automatic current self-recorders began to be used.

For a long time, the processes occurring at the underside surface of the ice, its relief, and structure remained completely unstudied. In 1956, sub-ice surveys were

(a)

(b)

Figure 1.6. (a) Aerologists Vasily Kanaki (on the left) and Igor Tsigelnitsky prepare for launching a radiosonde at the 'North Pole-3' drifting station (1954); (b) general view of the construction of the 'NP-3' drifting station (April 1954).
(a) Photo from V. Savostianov. (b) Photo from the RGMAA archives.

Figure 1.7. (inset) Head of the 'North Pole-3' drifting station Aleksey Treshnikov (July 1954); (main image) oceanographer Vladimir Shamontiev measures the ocean depth at the 'NP-3' drifting station.

(a) Photo from the RGMAA archives. (b) Photo from V. Savostianov.

conducted for the first time using two small size photo-theodolites placed in a waterproof box, and from 1969, underwater observations and studies using diving gear and photo-cinematic and video equipment were carried out at the drifting stations.

The living conditions during wintering (and not only at the NP stations) dictated an implicit law of behaviour to people during long Polar expeditions – the 'Polar Law'. Polar explorers do not like to speak about it – they are modest people and are shy of pompous words. Academic Aleksey Treshnikov, head of the NP-3 station in 1954–1955, the Second and Thirteen Antarctic expeditions in 1956–1958 and in 1967–1968, director of the Arctic and Antarctic Research Institute in Leningrad (St. Petersburg) for many years best summarized the Polar Law. It was Treshnikov who equipped and dispatched the NP expeditions on their dangerous journeys, thus earning the right to speak on behalf of all polar explorers saying of the Polar Law: 'Save your comrade even if you can perish at this. Remember that his life is always more valuable than yours'.

The Polar Law cannot be taught in the class, it is learned only during a difficult overwintering. This law was established at the NP stations by the authority of experienced polar explorers and was understood not as a dogma, but as a method helping to assess your own actions based on the interests of the entire station rather than your own. The primacy of collective over personal and individual is obvious here. This is the difference between life in ice and our everyday existence in towns on the mainland.

The Polar Law acts not only in that extreme form imparted to it by A. Treshnikov, but also in more standard situations (e.g., in emergencies connected with ice shears, break-ups of ice floes, storm winds ...).

A man stands on the roof of the radio shack with an antenna in his raised hands – it had been torn by the hurricane wind. He stands there for five, ten, twenty minutes blown through by the icy wind. He stands there as long as it is needed to transmit an urgent radiogram (he is a living antenna!). He has to be taken down from the roof, his frostbitten face is rubbed with snow and he is made to drink half a glass of cognac and something else hot and he is put to bed to sleep ... It is normal – it is the Polar Law.

The ice floe breaks up. One has to save the equipment, food supplies, a radio-station and drums with fuel ... Everyone participates in the emergency work including the head of the station, he has to have time to give orders and like everyone carry loads, roll drums, move huts on sledges frozen to the ice ... It is normal – it is the Polar Law.

It is summer, bright sunshine and peaceful stillness – no break-ups or storms – but the mechanic urgently needs an assistant to pour diesel fuel from a drum to the supply tank of the diesel engine. Of course he will ask anyone to help, and all will oblige him – they know that the life of the station and their own lives depend to a great extent on the regular operation of the power station ... It is normal – it is the Polar Law.

None of these presented cases were invented; they were all taken from the recollections of polar explorers and administration reports to the mainland.

But was the life really so ideal at the NP stations, were there no cases of breaking the comprehensive Polar Law at all, manifestation of egoism, refusal to work and mutual assistance, in general, 'deviant behaviour'? Of course there were such cases, but these were rare exceptions to the general rule. There were always people at the station who voluntarily or by duty took on themselves an uneasy job in order to unite a team of winterers, transforming them to a group of associates pursuing one common purpose – efficient and safe operation of the station. The primary role here always belonged to the head of the NP station.

Over the entire history of existence of the NP stations (1937–1991), 59 station heads took turns. The first of them became a legend of the Arctic and Antarctic expeditions a long time ago. These are real pioneers, prominent organizers and scientists, born polar explorers, who served as an example for all the next generations of investigators of the Earth's Poles: Ivan Papanin (NP-1), Mikhail Somov (NP-2), Aleksey Treshnikov (NP-3), Yevgeny Tolstikov (NP-4) and Pavel Gordienko (NP-4).

Nineteen heads were in charge of two or more wintering periods (shifts of polar explorers continuing for about a year). Among these, three wintering periods were spent at different stations by Nikolay Blinov (NP-8, NP-17, NP-26), Georgy Kizino (NP-21, NP-22, NP-23), Pavel Morozov (NP-16 for two shifts, NP-22) and Ilya Romanov (NP-8, NP-5, NP-18). Four wintering periods were spent by two heads – Yury Konstantinov (NP-10, NP-14, NP-16, NP–19) and Vasily Sidorov (NP-13, NP-25, NP-26, NP-31). Yury Tikhonov was the champion of the heads of drifting stations, far ahead of the others with seven wintering periods (NP-20, NP-22, NP-24, NP-25, NP-27 for two shifts, NP-31). NP-31 was the last drifting station in the 20th century (Figure 1.8).

Thus, a 'corps of heads' of the drifting stations was gradually formed in the Arctic. To be in charge of the stations (often for their opening) was their main job. They represented a contingent of polar explorers that was able to unite teams at the stations, direct their work toward a common goal and achieve the observation by all explorers of the holy of holies – the Polar Law.

All of them without exception were first-class specialists in the area of polar studies and good organizers, but the main thing was that they were good psychologists who could find an approach to each polar explorer, make out his good and weak points and correct him if he took a false step. Such a head knew whom and when he has to compliment, whom to pull and whom to let 'weep at a father's shoulder'.

People always wanted to get to the ice floe to work under such leaders, there was competition among polar explorers lining up to work at the stations, but the final word belonged to the head – often chosing a team of his own people, the core of the station with whom he drifted for several shifts.

In 2003, a known polar explorer, vice-speaker of the State Duma of the Russian Federation Arthur Chilingarov came forward with an initiative to resume organization of drifting stations in the Arctic Ocean. He found sponsors – well-to-do people

(a)

(b)

Figure 1.8. (a) The 'North Pole-31' drifting station (May 1990); (b) aeroplane at the airstrip of the 'NP-31' drifting station (May 1990).
(a) Photos from V. Volkov.

dealing with polar tourism who were not indifferent to Russian polar science. The initiative was supported by the Russian Federal Service for Hydrometeorology and Environmental Monitoring (Roshydromet). The scientific programme and organization of the first Russian drifting station the 'North Pole-32' were performed by the Arctic and Antarctic Research Institute together with the sponsor's representative 'Pole Center'.

The station was landed on drifting ice on 16 April, 2003. The official opening of NP-32 was on 25 April, 2003. For organization of the station, 5 flights of AN-12 aircraft were made and necessary equipment, instruments, transport vehicles and 12 polar explorers were transported (Figure 1.9).

A complicated ice situation and break-up of the airstrip did not allow completion of the supply of the station. In June 2003, missing fuel and general station equipment was dropped from the aircraft. In December 2003, additional drops of fuel were made.

During its operation from 16 April, 2003 to 6 March, 2004 the NP-32 station passed a total route of 2,306 km with an average speed of 7.5 km per day. The general drift comprised 772 km with an average speed of 2.2 km per day. During operation, 1078 weather data at standard synoptic hours were transmitted to the Global Telecommunication Network, 285 ozone layer observations were made, 172 ocean soundings from 500–3,460 m were conducted, 59,362 current speed measurements in different horizons were carried out and more than 1,000 ice and snow temperature measurements were made. New data on the ice dynamics and mechanics and hydrobiology were obtained.

It was planned to complete work at NP-32 on 15–20 March, 2004. People and cargoes were to be evacuated by means of nuclear-powered ice-breaker. However, nature had other plans. On 3 March, 2004 there was a break-up of the ice floe on which the NP-32 station was located. The break-up was under strong ice pressure conditions accompanied by extensive ridging. Ice ridges were 4–6 m high and several facilities of the station were covered under ice fragments. However, people did not suffer, but an urgent evacuation was necessary.

A. Chilingarov undertook leadership of the rescue expedition. After just 72 hours, two helicopters MI-26 and MI-8MTV had evacuated all the polar explorers and part of the cargo from the ice floe to Spitsbergen Island, where they were warmly met by Governor of Longyear (Norway).

However, this was not the end of the drift of the NP-32 station. On 16 August, 2004, the captain of the German research vessel *Polarstern* – Udo Domke – detected the station remains at coordinates of 82°16.1′N, 4°20.7′W. The ship's crew removed 304 drums of fuel and two tractors from the ice floe. This operation was terminated on 19 August, 2004 and the *Polarstern*, with a speed of 20 knots, continued westward to fulfill the research programme of the German Alfred Wegener Institute of Polar Research (AWI). Thus the first Russian drifting station the 'North Pole-32' was finally closed.

The Russian polar explorers warmly thanked the *Polarstern* crew and administration of the AWI for their generous help. It should be noted that during the last decade close scientific contacts were established between the Arctic

(a)

(b)

Figure 1.9. (a) Opening of the 'North Pole-32' drifting station; (b) Summer of 2003 at the 'NP-32' drifting station
Photo from the AARI archives.

and Antarctic Research Institute (AARI) and AWI in the area of studies of polar areas.

For the first time since 1991, the Government of the Russian Federation allocated dedicated funds for conducting a full-value Arctic expedition including organization of the next drifting station the 'North Pole-33' (NP-33).

The AARI of Roshydromet was entrusted with the organization of NP-33. Lead scientists of research institutions of Russia took part in the development of the research programme.

The question of the place and method of landing the drifting station was addressed on the basis of initially set conditions:

- the ice floe should drift for one and a half to two years;
- in spring of 2005, the drifting station should be near the North Pole;
- in summer of 2006, the NP should be evacuated by an ice-breaker at the approach to Fram Strait; and
- the costs for organization of NP should be minimized.

By analysing the current state of the ice cover, taking into account the climatic ice drift characteristics, the locality of NP landing was determined – in the area to the north of the New Siberian Islands. The optimal period of NP organization is September 2004 and the optimal method – ice-breaker + ship + helicopters.

Beginning in spring 2004, the AARI monitored the ice cover of the area of proposed for the landing of NP-33 in order to choose the most suitable ice floes. Data from 'Radarsat', Terra (MODIS) and ENVISAT satellites were used with data from the latter satellite being received through the Norwegian Nansen Environmental Remote Sensing Center with which the AARI has close scientific links.

In late August 2004, the research expedition ship *Akademik Fedorov* escorted by the atomic ice-breaker *Arktika* exited to the ice floe search area for organization of the NP-33 drifting station. The weather conditions were not favorable: the visibility was restricted and the low cloud boundary did not rise higher than 200 m. Several days of wearisome flights on board helicopter MI-8 did not yield any positive results. Unfortunately, it was not possible to identify any of the ice floes detected by the satellites.

On 3 September, 2004, in the area of 85°07′N, 155°05′E, the sun finally shone through clouds, the visibility increased to 10–20 km and the relief of ice floes became clear well recognizable for ice reconnaissance specialists. A helicopter hovered above the ice floe and flew lower. The chosen ice floe had the dimensions 3.5 × 4.2 km and a thickness of 2–4 m – quite suitable for organizing a drifting station for long-term habitation by a team of 11 polar explorers.

On 4 September, the R/V *Akademik Fedorov* moored to the ice floe and unloading began. About 400 tons of equipment and fuel–lubricants were unloaded on the ice, twenty living and laboratory complexes were built very quickly over several days (Figure 1.10).

On 9 September, the Russian flag and the flag of St. Petersburg were hoisted above the NP-33 drifting station. The first weather report was transmitted. On 10 September the R/V *Akademik Fedorov* and the atomic ice-breaker *Arktika*

(a)

(b)

Figure 1.10. (a) Biological sampling at the 'North Pole-33' drifting station (January 2005) by V. Kuznetsov (on the left) and A. Zhukovsky; (b) Organization of the 'NP-33' drifting station (September 2004).
Photos from V. Strugatsky.

departed from the ice floe where 11 people – remarkable polar explorers under the leadership of Aleksey Visnevsky and three dogs remained on polar watch.

1.4 GENERAL INFORMATION

During the period 21 May, 1937 to 31 December, 2004 (when this book was written), 90 teams on all the drifting stations have carried out scientific studies on the ice of the Arctic Ocean. The total drift duration of the 'North Pole' stations comprised 28,391 days. The ice floes on which the NP stations were located passed along a complicated meandering route measuring 152,519 km with an average speed of 5.4 km per day (Figures 1.11–1.18, see colour section). On this route, 2,032 people (polar explorers) participated in the yearly cycles of studies. For short-term studies under special programs, about 9,000 people visited the drifting stations episodically (without taking into account the crews of aircraft and ships). For the organization of drifting stations and maintaining their activity 24,170 tons of different cargoes were delivered to the ice, including: fuel, food products, domestic equipment, scientific gear and instruments. On the ice airfields of the drifting stations, created by polar explorers, 15,190 landings by different aircraft (including helicopters) were made: from R-5, TB-3, PE-8 to the post-war LI-2, AN-2, IL-12, TU-4, IL-14 and further back to AN-12, IL-18, AN-74 and helicopters MI-4, MI-6 and MI-8. Icebreakers and ice-breaking ships transported 4,960 tons of different cargoes to the stations.

Most of the drifting stations were organized in the spring period using aviation and only 7 stations of the 33 were opened in autumn (NP-10, NP-22, NP-24, NP-29, NP-30, NP-31 and NP-33).

During the break-ups of the NP ice floes (around 800) when it was impossible to prepare a runway, necessary cargoes were dropped on parachutes from large aeroplanes AN-12 and IL-76.

Table 1.1 and Table 1.2 present general data on all the stations.

The scientific observations that were carried out at the 'North Pole' drifting stations and in the framework of the high-latitudinal 'Sever' expeditions made a large contribution to understanding regularities of the natural processes of the Arctic. By the middle of the 20th century it was still believed that the seabed of the Arctic Ocean presented one cup with depths of 4–5 km. Due in many respects to the observation data of high-latitudinal expeditions, including NP, it is determined that the seabed of the Arctic Ocean has a complicated relief with high underwater mountains and ridges, alternating deep depressions and valleys. The greatest depth of 5,449 m was recorded in the Nansen basin and the underwater ridges recorded lesser depths of 900–100 m. The major underwater ranges were named after the Russian scientiests Lomonosov, Gakkel and Mendeleyev (*Regulation of Presidium of the USSR Academy of Sciences*, 1974).

The contribution of the 'North Pole' drifting stations to the study of the Arctic cannot be overemphasized.

During the period of NP operations, 214,000 synoptic hours were performed, 3,466 deepwater hydrological stations were made and tens of thousands of samples

Table 1.1. General characteristics of the drift of NP drifting stations.

		Drift coordinates				Drift extent (km)		Drift velocity (km/day)	
		Start		End					
Station	Drift duration (days)	Latitude (N)	Longitude	Latitude (N)	Longitude	By meandering trajectory	By the general course	By meandering trajectory	By the general course
NP-1	274	89°25′	78°40′W	70°40′	19°16′W	2,850	2,200	10.0	8.0
NP-2	374	76°03′	166°36′W	81°44′	163°48′W	2,600	640	6.9	1.4
NP-3	376	86°00′	178°00′W	86°00′	31°42′W	1,865	825	4.9	2.2
NP-4	1,108	75°48′	178°25′W	85°52′	00°00′	6,970	2,110	6.3	1.9
NP-5	536	82°10′	156°51′E	84°23′	63°19′E	3,630	1,080	6.7	2.0
NP-6	1,245	74°24′	177°04′W	82°06′	03°56′E	8,650	2,920	7.0	2.5
NP-7	721	82°06′	164°11′W	85°14′	33°00′W	3,520	1,240	4.9	1.7
NP-8	1,058	75°42′	163°10′W	83°15′	132°30′W	6,090	1,665	5.3	1.6
NP-9	335	77°23′	163°00′E	86°36′	176°00′W	2,660	1,340	8.0	4.0
NP-10	914	75°27′	177°10′E	88°32′	90°30′E	3,960	1,795	4.3	1.6
NP-11	373	77°10′	165°58′W	81°10′	139°34′W	2,400	675	6.5	1.8
NP-12	725	76°50′	165°34′W	81°06′	145°47′W	1,595	1,200	2.2	1.6
NP-13	1,099	73°50′	166°00′W	87°52′	02°00′W	3,545	2,670	3.2	2.4
NP-14	288	74°20′	175°20′W	77°00′	154°48′E	1,040	880	4.0	3.4
NP-15	710	78°50′	168°43′E	85°45′	10°30′W	2,330	1,880	3.3	2.6
NP-16	1,442	74°58′	171°40′W	86°00′	85°27′W	5,850	2,485	4.0	1.7
NP-17	546	80°28′	165°23′E	86°48′	25°47′E	1,750	1,500	3.2	2.9
NP-18	1,110	75°10′	165°02′W	86°06′	152°36′E	5,240	2,160	4.7	1.9
NP–19	1,256	74°34′	161°48′E	83°09′	15°17′E	6,705	3,055	5.3	2.4
NP-20	755	75°35′	179°04′E	81°44′	165°47′W	3,780	1,230	5.0	1.6
NP-21	748	74°42′	175°58′E	86°16′	143°36′E	3,605	1,580	4.8	2.1
NP-22	3,120	76°16′	168°31′W	86°10′	00°00′	17,069	5,762	5.5	1.8
NP-23	1,068	73°51′	178°25′W	87°40′	22°31′W	5,786	2,219	5.4	2.0
NP-24	931	76°45′	163°00′E	86°03′	29°40′E	5,652	1,916	6.2	1.9
NP-25	1,070	75°01′	168°35′E	85°50′	122°15′W	5,754	1,782	5.3	1.7
NP-26	1,044	78°30′	174°46′E	82°46′	170°31′W	5,380	685	5.1	0.6
NP-27	1,084	78°31′	160°30′E	86°28′	09°02′W	5,655	1,706	5.2	1.5
NP-28	978	80°40′	168°29′E	79°40′	03°09′E	7,634	600	7.5	2.3
NP-29	436	80°23′	112°59′E	84°43′	56°34′W	2,686	1,372	6.2	3.2
NP-30	1,255	74°18′	171°24′W	82°31′	126°26′W	7,675	1,611	5.9	1.3
NP-31	976	76°35′	153°10′W	73°33′	161°04′W	5,475	2,017	6.0	2.7
NP-32	325	87°52′	148°03′W	84°41′	03°33′W	2,418	794	7.4	2.4
NP-33		85°05′	156°31′W						

were collected and analysed for dissolved oxygen, silicon, phosphorus, pH value, presence of hydrocarbons, organochlorines, etc. About 800 samples of soil were collected; 48,000 ocean depths were measured; 120,000 series of measurements of the radiation balance components were made; and 32,859 radiosondes were launched. The upper air observations were carried out at 21 stations during 51 shifts. The Earth's magnetic field intensity and determination of the horizontal and vertical components with continuous registration at the magnetic-variation station were carried out at almost each of the drifting stations. At 12 stations, ionospheric observations were conducted during 31 shifts: vertical sounding, determination of absorption of radiowaves in the ionosphere and the intensity of the fields of signals of very low-frequency radio-stations.

A large complex of special studies was carried out at the drifting stations in addition to standard programmes. In particular, observations of the fine structure of oceanic waters in the 0–250-m layer were made; the heat exchange characteristics between the ocean and the atmosphere through ice of different thicknesses were determined; topographic underwater surveys of the lower ice surface including ice ridges were made; the dynamics of the sub-ice water layer was determined; and wave processes in the upper thermocline and in ice were investigated. Conditions for propagation of electromagnetic emission of the optical spectrum region in the ocean, optical characteristics of water masses, of different types of sea ice and a dependence of electrical ice parameters on its physical characteristics were studied. Investigations into the fine structure of the fields of sonic speed propagation in water were undertaken. A large number of measurements were performed to determine the prognostic characteristics of ice, thermal properties, texture, crystalline structure; natural fluctuations and parameters of stresses in ice and its deformations were also determined.

At the NP-22 drifting station, complex studies of the aerosol–optical parameters of the atmosphere were carried out for the first time in the framework of the Global Atmospheric Research Program (GARP). Unique results on polar aerosols, cloudiness and radiation fluxes and their interaction in the Arctic atmosphere were obtained. These observations were later continued at NP-28 under the extended program including spectroscopic measurements of carbon oxide and methane.

The major results of the studies at drifting stations are contained in such basic editions as the *Atlas of the Oceas. The Arctic Ocean* (1980) and the *Atlas of the Arctic* (1985), in a large number of monographs and handbooks, and are also presented in the next chapters of this book.

Table 1.2. Dates of the drift of the stations and their drift patterns.

Station	Drift period	Platform	Transportation Means	Closed due to:
NP-1	21 May 1937–19 February 1938	Ice floe	Aircraft	Complete break-up, demobilization by ice-breaking vessels.
NP-2	2 April 1950–11 April 1951	Ice floe	Aircraft	Break-up, the ice floe was changed.
NP-3	15 April 1954–19 April 1955	Ice floe	Aircraft	The station approached Fram Strait.
NP-4	3 April 1954–19 April 1957	Ice floe	Aircraft	Complete break-up.
NP-5	21 May 1955–08 October 1956	Ice floe	Aircraft	The station drifted towards the Siberian shelf.
NP-6	15 April 1956–14 September 1959	Ice floe	Aircraft	The station approached Fram Strait.
NP-7	23 April 1957–11 April 1959	Ice floe	Aircraft	The station approached Canada.
NP-8	15 April 1959–19 March 1962	Ice floe	Aircraft	Complete break-up.
NP-9	28 April 1960–28 March 1961	Ice floe	Aircraft	Significant break-up.
NP-10	17 October 1961–29 April 1964	Ice floe	Ice-breaker	Significant break-up.
NP-11	12 April 1962–20 April 1963	Ice floe	Aircraft	Significant break-up.
NP-12	30 April 1963–25 April 1965	Ice floe	Aircraft	The station approached Canada.
NP-13	22 April 1964–17 April 1967	Ice floe	Aircraft	Complete break-up. The personnel moved to the ice floe of former subsidiary NP-13 (3 April 1965–14 October 1966).
NP-14	1 April 1965–11 February 1966	Ice floe	Aircraft	The station collided first with Zhokhov Island, then with Genrietta Island, complete break-up.
NP-15	29 March 1966–22 March 1968	Ice floe	Aircraft	Significant break-up, transfer to new ice floe.
NP-16	9 April 1968–22 March 1972	Ice floe	Aircraft	Complete break-up, two transfers to new ice floes.
NP-17	29 April 1968–16 October 1969	Ice floe	Aircraft	The station was approaching Fram Strait.
NP-18	9 September 1968–24 October 1971	Ice floe	Aircraft	From 9 September 1968 to 18 May 1969, the station was on the ice island, where NP–19 was set up.
NP–19	7 November 1969–14 April 1973	Ice island	Aircraft	In January 1970 the ice island grounded and broke-up. Transfer to a larger remaining part of the ice island.
NP-20	11 April 1970–10 May 1972	Ice floe	Aircraft	The studies were completed.
NP-21	1 May 1972–25 May 1974	Ice floe	Aircraft	The studies were completed.
NP-22	13 September 1973–8 April 1982	Ice island	Ice-breaker	The station was approaching Fram Strait.

NP-23	5 December 1975–1 November 1978	Ice island	Aircraft	The station approached Greenland.
NP-24	23 June 1978–19 November 1980	Ice island	Aircraft plus ice-breaker	The station was calculated to enter Fram Strait in polar night.
NP-25	16 May 1981–20 April 1984	Ice floe	Aircraft	The station drifted away to the Beaufort Sea.
NP-26	21 May 1983–9 April 1986	Ice floe	Aircraft	The station approached Fram Strait.
NP-27	2 June 1984–20 May 1987	Ice floe	Aircraft	The station approached Fram Strait.
NP-28	21 May 1986–23 January 1989	Ice floe	Aircraft	The station approached Fram Strait.
NP-29	10 June 1987–19 August 1988	Ice floe	Ice-breaker	The studies were completed.
NP-30	9 October 1987–4 April 1991	Ice floe	Ice-breaker	The station drifted away to Canada.
NP-31	22 October 1988–25 July 1991	Ice floe	Ice-breaker	The ice floe melted.
NP-32	25 April 2003–6 March 2004	Ice floe	Aircraft	Complete break-up, demobilization by ice-breaking vessels.
NP-33	9 September 2004	Ice floe	Aircraft	Break-up, the ice floe was changed.

2

Meteorological investigations

2.1 METEOROLOGICAL OBSERVATIONS

2.1.1 Methods and instruments

Meteorological observations at the drifting stations were made using standard methods in compliance with the 'Manual' (*The Manual for Hydrometeorological Stations and Posts*, 1969). Some deviations connected with the specifics of observations on drifting ice were performed in accordance with Arctic and Antarctic Research Institute (AARI) instructions. Data of the meteorological observations at all 'North Pole' drifting stations were published in digital form in *Arctic Ocean Snow and Meteorological Observations from Drifting Stations* (1996).

Usually, meteorological observations were performed eight times daily (at 0000, 0300, 0600, 0900, 1200, 1500, 1800, and 2100 Moscow Local Time (MLT; which is GMT + 0300). However, sometimes observations were performed four times daily (at 0300, 0900, 1500, and 2100 MLT) as a rule during periods of ice breakup and during arrival of new personnel (Table 2.1) – so that some stations had observations at both three and six hourly periods.

The location for the meteorological site was chosen on level ice at a distance of not less than 70–100 m from the ice ridges and living premises. At a site 26 × 26 m in size, instruments and equipment were arranged in a standard manner: psychrometric shield with thermometers and hygrometers, shield for self-recorders with thermograph and hygrograph and also masts with weather vanes and a mast with wind indicator. Near the psychrometric shield, thermometers for measuring snow surface temperature were located and three snow line measurement staffs were established at different angles to the meteorological site. In the dark time of the year, the meteorological site had electrical lighting, and in emergency, a pocket flashlight was applied.

For atmospheric pressure measurement, a cistern barometer and a barograph that were placed in the meteorological room were used. If necessary, a control inspection barometer was used with the readout accuracy of 0.05 hPa and

Table 2.1. Periods during which observations were made four times a day at the drifting stations.

Station	Start	End
NP-8	6 December, 1961	18 March, 1962
NP-12	30 April, 1963	11 April, 1964
NP-13	1 May, 1964	20 May, 1964
	1 November, 1964	31 March, 1965
	21 May, 1966	25 May, 1966
	1 November, 1966	10 May, 1967
NP-14	1 December, 1965	12 February, 1966
NP-15	15 April, 1966	1 May, 1966
NP-16	22 April, 1970	31 May, 1971
NP-17	1 April, 1968	31 December, 1968
NP-18	1 January, 1969	11 March, 1969
	15 March, 1969	10 June, 1969
	7 November, 1969	30 April, 1970
	14 October, 1970	1 April, 1971
	25 April, 1971	20 October, 1971
NP-23	1 January, 1976	30 April, 1976

maximum error of not more than ±0.2 hPa. During technical failure of the barometers, the atmospheric pressure was measured using an aneroid altimeter (NP-5, NP-8, NP-9, and NP-16) with a readout accuracy of 1 mm mercury and the maximum measurement error of not more than 0.8 mm mercury (i.e., ~1.1 hPa).

The air temperature was determined by means of mercury and alcohol thermometers installed at the psychrometric shield at a height of 2.0 m above the ice surface. For temperatures higher than −30°C, mercury thermometers were used while for temperatures lower – alcohol thermometers were used.

The snow cover surface temperature was measured from September to April only – during the period with stable below zero temperatures. For measuring snow surface temperatures and air temperatures higher than −30°C, a mercury thermometer TM-3 was used (it is intended for surface temperature determination with a scale factor of 0.5°C and a measurement error of ±0.5°C in the range up to −20°C, and at lower temperatures – with an error of ±0.7°C). At air temperatures below −30°C, alcohol thermometers with a scale factor of 0.5°C were applied. The measurement errors for these were as follows:

Temperature range	Error not higher than
Up to −20°C	±0.5°C
Up to −30°C	±0.8°C
Up to −40°C	±1°C
Up to −50°C	±1.5°C
Up to −60°C	±2°C

At active ablation (a combination of processes such as sublimation, fusion or melting, and evaporation, which removes snow or ice from the surface of ice or from a snow field) from May to August, the measurement methodology does not allow the obtaining of reliable information on the surface temperature of melting snow and ice – measurements during this period were therefore stopped.

The air humidity was measured by means of psychrometers in the warm time of the year and a hygrometer in the cold season.

The cloud amount was determined visually using a 10-point scale. For measuring the cloud height, one used a cloud searchlight (until 1963), pilot-balloons and radio-sounding data. From 1969, the ceilometer IVO-1 was applied. When it was impossible to carry out instrumental observations, the cloud height was determined visually.

The precipitation amount at NP-1, NP-2 and NP-3 was determined using a rain gauge with a Nipher shield. From NP-4 (1954) onwards a Tretyakov precipitation gauge was used. The instrument mounting height was 2 m above the ice surface. From 1966, a correction for measurement error due to wetting of the precipitation gauge bucket was introduced to each measurement. A correction of +0.2 mm was introduced for liquid and mixed precipitation and of +0.1 mm for solid precipitation. The precipitation measurements were made twice in 24 h.

The wind speed and direction were determined until 1962 by a Wild weathervane with light and heavy boards mounted at a height of 6–10 m and by remote meteorological stations DMS-49 and DMS-53. From 1963, wind indicators M-7, M-49, M-63, M-63M and anemorumbographs M-12 and M-64 were installed. The orientation of the wind-measuring instruments was systematically checked in connection with the ice drift and the altered orientation of the meteorological site with time.

The accuracy of measurements of meteorological parameters is presented in Table 2.2.

Table 2.2. Accuracy of measurements of meteorological parameters.

Parameter	Instrumental method (measurement units)	Error: Single observation Avg.	Error: Single observation Max.	Error: Daily values Avg.	Error: Daily values Max.
Pressure	Mercury barometer (mb)	0.2	0.5	0.2	0.4
Wind speed	Wind indicator (m/s)	1–2	4–6	1–2	2–4
Wind direction	Compass points (22.5°)	1	2	0.5	1
Air temperature	Thermometer: psychrometric (°C)	0.3–0.4	1.0	0.3–0.4	0.6–0.8
	Maximum (°C)	0.4–0.6	2.0	0.4	0.8–1.0
	Minimum (°C)	0.5	2.0	0.5	1.0
Air humidity	Psychrometer, self-recording hygrometer (mb)	0.2	0.4	0.2	0.4
Liquid precipitation	Tretyakov precipitation gauge (% of the measured amount)	15–20	40	–	–
Solid precipitation	Tretyakov precipitation gauge (% of the measured amount)	45–55	150	–	–

2.1.2 Air temperature

The main peculiarity of the temperature regime of the Arctic basin is a below zero mean monthly air temperature in all months. The mean yearly air temperature in the Pole area is −19.4°C. The area of the lowest temperatures almost throughout the entire year is located near the Canadian Arctic archipelago and Greenland. In February, the coldest winter month, mean monthly temperatures up to −36°C are recorded here. The period exhibiting above zero temperatures in the Pole area is short. Transition of mean daily air temperature across 0°C from below zero to above zero values occurs on 15 July, on average, and back to below zero mean daily temperatures – on 21 July (Aleksandrov *et al.* 2004; Aleksandrov *et al.* 1999b).

Yearly variations. The peculiarities of the thermal regime of the Arctic basin are determined by a complex interaction of many factors. The most important are the following: the incoming solar radiation and its absorption by the underlying surface, heat and moisture transfer in the atmospheric circulation system, heat re-emission by the underlying surface and the atmosphere and the heat flux from the ocean.

The complicated character of the impact of these and other factors determines a spatial non-uniformity of the air temperature field in the Arctic basin. It is especially pronounced in its Atlantic sector in wintertime. Different regions of the Arctic basin have their own peculiarities in the yearly air temperature variations expressed in the dates of manifestation of temperature extremes.

In general in the Arctic basin, in spite of ice cover presence, features of the marine climate are clearly expressed. These are a lag of the yearly temperature extremes, smaller amplitude of its yearly variations compared with the Arctic continental regions and asymmetry of the curve of yearly variations reflecting a more intense temperature increase in spring as compared with its decrease in the autumn months (Figure 2.1) (*The Climatic Atlas of the Arctic,* 1963; Prik, 1961; Leontyeva, 1947; Prik, 1965).

The highest mean monthly air temperatures in the Arctic basin are observed in July. In summer, under the conditions of substantial heat losses due to snow and ice melting, the convective and turbulent heat exchanges of the ocean with the atmosphere are practically absent here and the temperature changes according to the continental type, with an air temperature maximum in July. Unlike the Arctic basin, the temperature maximum in the marginal seas is displaced to August.

The yearly amplitude of mean monthly temperature fluctuations above the Arctic basin is within 32–38°C. The maximum value of the yearly amplitude of 40°C was recorded in 1977 at NP-22. For coastal and continental stations, the amplitude values are higher. Thus, at Tiksi station, the yearly amplitude of mean monthly temperatures in 1946 was 48°C and at Verkhoyansk station in 1908, it comprised 73°C.

In some years, the mean monthly air temperature can differ significantly from a multiyear average. In winter, the amplitude of mean monthly temperature changes from one year to another comprises 8–10°C, and sometimes it can be as high as 15°C.

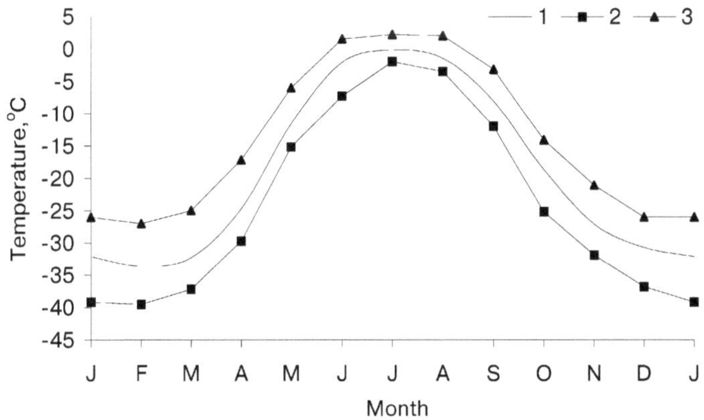

Figure 2.1. Yearly variations of air temperature. 1 – mean monthly; 2 – mean monthly minimum; 3 – mean monthly maximum.

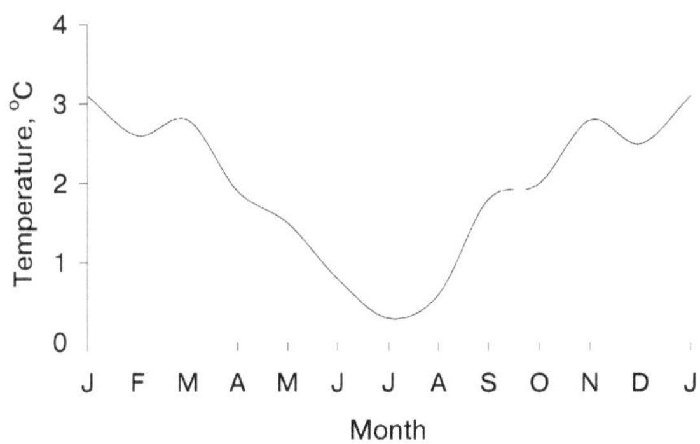

Figure 2.2. Yearly variations of the RMSD of mean monthly air temperature in the Arctic basin.

In summer, the interannual variability of air temperature is much smaller and the amplitude in July in the Arctic Basin is not higher than 1°C.

Correspondingly, the root-mean-square deviation (RMSD) of mean monthly temperature from a multiyear average in the cold half of a year is 3–4 times greater compared with summer months. In the yearly variations of RMSD of air temperature in the Arctic basin, the maximum falls on January and the minimum variability is observed in July (Figure 2.2). The largest variability of winter temperatures is recorded in the western Arctic basin.

Spatial distribution. The spatial structure of mean air temperature fields in winter months is significantly different in character from summer months. In winter, the heat flux to the Arctic basin with Atlantic cyclones and oceanic currents intensifies. As a result, comparatively high air temperatures form in the western Arctic basin. The horizontal temperature gradients in winter months here

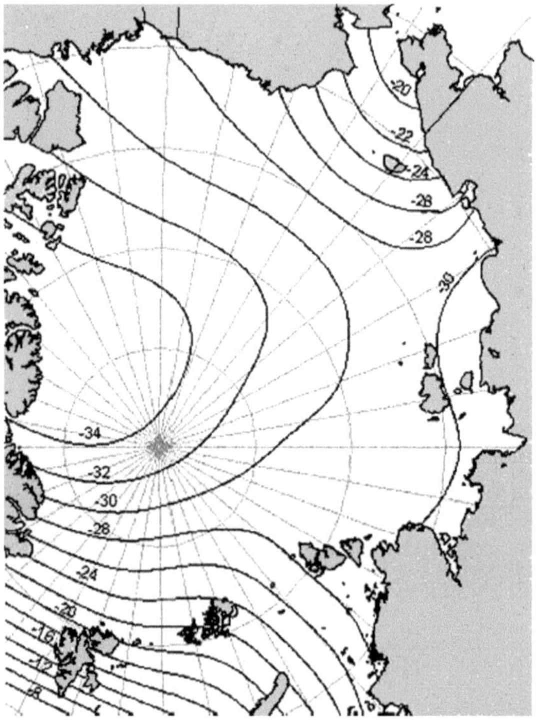

Figure 2.3. Mean monthly air temperature in January (°C).

comprise 2°–3° per 100 km. Average January temperatures (Figure 2.3) range from −18°C at 80°N (at the northern tip of Spitsbergen) to −34°C in the Pole area.

The air temperature distribution has a similar character in the other months of the cold period of the year (November–April), with only absolute values changing. The summer type of air temperature distribution (June–August) is characterized by the presence of an extensive low-gradient closed area of temperatures close to 0°C, the boundaries of which are close to the location of the multiyear ice edge. In July (Figure 2.4), all incoming heat in the enormous territory of the near-pole regions, with an area of several millions of square kilometers, is lost due to ice melting and a temperature around 0°C is preserved everywhere (slightly below the freezing point, on average for a month).

The entire Arctic basin is an area of absolute dominance of below zero air temperatures. While along the northern boundary of the Arctic Seas, the number of days with a positive mean daily temperature comprises 20–30 over the year, the period with a stable temperature above 0°C is absent in the near-Pole area.

Temperature extremes. The range of possible air temperature fluctuations characterizes its absolute extremes. The absolute yearly amplitudes achieve the greatest values in the coastal zones of the Arctic Seas where they are equal to 70–80°C. Approaching the pole, the absolute amplitudes decrease everywhere to 60–65°C.

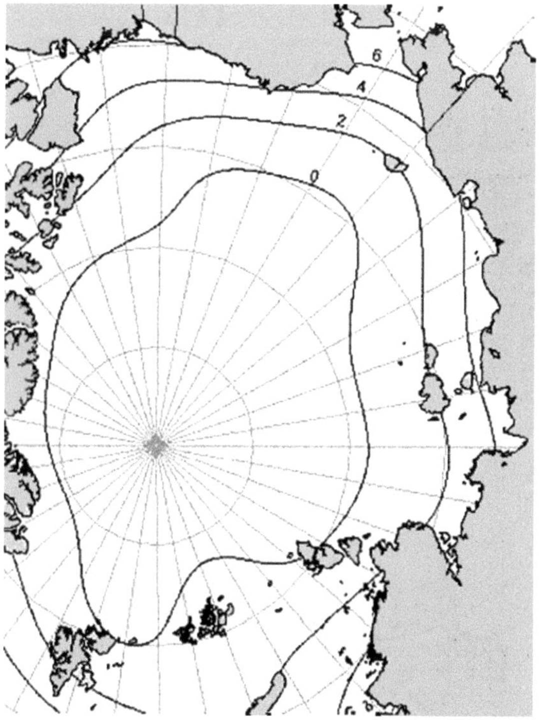

Figure 2.4. Mean monthly air temperature in July (°C).

The spatial distribution of the absolute minimum temperatures from October to April in the Arctic basin is characterized by an extensive low-gradient area. The absolute minimum temperature in the middle of winter can drop to −55°C (Figure 2.5). An analysis of the observation data of drifting stations revealed that the yearly minimum air temperature in the centre of the Arctic basin differed from the absolute extreme by only 8–10°C.

The spatial distribution of the absolute yearly maximums of air temperature in the Arctic basin is characterized by their rather uniform decrease towards the Pole. In July, the character of the maximum temperature distribution is close to yearly (Figure 2.6). In the near-Pole area, the maximum air temperature in July is not higher than 5–6°C, and above the northern offshore of the Arctic Seas, it can comprise 14–16°C. The distribution of extremes in July and August is similar.

2.1.3 Atmospheric pressure

Before the beginning of observations at the drifting stations, there was an idea about the dominance of a constant anticyclonic cell above the Arctic (Prik, 1965). However, the data of observations even at the first drifting stations showed cyclonic activity in the central part of the basin to be a common phenomenon. In Prik (1974), a rather frequent appearance of cyclones is noted in winter in the

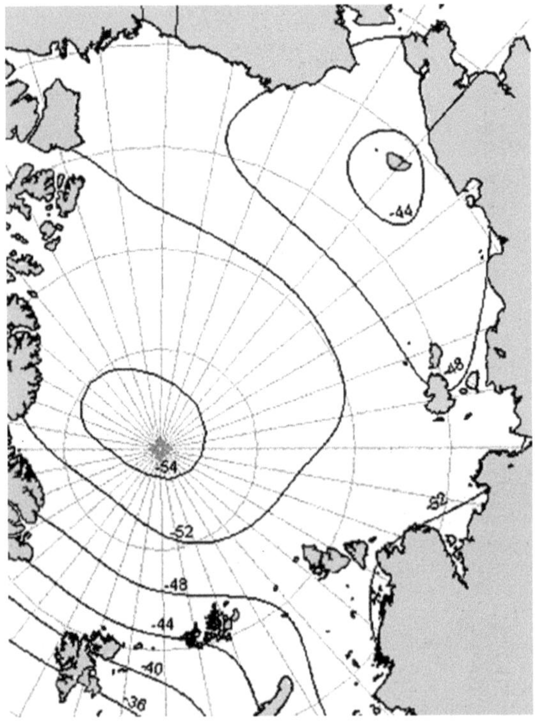

Figure 2.5. Minimum air temperature in January (°C).

Atlantic sector of the basin, whereas in summer, an intense cyclonic activity is also observed in the Pole area.

Yearly variations. The air pressure variability within a year in the Pole area and in the western and eastern parts of the basin is presented in Figure 2.7 (Aleksandrov *et al.* 2004).

The main maximum in yearly variations is in March–April, when thermal and pressure gradients begin to decrease and the Icelandic Low is filled. Due to this, the air pressure also increases in the western part of the basin. In the eastern part of the basin, a pronounced high is preserved. A minimum in the yearly pressure variations is observed in July when an extensive depression forms in the western area and an extensive but poorly expressed anticyclone is generated in the eastern area.

The pressure changes from day-to-day in the Arctic basin are large varying in winter months on average over the basin around 7–8 hPa, while in some regions, the values can be even higher. In some cases, the changes from day-to-day comprise several tens of hPa.

The mean monthly pressure variability in the Arctic basin is 4–5 hPa in summer months and 6–8 hPa in winter months (Figure 2.8).

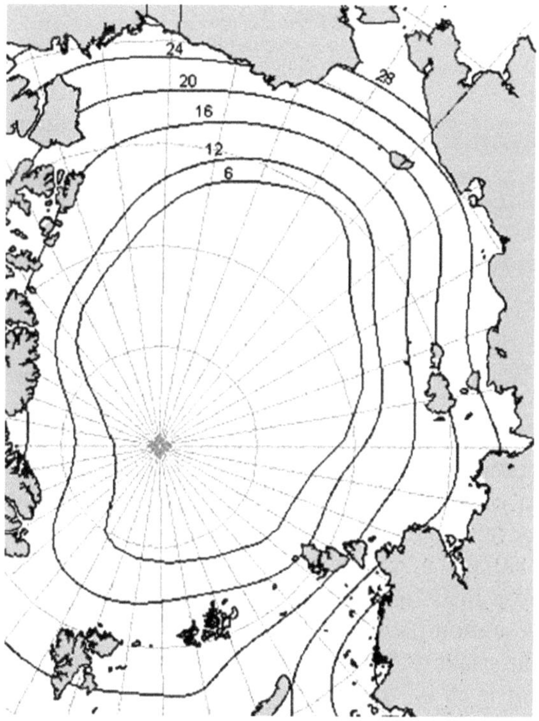

Figure 2.6. Maximum air temperature in July (°C).

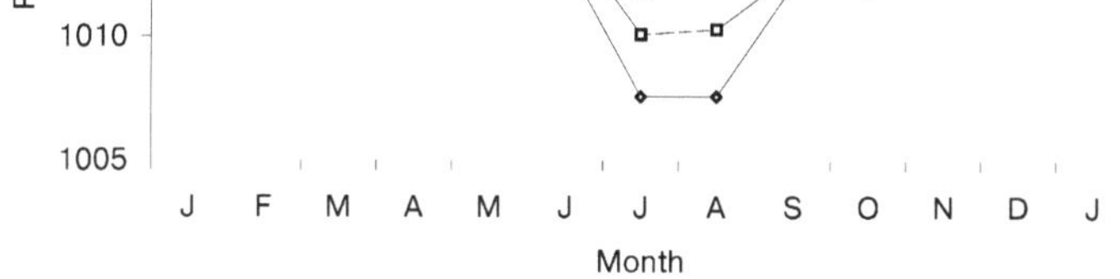

Figure 2.7. Yearly variations of atmospheric pressure in the Arctic basin. 1 – the Pole; 2 – western area; 3 – eastern area.

An absolute pressure maximum in the Arctic basin achieved 1,063 hPa (April 1956), and the absolute minimum comprised 946 hPa (June 1971).

Spatial distribution. The winter distribution of the air pressure fields from November to March is characterized by the presence of a large trough extending

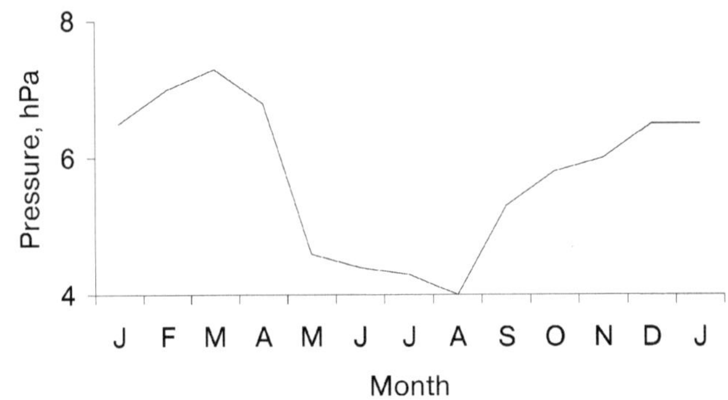

Figure 2.8. Yearly variations of RMSD of mean monthly atmospheric pressure in the Arctic basin.

from the Icelandic Low across Greenland and the Barents Sea up to the North Pole along an isobar of 1,012–1,016 hPa (Figure 2.9). The location of this trough from Spitsbergen to the Pole is determined by the trajectories of cyclones. On mean air pressure charts, the Aleutian Low trough, with a small area directed to the Arctic basin across the Chukchi Sea, is also quite clearly delineated. The next typical feature of these months is a high-pressure bridge extending from the north of Canada across the eastern Arctic basin to the Siberian High. The pressure in this part of the basin is equal to 1,020–1,022 hPa. The pressure gradients are the highest in the western Arctic basin.

From June to August, a low-pressure area extends to the north of Greenland with its centre above the Pole (Figure 2.10). Its occurrence is determined by cyclones of low intensity penetrating the Arctic basin up to the near-Pole area from the territory of Canada and Eurasia. Warmer waters of the North Atlantic freely flowing to the Arctic basin also contribute to the formation of this depression.

2.1.4 Wind

The wind regime in the Arctic basin is determined by the location of mean multi-year fields of atmospheric pressure and seasonal peculiarities of the circulation conditions.

In winter, most of the basin is covered with intense cyclonic activity. Simultaneously, a high air pressure is located above the Asian continent and a similar area is observed above north Canada. This determines the dominance of the air flows mainly directed from the Eurasian continent towards the Arctic basin. However, the winds here are unstable due to interaction of the aforementioned high pressure areas and the Aleutian Low trough. In the west of the Arctic basin, the direction of air flows depends on the location of the Icelandic Low of atmospheric pressure.

In summer the pressure field is characterized by the location of the increased pressure area in the west of the region and in the Beaufort Sea, and of the depression in the near-Pole area and the decreased pressure area above the continents. This creates conditions for changes in wind direction in the Arctic basin.

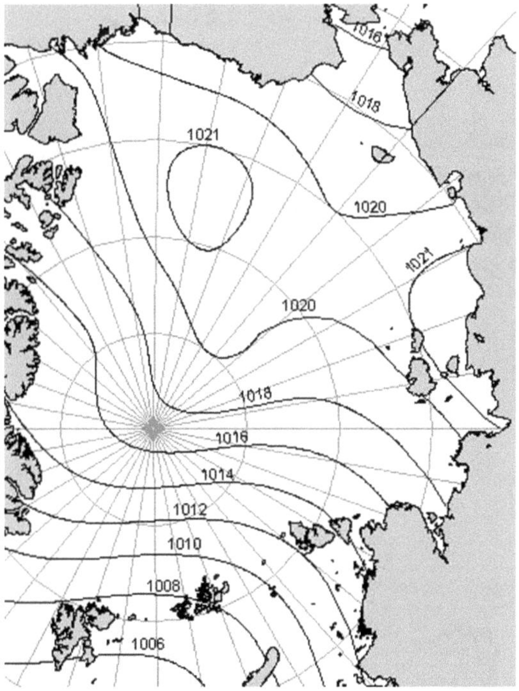

Figure 2.9. Atmospheric air pressure at sea level in January (hPa).

The wind speed regime depends little on the season (Aleksandrov *et al.*, 2004). Two figures below present data for the point with coordinates 82°N, 175°E. The yearly variations of the wind speed do not have pronounced maximums and minimums (Figure 2.11). In the summer months, the mean monthly wind speed is equal to 4.5 m/s and in winter to 4.7 m/s.

The distribution of wind speeds by gradations in January and July is shown in Figure 2.12. The most frequently recorded are weak winds with a speed of 2–5 m/s. Their frequency of occurrence is 50–60% both in winter and in summer. However, the winds with a speed of 2–3 m/s are more frequent in winter while those with a speed of 4–5 m/s are more frequent in summer. The maximum recorded speed in winter was equal to 25–28 m/s (frequency of occurrence of 0.01%). In summer, the maximum comprised 16–17 m/s (frequency of occurrence of 0.01%).

Histograms of the wind speed distribution above the Arctic Ocean are distinguished by significant asymmetry. The greatest asymmetry is observed in the central region of the Arctic basin, where the wind speed is less variable and the frequency of occurrence (55–70%) of weak winds (1–5 m/s) is high. The least asymmetrical is the distribution of wind speeds in the Atlantic sector. Intense cyclonic activity in this region leads to predominating (frequency of occurrence of 30–50%) moderate winds (6–10 m/s). However, in the eastern part of the Arctic basin the frequency of occurrence of moderate and weak winds is close.

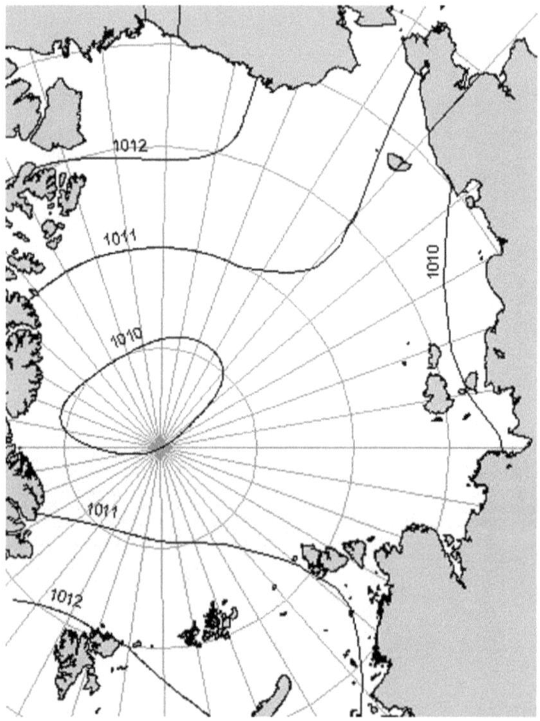

Figure 2.10. Atmospheric air pressure at sea level in July (hPa).

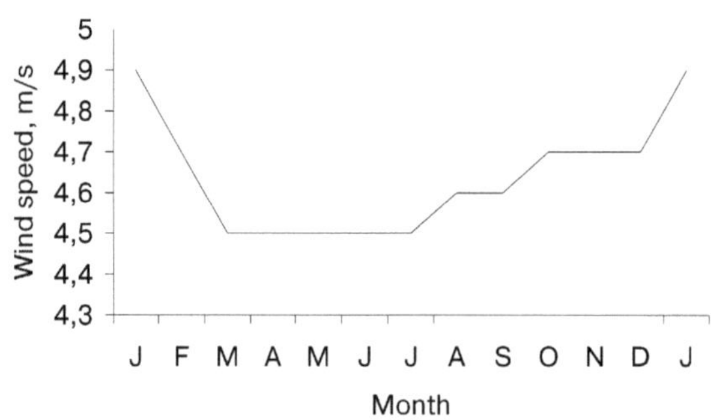

Figure 2.11. Mean monthly wind speeds.

2.1.5 Cloudiness

According to ground observation data, the most widespread form of clouds in the Arctic is stratus clouds (St), typical of the stable air mass (Zavyalova, 2000). They occur during the whole year. In connection with the fact that uniform air masses with stable thermal stratification encompass extensive Arctic regions, the horizontal extent of these clouds can comprise hundreds and even thousands of kilometres. However, most frequently it is not greater than 600 km. Since the indicated thermal

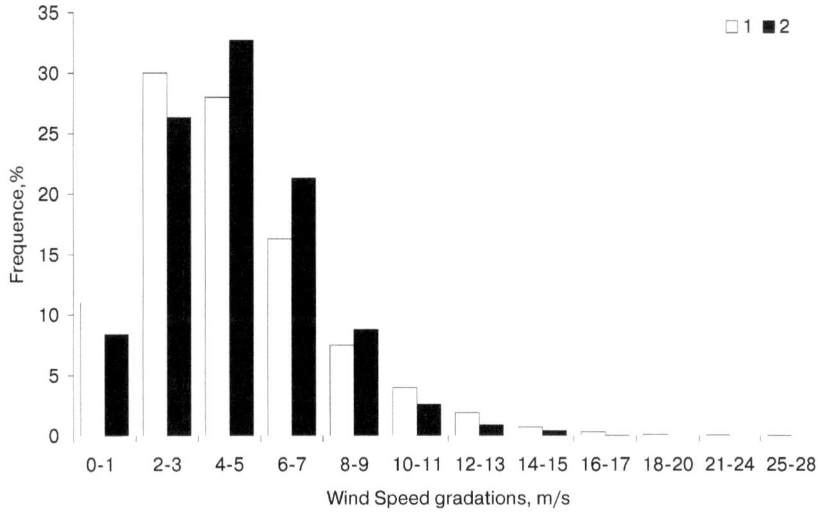

Figure 2.12. Distribution of wind speeds by gradation. 1 – January; 2 – July.

conditions in the Arctic are more typical of the summer period, the frequency of occurrence of these clouds significantly increases from the cold period to the warm one. Regardless of the time of the year, the frequency of occurrence of the given cloud form is higher in the regions remote from the main centres of atmospheric action. At some points it comprises up to 40% of all observations. In 70% of the cases, stratified clouds are observed at 8–10 tenths.

The second most widespread forms are stratocumulus clouds (Sc), which also form as a rule in a stable air mass but as a result of wave motions in inversion layers located in the lower troposphere. The frequency of occurrence of this form, although close to the frequency of occurrence of stratified clouds, is however, often higher – especially in the summer period. The minimum frequency of occurrence of Sc is noted in the cold period of the year (12–18%), but already in May their frequency of occurrence comprises 28–34% achieving 47% at the beginning of autumn. As a rule, in 60% of the cases, this cloud form is recorded at 8–10 tenths.

The height of the lower cloud boundary significantly varies depending on the season and region. In winter, the lower boundary of low-level clouds is located higher than in the other seasons. From December to February, the frequency of occurrence of the heights of the lower boundary in the Arctic basin less than 300 m comprises approximately 15–20% and up to 400 m – 30%. The heights up to 600–700 m comprise about 50% of all cases of location of the lower cloud boundary. In March–April, the frequency of occurrence of lower heights of the lower boundary of clouds increases and achieves 60–65% by summer for heights less than 300 m.

The heights of the lower cloud boundary up to 150–200 m comprise about 30% of all cases. From September, the height of the lower cloud boundary increases and their frequency of occurrence increases as well.

Table 2.3 contains the frequency of occurrence of different ranges of the lower cloud boundary height in the Arctic basin by months.

Table 2.3. Frequency of occurrence of different heights of the lower cloud boundary during a year (%).

Height (m)	Jan	Feb	Mar	Apr	May	Jun	Jul	Aug	Sep	Oct	Nov	Dec
<100	0	1	3	2	4	7	20	15	8	2	0	0
101–300	14	22	21	24	44	52	57	60	45	26	20	16
301–600	42	50	45	50	79	78	75	78	69	50	46	40
601–1,000	67	70	66	66	91	87	82	87	81	70	69	63

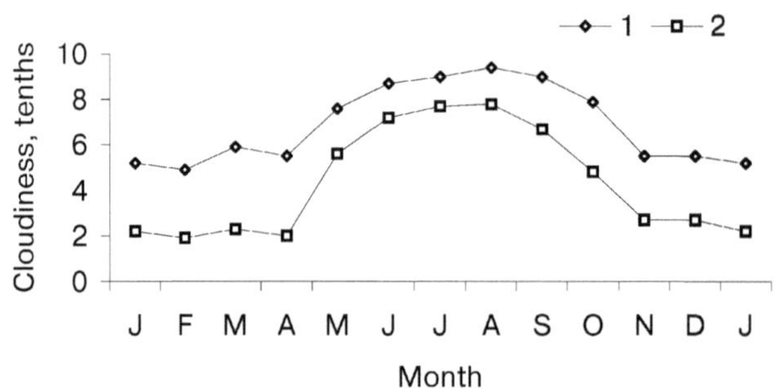

Figure 2.13. Total and lower cloud amount near the Pole. 1 – total; 2 – lower.

Yearly variations in cloudiness (Figure 2.13) are characterized in general by its greatest values during the period May–October and its least – from November–April. One should note that at transition from the winter regime to the summer one especially from April–May, a very intensive cloud cover increase is observed.

The amplitude of the yearly variations of cloudiness above the central part of the Arctic Ocean is 5 tenths.

A typical peculiarity of the spatial–temporal structure of the fields of average cloudiness is that during the period May–October, the largest values of the average total cloudiness are observed in the near-Pole area comprising 9–9.5 tenths. Such significant and stable cloudiness is connected both with intense cyclonic activity and with the impact of a cold ocean covered with melting ice above which the air masses rapidly transform contributing to the formation of frequent inversions, fogs and low stratified clouds. By winter, the amount of clouds above much of the Arctic Ocean dramatically decreases comprising 4–5 tenths in January. The least average cloudiness is observed at this time to the north of the East Siberian Sea, to which a tendency toward the anticyclonic circulation in the given region and also its remoteness from the sources of moistening contribute.

A special kind of cloudiness is noted in the Arctic, which belongs to the type of stratified clouds, the so-called cloud veil. It presents a thin whitish layer of condensation products forming beneath the inversion layer at the cold time of the year. The height of the lower veil boundary varies between 30–200 m. The cloud veil is a rather frequent phenomenon during the polar night period in the Arctic, especially above drifting ice.

To correctly take into account the medium-level clouds based on ground observation data is possible only in the absence of low-level clouds or other weather phenomena deteriorating visibility in the surface layer. In summer, study of medium-level clouds is difficult due to an exclusively high frequency of occurrence of fog and low clouds, and in winter – due to the polar night and snow storms. However, the results of observations of the Flying Meteorological Observatory (FMO) indicate a significant frequency of occurrence of medium-level clouds. Thus, of 183 flights in the summer–autumn period made higher than the upper boundary of low clouds, altocumulus clouds (Ac) and altostratus (As) clouds were noted in 96% of the cases. In spring, the frequency of occurrence of these cloud forms was only 35% (Burova and Voskresensky, 1962).

The convective cloudiness is not typical of the Arctic basin, only single cases of cumulus clouds were recorded, their frequency of occurrence being not greater than 2% in summer months and equaling 0–0.2% in winter. Of the other forms of clouds, one can note the nimbostratus clouds (Ns), their frequency of occurrence during the year varying from 1–7%.

2.1.6 Precipitation

In terms of the amount of precipitation received we arrive at about 150 mm for a year in the central area and up to 200 mm at its boundaries (Bryazgin, 1976; Bryazgin, 1983; Bryazgin, 1998; Bryazgin and Voskresensky, 2000) – the Arctic basin presents a polar desert (Korotkevich, 1972). Such insignificant amounts of precipitation fallout is related to the small moisture content of cold air masses at high latitudes. The moisture content of the atmosphere above the Arctic Ocean comprises 6.5 mm on average for a year, which is 4–5 times less compared with the other oceans (Burova, 2000). In general, the precipitation fallout is associated primarily with the atmospheric fronts and the vertical motions in the atmosphere. In this part of the Arctic, the intensity and frequency of manifestation of both phenomena is much less than in other World regions.

Liquid precipitation here falls out only in July–August and rarely in June and September (predominantly in the form of drizzle). Solid precipitation falls out all the year round, in winter there are mainly snowflakes and snow grains. The yearly variations of their monthly totals are shown in Figure 2.14. The mean multiyear precipitation total for a year is close to 150 mm. The maximum of the monthly precipitation total of 22 mm is observed in July and the minimum of 6 mm in April. From October–May, only solid precipitation is observed and from June–September – all forms of precipitation: liquid, solid and mixed. The contribution of liquid precipitation is estimated from data given in Figure 2.14 as a difference between their total amount and a sum of solid and mixed precipitation. In July, the contribution of liquid precipitation is at its maximum and exceeds 40% of the total amount for the month.

The so-called trace precipitation, forming at the surface and on different objects due to condensation or sublimation of water vapor and also deposition of water drops or crystals at fogs, are not recorded by the precipitation gauge. At well below

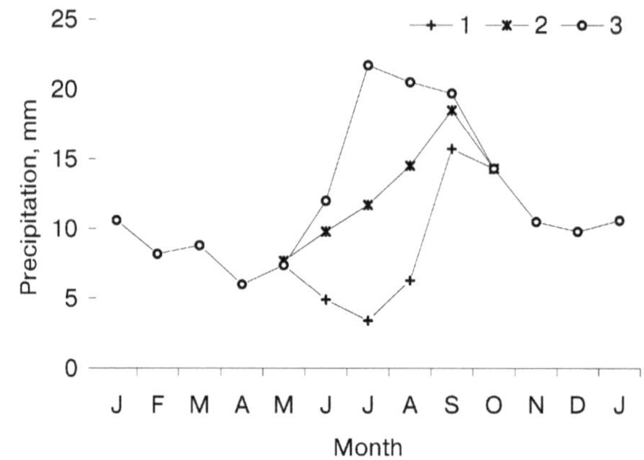

Figure 2.14. Yearly variations of precipitation amount. 1 – solid; 2 – liquid and mixed; 3 – total.

zero air temperatures, super-saturation over ice exists up to 120% in cases of a sufficient cooling of ice compared with the air temperature, mostly by nocturnal radiation (outgoing radiation). As a result, one observes a high frequency of occurrence of hoar-frost, rime, fog, ice needles and ice crystals formed by sublimation (i.e., the direct transition of water vapour to a crystal phase on ice surfaces and objects). According to the observation data of the drifting stations, in each of the months from November to April, the number of days with this indicated atmospheric phenomena varies from 20–31 days. The average frequency of occurrence of the number of days with trace precipitation over winter changes from 82–90%. The intensity of trace precipitation is low while the duration and frequency of occurrence of these phenomena is significant. As a result, their amount in total for a year comprises 20 mm. In summer, the fog drops depositing at the surface make an additional contribution (of 2.5 mm) to the increased water equivalent of the snow cover. Table 2.4 presents average variations of their amount within a year. In general for a year, they add 15%, on average, of the total precipitation measured by the precipitation gauge with up to 25% in some individual years (Bryazgin *et al.*, 1967; Dolgin, 1971).

Along with snow accumulation on the ice of the central Arctic basin, the process of snow evaporation also takes place. Evaporation has pronounced yearly varia-

Table 2.4. Average amount of trace precipitation in the Arctic basin (mm).

Type of precipitation	Jan	Feb	Mar	Apr	May	Jun	Jul	Aug	Sep	Oct	Nov	Dec	Oct–May	Jun–Sep	Year
Rime	0.3	0.2	0.2	0.3	0.4	1.2	1.8	3.0	4.2	3.0	0.9	0.5	5.8	10.2	16.0
Ice needles	0.6	0.7	0.3	0.2	0.1	0.0	0.0	0.0	0.0	0.2	0.3	0.4	2.8	0.0	2.8
Hoar-frost	0.2	0.2	0.2	0.1	0.1	0.0	0.0	0.0	0.0	0.1	0.1	0.2	1.2	0.0	1.2
Water drops	0.0	0.0	0.0	0.0	0.0	0.4	0.9	0.7	0.5	0.0	0.0	0.0	0.0	2.5	2.5
Total	*1.1*	*1.1*	*0.7*	*0.6*	*0.6*	*1.6*	*2.7*	*3.7*	*4.7*	*3.3*	*1.3*	*1.1*	*9.8*	*12.7*	*22.5*

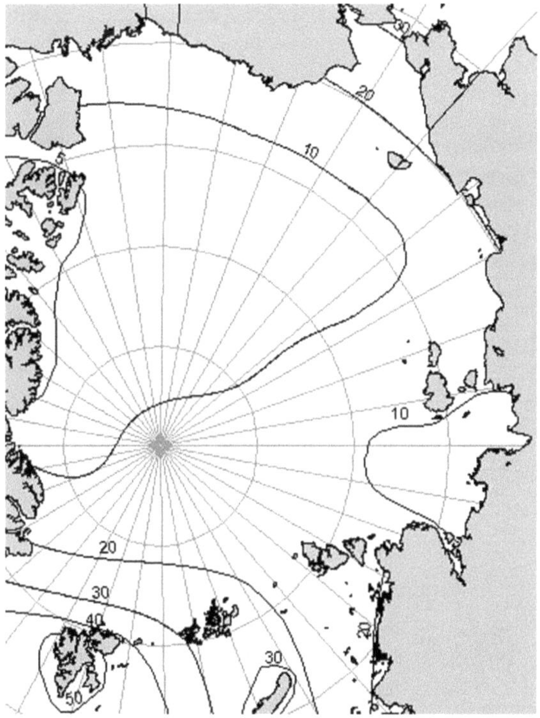

Figure 2.15. Precipitation totals in January in the Arctic basin (mm).

tions. From October–March, less than 2 mm of water equivalent evaporates. In spring, in clear sunny weather even at below zero air temperatures, the evaporation intensity from the snow cover surface significantly increases. The rate of this process depends primarily on the air moisture deficit. The most favourable conditions for this are realized in April–May. The average evaporation intensity comprises 0.012 mm/h, and the maximum comprises 0.035 mm/h in one equivalent according to data of observations at the NP-6 and NP-22 drifting stations. An assessment of the yearly snow thickness decrease due to its evaporation is close to 50 mm (Bryazgin, 1966; Bryazgin, 1996).

Spatial distribution of precipitation. The distribution of the monthly precipitation totals in the Arctic basin in summer and winter differs. In winter in the Arctic basin, the amount of monthly precipitation is only about 10 mm with the least precipitation (about 5 mm a month) falling out near the Canadian archipelago (Figure 2.15).

The summer distribution of precipitation is most close to being zonal. At this time, the largest amount of precipitation falls out during the year. In the Arctic basin in July–August, 20–25 mm falls out. A comparatively large precipitation amount for the central Arctic is connected with the increased moisture content of the Arctic air and a frequent exit of cyclones from the south. The spatial distribution of the yearly

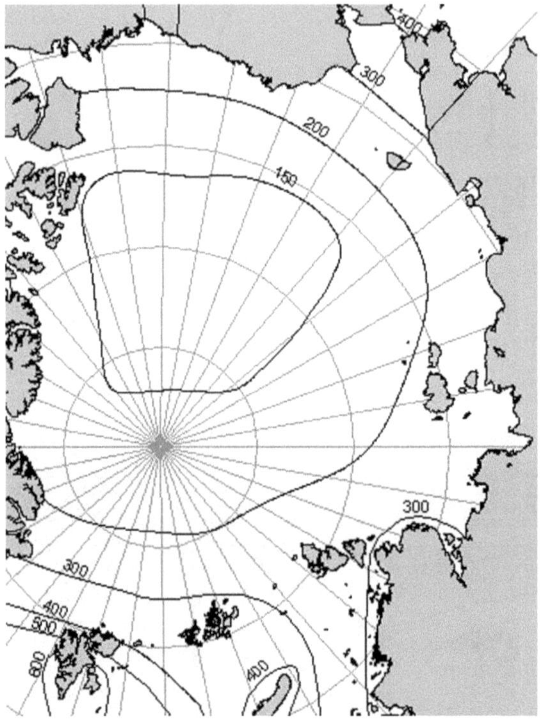

Figure 2.16. Yearly precipitation totals in the Arctic basin (mm).

precipitation total is similar to the winter distribution. Its least amount is noted in the Canadian sector of the Arctic basin (150 mm). In the Arctic Seas, the precipitation amount increases up to 200–300 mm (Figure 2.16).

The distribution of the yearly totals of solid precipitation lacks the latitudinal character of the distribution of isolines. The largest solid precipitation totals are noted in the sub-Arctic zone of Eurasia rather than in the Arctic basin. The smallest amount of solid precipitation for a year falls out in the Arctic Seas and in the near-Pole area comprising 80–120 mm. In the yearly variations of solid precipitation in the Arctic basin, the maximum is recorded in September–October and the minimum in July (see Figure 2.14) (Bryazgin, 1976).

The maximum monthly precipitation totals in the Arctic basin from measurement data in different years at the drifting stations are presented in Table 2.5.

2.1.7 Snow cover on drifting ice

The drifting ice surface in the Arctic basin on which the snow cover is established is characterized by a sufficiently large diversity of relief features of different sizes. Young ice has the most even surface. The multiyear ice floes are distinguished by a strongly developed relief: ice hillocks alternate with level segments of puddles frozen in autumn and isthmuses elevating between them. The ice floes are usually

Table 2.5. Maximum monthly amount of precipitation from drifting station data (mm).

Month	Jan	Feb	Mar	Apr	May	Jun	Jul	Aug	Sep	Oct	Nov	Dec
Monthly total	36	22	25	24	26	40	62	73	56	33	42	35
Observation year	1952	1956	1962	1963	1989	1989	1989	1958	1982	1969	1965	1965

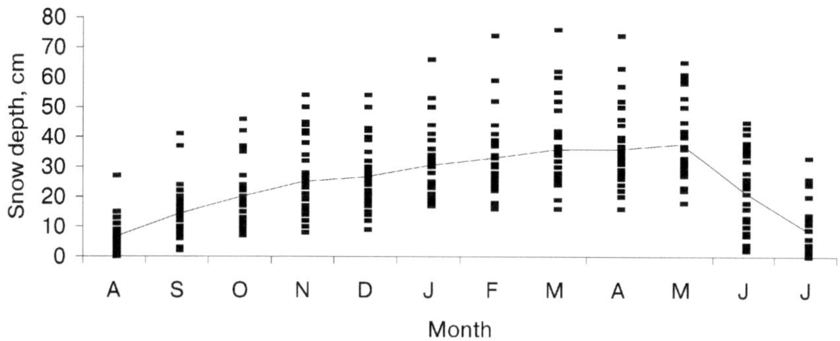

Figure 2.17. Snow cover depth at drifting stations in different years.

contoured by chaotic ice ridges. Due to relief irregularities, the snow cover beds non-uniformly. The snow layer on the ice irregularities in the form of gently sloping hillocks up to 40–50 cm in height repeats the ice relief shape. Wind blows away the fallen snow from the tops of higher ice hillocks and ice ridges and the ice surface becomes bare with snowdrifts forming between the ice ridges (Aleksandrov *et al.*, 2004; Radionov *et al.*, 1996).

The seasonal cycle of snow cover depths in different years at different stations and their mean multiyear values for each of the months are given in Figure 2.17. The range of variability of standard deviation values is within 4–20 cm.

In general, sufficiently rapid snow accumulation is recorded in early autumn, since the maximum snowfalls are practically always observed in September. In December and January, the snow accumulation rate is minimal. In subsequent months, the snow thickness slowly increases up to maximum values: in May with a mean multiyear thickness of snow cover comprising 35 cm. The most intense snow melting was from mid-June–mid-July, but its rates strongly differed from year to year.

In autumn, snow most rapidly accumulates on ice to the north of Greenland and the Canadian archipelago islands (Figure 2.18(a)). Snow accumulation in the Alaskan and Siberian sectors of the Arctic basin occurs more intensely during the next months compensating the autumn lag. By February, the snow cover thickness becomes sufficiently uniform over the basin territory and is close to 30 cm (Figure 2.18(b)).

Snow accumulation in spring occurs again more intensely to the north of

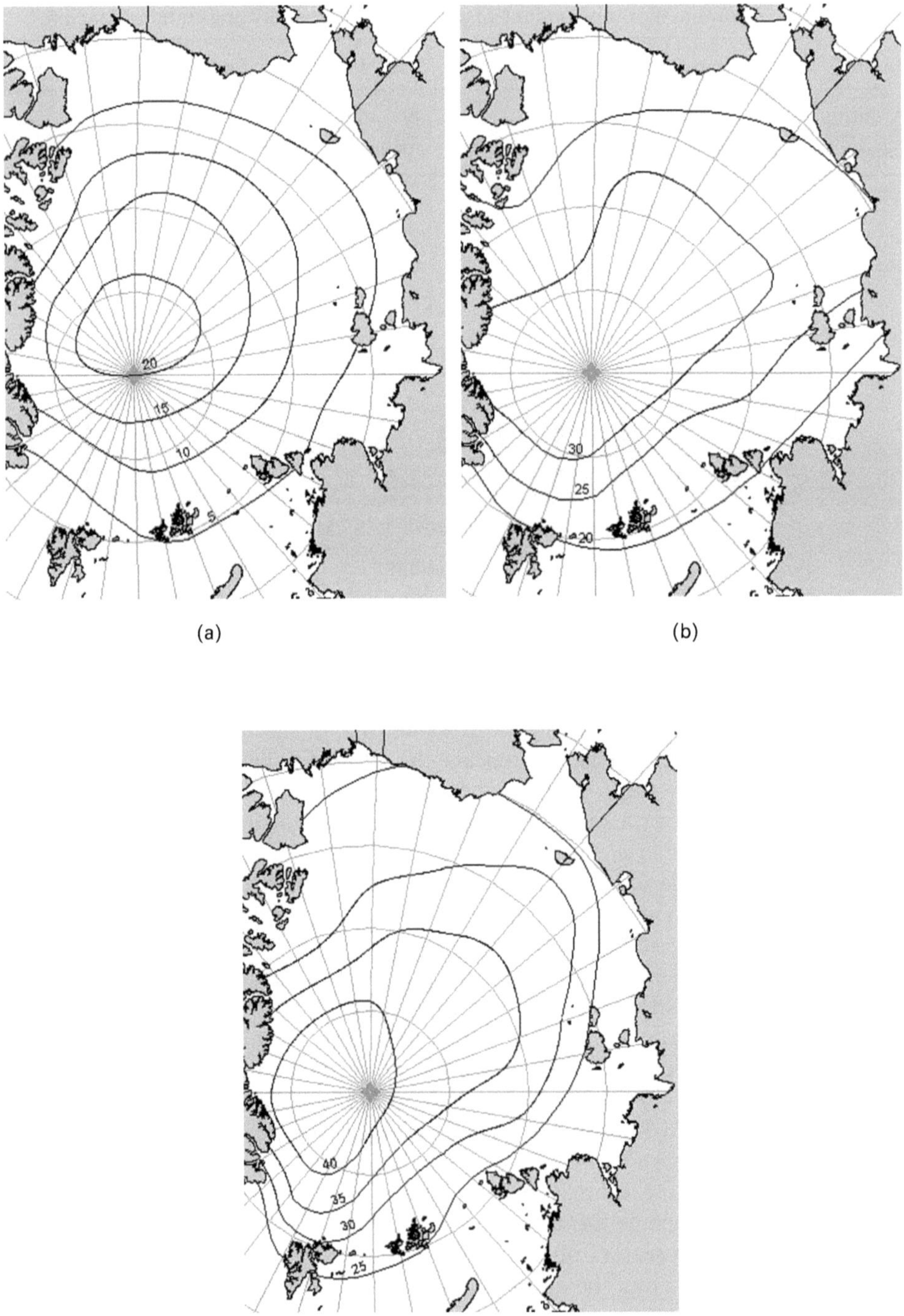

Figure 2.18. Snow cover depth on the drifting ice in (a) September, (b) February and (c) May.

Greenland and Canada. Figure 2.18(c) presents the main features of the geographical snow distribution during the maximum snow accumulation. The regional differences of the snow cover thickness are maximal in June due to a rapid decrease of its thickness in the regions to the north of the coast of Alaska and Siberia throughout May–June, while at high latitudes snow accumulation near Greenland continues during this period. In July, snow melting is observed everywhere. During the second half of July and practically all August, snow is absent over much of the Arctic basin, except for some years when it does not melt completely.

The snow depth in ridged ice is greater compared with level ice. Snow is transported here by wind and snow storms and accumulates between ice ridges. The snow depth depends on the height of ice ridges and the distances between them. On average, it is equal to (82 ± 20) cm. Near Greenland and in the central Arctic basin, the maximum snow cover depth in ice ridges at the end of winter is equal to 120–140 cm, whereas near the coast of Greenland, one observes snow depths up to 160–180 cm. In the other areas, it varies within 50–100 cm (Radionov *et al.*, 1996; Yakovlev, 1960; Warren *et al.*, 1999).

During the snow-line survey measurements, snow density is measured simultaneously with its depth. This makes it possible to calculate the water equivalent of the snow cover, which is extremely important for freshwater balance estimates both in the Arctic basin, and in the Arctic Ocean in general. The least density is noted for freshly fallen snow comprising 50–90 kg/m^3. In snow storms it is compacted to 170 kg/m^3. The maximum density is observed for melting snow and for snow saturated with meltwater comprising 500–550 kg/m^3.

The water content in snow or its water equivalent is calculated using simple multiplication of the snow depth by its density. Therefore, the pattern of the snow water content distribution throughout the year in the Arctic basin repeats in general the main peculiarities of the snow cover depth distribution. At the peak of the seasonal cycle in May, the average water content in snow over the Arctic basin comprises 11 g/cm^2 or a liquid water layer with a thickness of 11 cm.

Melting of snow in spring begins under the action of solar radiation already at the below zero air temperature. In the Arctic basin, melting of the upper snow layer occurs at -4°C if the total solar radiation fluxes at that time comprise 600 W/m^2. The daily rate of the snow cover disappearance (ablation) – melting evaporation, in the Arctic basin changes between 0.3–1.2 cm/day, and in June, July and in the first 10 days of August, it comprises 0.6–0.7 cm/day, on average (Radionov *et al.*, 1996).

The formation of stable snow cover begins again in the Arctic basin in the third 10-day period of August rapidly spreading southward. On 8 September, it is already established at 75°N and reaches the islands, while at the coast of the Arctic islands, the snow cover forms in the first 10 days of October. The snow cover decay begins from the south – at the coast of the Arctic Seas in the first 10-day period of July and only in the second 10-day period of July in the Arctic basin. The map of the distribution of snow deposition duration is presented in Figure 2.19.

Duration of snow deposition comprises 350 days in the Pole area and 300 days and less at the coast of the Arctic Seas. In 25% of the cases of year-round observations, snow on drifting ice did not melt completely (Radionov *et al.*, 1996).

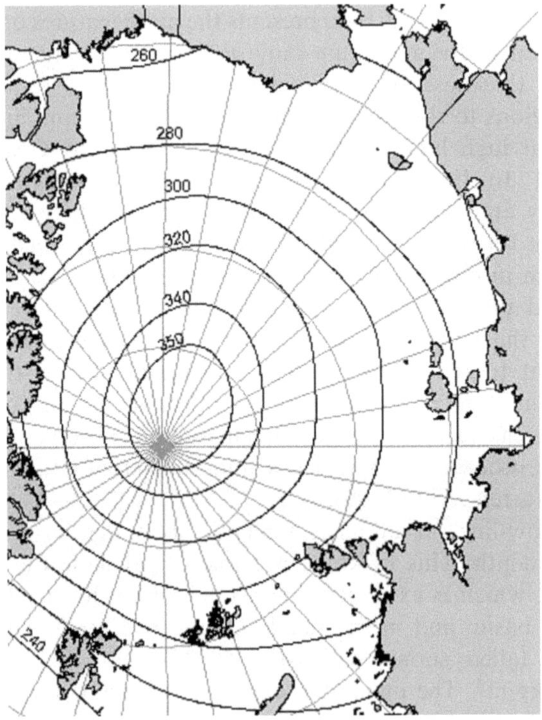

Figure 2.19. Number of days with snow cover.

2.2 SOLAR RADIATION MEASUREMENTS

First measurements of direct solar radiation in the Arctic were carried out by Westman (Westman, 1903) in 1899–1900 in Ttreurenberg (Spitsbergen) during the expedition measurements of an arc of meridian. In the Soviet Arctic, first expedition solar radiation observations were made by N. N. Kalitin on Novaya Zemlya and on Vaigach Island in 1920–1923.

Of great importance for investigating the radiation regime of the East Siberian Sea were observations of the Norwegian Polar Expedition on board the ship *Maud* in 1922–1925 (Mosby, 1932). Among the studies of this period, it is also necessary to note the work of V. S. Samoilenko in the Barents Sea in 1926 and of V. A. Berezkin on Wrangel Island in 1929 and 1934. Stationary observations of direct solar radiation at the Matochkin Shar Observatory began in 1928 and were continued in 1931–1933. Until 1932, solar radiation observations in the Arctic were predominantly conducted in summer at the time of navigation.

Complex solar radiation observations were only organized during the period of the Second International Polar Year (1932/1933) in Tikhaya Bay, at Matochkin Shar, in Tiksi Bay, at Cape Dezhnev and in Lawrence Bay. The same year, V. V. Shuleikin on board the ice-breaking ship *Taimyr* in the Kara Sea carried out investigations of the heat balance components of the sea devoting a great deal of attention to radiation measurements. In 1934/1935, solar radiation observations were made at Uyedineniya Island and in 1935 a polar station in Tiksi Bay was opened. In 1936, there were already six operating stations (Tikhaya Bay, Tiksi Bay, Cape Cheluskin, Cape Schmidt, Cape Zhelaniya and Uyedineniya Island).

Solar radiation measurements on ice in the Arctic basin were made for the first time in the area of the Relative Inaccessibility Pole in April 1941 (Chernigovsky, 1946). From 1950, systematic solar radiation observations began at the drifting stations in the central Arctic. One of the first generalizations of the observation data on the net radiation components in the Arctic, including observations at the first drifting stations was made in the monograph of Chernigovsky and Marshunova (1965).

2.2.1 Methods and instruments

All types of observations and registration and their processing and control were performed in compliance with the 'Manual' (*The Manual*..., 1971) and the AARI methodological directions and instructions. A set of standard solar radiation observations included measurements at fixed intervals of the fluxes of direct, diffuse and global radiation, reflected radiation or albedo and net radiation of the underlying surface. All measurements of the radiation fluxes were accompanied by the determination of solar elevation and detailed meteorological characteristics of the observation conditions. Measurements of the radiation fluxes were taken from four to eight times daily, measurements were always taken at noon and midnight local solar time, with the remaining measurements distributed evenly between noon and midnight. In addition, global radiation was measured continuously at all drifting stations and, beginning in 1965, the diffuse and reflected radiation and net radiation were also measured. As a result, daily variations of the radiation regime parameters were obtained in the form of hourly, daily, monthly and yearly sums of the radiation characteristics. In the cases of absence of the possibilities of continuous registration, the daily totals of the corresponding radiation elements were calculated from the results of observations at fixed times by linear interpolation for the whole hours between the measurement times.

The radiation observation sites were chosen so that the underlying surface was characteristic of most of the ice floe and so that the sites would not be flooded during the summer snowmelt.

To investigate the underlying surface reflectivity, observations of albedo were carried out one to two times each month and in certain years the net radiation was also sampled in the vicinity of the stations over a spatial grid. During the summer melt, the number of these spatial sampling observations was increased to five or six times per month.

The following standard thermoelectric instruments were used to obtain the radiation values: AT-50 actinometer, M-80 pyranometers (or GP 3×3 pyranometer sensor heads) and M-10 balance meters. GSA-1 galvanometers were mainly used to record the signal voltages during observations at fixed time intervals. In the last years of the operations of drifting stations, PP-63-type portable recorders were also used.

Until 1965, the data recording was carried out using disk galvanometers or self-recording millivoltmeters. In 1959, the EPP-09 multi-channel electronic potentiometer was used for the first time to record global and diffuse radiation. After 1965 the potentiometer was adapted for continuous recording of the global, diffuse and reflected radiation as well as net radiation.

In 1970 significant modifications in standard solar radiation measurements were made to allow continuous automatic recording of all net radiation elements (except for direct solar radiation) without observations at fixed hours. Because of the lack of a suitable guidance system to point an instrument at the Sun's disk, the direct radiation was recorded for a few minutes at defined intervals four times a day. The transfer to continuous recording was accomplished following a series of parallel measurements of solar radiation elements in the recording mode and at the specific intervals and special methodological studies. This made it possible to choose the appropriate method for processing continuous registration records (tape of self-recorders) and the presentation format of the observation materials, and to provide homogeneity of the series of observations. To check the calibration of the recorded data, comparative observations were made using supplementary instruments.

All thermoelectric sensors were calibrated at the Central Calibration Bureau of the Soviet Hydrometeorological Service, located at that time in Leningrad, before and after deployment at the drifting stations. During the operation at the drifting stations, the sensors used for registration and the sensors for observations at fixed hours and comparative observations were calibrated periodically against a standard actinometer during the day or a net radiation meter at night. This procedure of calibration and data control ensured the reliability of all the components of short-wave radiation: direct radiation fluxes to the perpendicular and horizontal surfaces and scattered and global radiation fluxes. The measured long-wave radiation and the net radiation values were less reliable because of significant systematic errors (up to 30%) in the measurements made with the net radiation meters.

Generalized results of the measurements of radiation regime characteristics in the central Arctic basin obtained from data of solar radiation observations from 1950–1991 at the drifting stations beginning from NP-2 and until the end of operation of NP-31 are presented in Marshunova and Mishin (1994).

2.2.2 Conditions of the formation of radiation regime components

The first summaries of the observation data obtained at the drifting stations showed that the radiation conditions in the Arctic basin were characterized by distinctive features compared with all other regions on Earth. They result from the underlying surface homogeneity and are characterized by high albedo values throughout the year, high transparency of the atmosphere, structure of the atmosphere and cloudiness.

The polar day and polar night phenomena in the region under consideration govern the non-uniform incoming solar radiation throughout the year. The moment of the upper Sun's disk edge appearance (disappearance behind the horizon) is

Table 2.6. Dates of the start and end of the polar day and polar night.

	Polar day		Polar night	
Latitude (°N)	Start	End	Start	End
68	27 May	17 July	9 December	4 January
70	17 May	27 July	26 November	17 January
72	9 May	5 August	16 November	26 January
74	2 May	12 August	9 November	2 February
76	25 April	18 August	3 November	9 February
78	19 April	24 August	27 October	15 February
80	14 April	30 August	22 October	21 February
82	8 April	4 September	16 October	26 February
84	3 April	9 September	11 October	3 March
86	29 March	15 September	6 October	8 March
88	24 March	20 September	30 September	13 March
90	19 March	25 September	25 September	19 March

assumed to be the sunrise (sunset) in meteorology. Table 2.6 contains the dates of the beginning and end of the polar day and night (i.e., solar elevation $H_{\text{Sun}} \leq -50'$ at noon and midnight). It was taken into account that the angular radius of the Sun was $16'$ and regular refraction was $34'$ (Marshunova and Mishin, 1994).

Due to refraction, the polar day duration is slightly greater compared with the polar night. During the polar day, the noontime elevation of the Sun decreases with increasing latitude, but its elevation at midnight increases. As a result, the daily mean elevation of the Sun at different latitudes within the Arctic remains about the same. This governs the essential effect on the geographical distribution of daily and monthly radiation totals.

The character of the radiation processes depends to a great extent on the conditions of cloudiness and the transparency of the atmosphere, their variations being substantially determined by atmospheric circulation. A typical feature of the spatial–temporal structure of the medium-level cloud fields in the entire Arctic is that the highest average total cloudiness values are observed during the period May–October above the Arctic Ocean (more than 8 tenths), comprising 9–9.5 tenths in the near-Pole area. Only approaching the coastline does the cloud amount decreases to 7 tenths. Such significant and stable cloudiness is related both to the intense cyclonic activity and the impact of the cold ocean covered with melting ice above which the air masses rapidly transform contributing to the formation of frequent inversions, fogs and low-level stratus clouds.

By winter, the amount of clouds above much of the Arctic Ocean sharply decreases and comprises 4–5 tenths in January. The least average cloudiness is observed at this time to the north of the East Siberian Sea, to which anticyclonic circulation in the region and its remoteness from sources of moisture contribute predominantly. The area of maximum cloudiness (about 9 tenths) is displaced from the near-Pole area to the southern Barents Sea.

Small moisture content and relatively low levels of aerosol pollution govern the high transparency of the atmosphere in the Arctic.

2.2.3 Intra-annual variability of the radiation regime components

All these specific features are manifested in the character of the intra-annual variability of the radiation regime components: direct (S), diffuse (D), global (Q) and reflected (R) solar radiation. Their variations determine in turn the variability of net radiation (B) (Marshunova, 1983; Marshunova and Mishin, 1994; Marshunova and Chernigovsky, 1971; Radionov, 1997; Smetannikova, 1983; Chernigovsky and Marshunova, 1965).

Variations of the radiation heat fluxes in the Arctic throughout the year are primarily related to the presence of the polar day and night (Marshunova, 1983; Timerev, 1981; Chernigovsky and Marshunova, 1965).

During the light period, the incoming solar radiation $Q = S' + D$ to the surface is determined by the atmosphere transparency and the characteristics of cloudiness. The transparency variations are mainly influenced by the aerosol extinction and the solar radiation absorption by water vapor. They are small in the Arctic compared to the other regions of Earth and their maximums are observed during different periods of the light time. The maximum of aerosol turbidity of the Arctic atmosphere is recorded in late March and of the water vapor content in July (Marshunova and Mishin, 1994). As a result, the direct solar radiation fluxes to horizontal surface S' in the Arctic are 20% greater than similar fluxes at temperate latitudes at the same solar elevations.

The diffuse radiation flux values depend on the elevation of the Sun, amount and form of clouds, transparency of the atmosphere and albedo of the underlying surface (A). The cloudiness in the Arctic is maximum in the summer while the cloud conditions in the winter strongly vary from one part of the Arctic to another. The highest diffuse radiation values are observed at the medium and low-level clouds above the snow covered surface.

Snow and ice are the prevailing types of surface in the Arctic much of the year. At the time of stable snow cover from October–May, snow cover albedo is equal to 80–85% and there is little change from year-to-year over the entire Arctic. On drifting ice during the period of melting, the albedo values change over a wide range: from 10–70% (Radionov *et al.*, 1996).

A combined action of variations of astronomic factors, cloud conditions, characteristics of the atmosphere and the underlying surface albedo result in significant differences in the values and intra-annual variations of the net radiation components (Marshunova and Radionov, 1995; Radionov *et al.*, 1996). Figure 2.20 presents mean monthly values of monthly totals in the Arctic basin at latitude 80°.

In the yearly variations of the monthly totals of direct radiation, the maximum can be observed not only in June, when the duration of the day and the solar elevation are largest, but also in other months (May, July) depending on the cloud conditions.

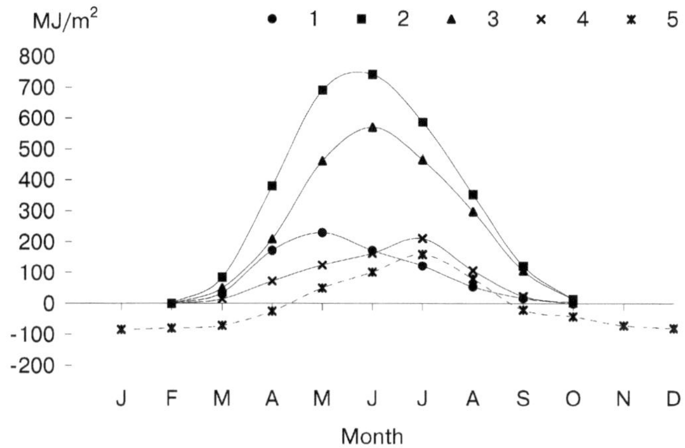

Figure 2.20. Monthly totals of direct (1), global (2), diffuse (3) and absorbed (4) radiation and the net radiation (5) at 80°N.

2.2.4 Direct solar radiation

From November to January the direct solar radiation is not observed to the north of 70°N. From late February when the polar night boundary moves to 80°N, the incoming solar radiation in the Arctic basin begins to increase.

In March with increasing day duration, the monthly total of direct solar radiation increases to 50 MJ/m^2 near the circle of latitude 75°N. In April, a sharp increase of day duration occurs. There is the increased number of clear days and the monthly total of direct solar radiation increases to 150 MJ/m^2. Beginning in May, the number of fogs and low-level stratified clouds increases over much of the Arctic. Compared with April, the radiation above the ocean does not significantly increase.

In June and July, the spatial distribution of the monthly totals of direct radiation to the north of 70°N corresponds to the distribution of clouds. Especially significant cloudiness is observed in the Arctic basin. As a result, in these months similar to April, the monthly totals in the central Arctic comprise only about 150 MJ/m^2. In August, the direct radiation decreases to less than 100 MJ/m^2.

In September, solar radiation continues to decrease due to the reduced duration of the day and elevation of the Sun. To the north of 80°N the polar has already ended.

In October to the north of 85°N, the polar night begins. The isoline 25 MJ/m^2 passes along the Arctic Circle latitude. In November, the polar night reaches 70°N, while in December no solar radiation is observed to the north of the Arctic Circle.

In general over the year, the direct solar radiation to the Arctic Basin comprises 800–1200 MJ/m^2.

2.2.5 Global radiation

It was believed practically until the early 1960s that the monthly radiation totals decrease toward the Pole during the entire light time. In reality, generalization and analysis of the results of solar radiation observations at the drifting stations showed that this did not occur during the polar day. Only in spring and autumn does the distribution of global radiation have a latitudinal character.

The distribution of the monthly global radiation values in spring and autumn, similar to direct radiation, is mainly determined by astronomical factors: with increasing latitude the radiation decreases. During the polar day when the mean daily solar elevation does not depend on the latitude, the radiation field is specified by the cloud cover and the underlying surface properties (Marshunova and Mishin, 1994; Chernigovsky and Marshunova, 1965).

In February–March and September–October, the distribution of the radiation values has an almost regular latitudinal character. In February, the polar night in the southern Arctic already comes to an end. The global radiation changes from 0 MJ/m^2 at 78°N to 50 MJ/m^2 at the Arctic Circle's latitude. The monthly radiation total in February comprises 0–2% of the yearly total being very close to the monthly values of possible global radiation as the cloud cover is at its minimum in February.

In March, with increasing solar elevation and the length of the day, a rapid increase of global radiation occurs: its totals are 5–10 times greater compared with February. The latitudinal distribution of isolines is preserved: from 50 MJ/m^2 in the near-Pole area to 250 MJ/m^2 at the Arctic Circle, with a small decrease towards the Barents and the Norwegian Seas. The cloud cover over much of the Arctic is small (4 to 6 tenths) and the monthly values of global radiation are as close as is possible to values of global radiation in cases of cloudless conditions. The radiation in March is 4–8% of the yearly total.

In April, the radiation continues to increase, this increase being particularly strong in the northernmost regions where the polar day has already begun. In the near-Pole region, the monthly total comprises 350–400 MJ/m^2 instead of 50 MJ/m^2, recorded in March (i.e., the radiation increases 7–8-fold). In more southern areas, it increases only 2–2.5-fold. The clear latitudinal radiation variations are significantly broken. The isolines elongate towards the regions with a larger cloud cover – the Barents and the Norwegian Seas.

From May to August, the location of global radiation isolines in the Arctic does not have pronounced latitudinal variations, since the influence of astronomical factors during the polar day is smoothed. The formation of individual regions of decreased and increased radiation depends on the distribution of cloud cover and ice.

In May, the global radiation increases 1.5-fold compared with April and becomes higher than at temperate latitudes. The global radiation fraction over May in the yearly incoming radiation increases to 20%.

June is a month of maximum incoming radiation. The increase of global radiation in the Arctic Basin occurs due to the increased fraction of diffuse radiation at significant cloud cover (8–8.5 tenths) and high albedo. In June, the incoming radiation amount is similar to May comprising 20% of the yearly total.

In July, with decreasing solar elevation, the incoming radiation decreases 1.2–1.4-fold compared with June. The contribution of July to the yearly incoming radiation decreases to 17–19%. In August, the radiation value decreases 1.5–1.7-fold compared with July. A reconstruction of the latitudinal character of the spatial distribution of radiation is outlined again. The incoming radiation in August comprises 10–12% of the yearly value.

In September, the distribution of global radiation in the Arctic again attains almost regular latitudinal variations changing from 80 MJ/m^2 at the Pole to 200 MJ/m^2 at the Arctic Circle's latitude. The fraction of September contribution to the yearly incoming radiation comprises only 4–5%.

In mid-October, the polar night has already begun to the north of 82°N. At the Arctic Circle's latitude, the radiation comprises 100 MJ/m^2.

In the yearly incoming global radiation, the Arctic basin is especially distinguished by its high values in spite of the large cloud cover presence particularly in summer (the average cloud amount for April–September is 8.4 tenths). The yearly incoming global radiation in the Arctic basin is also high, being equal to more than 3,000 MJ/m^2, like at the coast of the East Siberian and the Chukchi Seas, and comprises 70% of the possible radiation values. High global radiation values in the Arctic basin are governed by the increased contribution of diffuse radiation to global radiation due to a large reflectivity of the underlying surface throughout the year (the average albedo for a year is 78%). This is connected with the fact that the reflected radiation diffuses in the atmosphere again. The diffuse radiation increases even more significantly in the presence of cloud cover.

2.2.6 Net radiation

Before the beginning of direct measurements of net radiation components at the drifting stations it was believed that the net radiation of the snow–ice surface of the Arctic basin was positive (Yakovlev, 1957). Of a large set of the factors influencing the net radiation variability, it is mainly regulated by cloud cover and underlying surface properties. In the light time, the influence of absorbed radiation increases, its value and variability being determined by the global radiation values and predominantly by the underlying surface albedo.

From October–March, during the polar night when the incident global radiation is small, the net radiation is determined only by the outgoing radiation, which depends on the stratification of the atmosphere and the temperature of the underlying surface. In these months, its average value is negative changing from -50 to -90 MJ/m^2. In April and September, the net radiation is close to zero being positive only for four months (from May to August).

The yearly net radiation of the surface in the Arctic basin changes within a wide range: from the values close to zero at the boundary of multiyear ice to -140 MJ/m^2 in the area of the Pole. A zero isoline coincides with the boundary of multiyear ice.

Multiyear observations at the drifting stations revealed that at an average albedo value for a year of more than 70%, the yearly net radiation is always negative no matter how high the global radiation was. Only in some years, when the albedo in July–August on some surface segments decreased to 50–60%, can the net radiation for a year be close to or slightly higher than zero (Marshunova and Radionov, 1995). The lowest yearly net radiation values were obtained at the NP-6 station in 1957. ($B = -340$ MJ/m^2) and NP-16 station in 1970 ($B = -335$ MJ/m^2), when a very insignificant cloud cover was observed throughout the winter with a practically continuous cloudiness throughout the summer.

2.3 ALBEDO

The Earth's surface albedo is one of the important climate-forming factors. Its value is determined by the ratio of the absorbed to reflected solar radiation. In that way, the albedo directly influences the formation of the radiation regime of the Earth–atmosphere system. The snow cover albedo depends on the structure and state of the snow cover, the degree of its contamination and the snow density and moisture content. All these characteristics change continuously under the impact of meteorological conditions and solar radiation, resulting in a large variety of the types of snow surface and snow cover relief and, hence, in significant albedo variations.

In principle, the integral albedo variations of snow can also be related to the illumination conditions of clear or overcast skies and to changes in the spectral composition of incident solar radiation depending on the elevation of the Sun.

Measurements of the integral snow cover albedo were carried out at all drifting stations beginning from NP-2, as one of the standard types of meteorological observations. The results of the analysis and generalization of data of these measurements were partly published in Marshunova and Mishin (1994), Timerev (1976) and Chernigovsky and Marshunova (1965).

2.3.1 Methods and instruments

The snow cover albedo values were calculated from the results of measurements of the global and reflected solar radiation at the meteorological site using a universal M-80 pyranometer (Yanishevsky, 1957). The receiving surface of the pyranometer was located at a height of 1.5 m from the snow cover surface. The spectral range of recorded solar radiation was 0. 4–4.0 μm.

The absolute error of single measurements of solar radiation fluxes is not greater than 7 W/m^2. In this case, the relative error of a single albedo measurement under the real conditions was not more than ±5%.

A pyranometer is a relative instrument. Its calibration is made by the Yanishevsky AT-50 actinometer (Yanishevsky, 1957). In the process of measurements, the control of the sensitivity of the instruments was performed on a monthly basis by the control actinometer used only for this purpose. If the sensitivity change of the pyranometer between two successive calibrations was not more than 5%, then the conversion coefficient for calculating the recorded solar radiation by the measured electrical signal value was assumed to be constant during the month until the next calibration.

A careful selection and preparation of the instruments before the expeditions, and their accurate handling, ensured their unchanged characteristics for several years. Under extraordinary conditions when the instrument sensitivity changed by more than 5% between the two intervals of control measurements, the instrument was replaced by a reserve. These measures provided the necessary accuracy of measurements and uniformity of the observation series.

2.3.2 Integral albedo

The state of the active surface of the site where the observations were made was described verbally by visual assessment. The limits of the variability of albedo for different types of snow–ice surfaces are presented in Table 2.7. This snow-free period in the Arctic basin is very short, so the mean monthly albedo is less than 80% only from July to August (Radionov *et al.*, 1996).

The snow cover structure was determined by its age (fresh, old), origin (fallen out in the form of precipitation or by wind transport) and the degree of contamination (clean, contaminated, dirty). The site coverage by snow was characterized by two gradations: completely snow covered and snow covered in places. The type of snow cover was determined thus: dry, moist, with a snow crust (thin crust of ice over snow) and with an ice crust.

The values characterizing the albedo of different snow surface types are presented in Table 2.8.

Table 2.7. The albedo of different surfaces in the Arctic basin.

Surface	Albedo (%)
Fresh dry snow	80–90
Melting snow	60–70
Dry ice	50–60
Puddle	30–40
Water	10–20

Table 2.8. Albedo of different types of snow surface (%).

Snow state	Moisture content and color	Avg.	Max.	Min.
Fresh fallen	Dry, dazzling white, clean	88	98	72
	Moist, dazzling white	80	85	80
Fresh blown, weakly compacted	Dry, clean	85	96	70
	Moist	77	81	59
Fresh fallen or blown 2–5 days ago	Dry, clean	80	86	75
	Moist, grey–white	75	80	56
Compacted	Dry, clean	77	80	66
	Moist, grey–white	70	75	61
Recrystallized	Moist	63	75	52
Snow and ice	Dry, grey–white	65	70	58
Water saturated	Light green	35	–	28

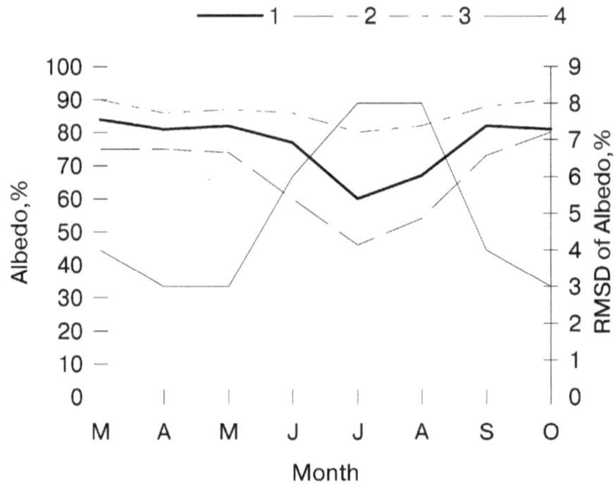

Figure 2.21. Yearly variations of albedo at the drifting stations. 1 – mean monthly; 2 – mean monthly minimal; 3 – mean monthly maximal; 4 – standard deviation.

As can be seen from the values presented in Table 2.8, the snow surface albedo varies over a very wide range depending on the snow structure, age and degree of moistening.

It is natural that a visual description of the snow surface state is very subjective. To assess the changes of physical snow parameters on this basis is impossible. This makes it rather difficult to reveal the causes of variability of snow reflectivity from the analysis of earlier made standard measurements.

The yearly variations of mean monthly albedo values, bounds of their variability and standard deviations derived by averaging for the entire set of observations at all drifting stations over the period 1950–1991 are shown in Figure 2.21 (Radionov *et al.*, 1996).

In general from October–April, the mean multiyear albedo of snow on ice of the Arctic basin changes from 80–86%. The main albedo changes occur during the periods of intense decay in late May–June and establishment of the new snow cover in September.

The maximum albedo values of 96–98% recorded at specific intervals are typical of freshly fallen snow. During the period of intense snow melting, especially when it is saturated with water, the minimum albedo values are close to 30%.

Dependence of the integral albedo on the illumination conditions in the cloud cover presence or absence is presented in Figure 2.22. As can be seen in this figure, snow surface albedo for an overcast sky is 3–6% greater than for a clear sky. A careful data analysis also showed that under the continuous cloud conditions, the albedo values measured at the low-level clouds are systematically higher by 2–3% than the albedo measured at the medium-level clouds (Timerev, 1976). In general taking into account the accuracy of albedo measurements of ±4–5%, it can be concluded that the influence of clouds on mean multiyear albedo values is insignificant and is not greater on average than the accuracy of the observations.

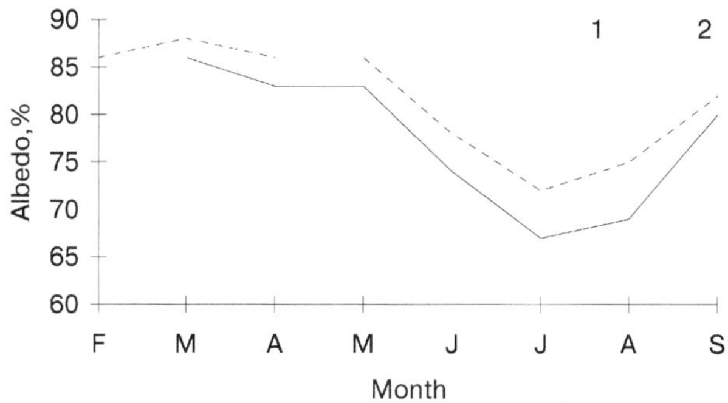

Figure 2.22. Mean multiyear albedo values at clear (1) and overcast (2) sky.
From data in Timerev (1976).

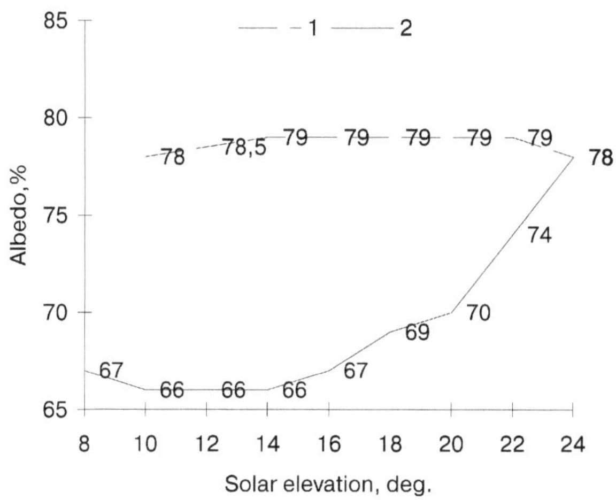

Figure 2.23. The dependence of albedo on the Sun's height at the beginning of snow cover melting (NP-4, May 1957). 1 – before noon; 2 – after noon.

The diurnal variations of the snow cover albedo (i.e., a dependence of the albedo value on the solar elevation) was typically manifested according to data of observations at specific intervals on clear days in spring. The albedo maximum was more frequently recorded at noon. As noted above, the reflectivity of the snow surface depends on its structure and type and the snow cover relief at the meteorological site. These parameters are determined by the preceding and current meteorological conditions. In spring, clear days prevail in the Arctic and they are characterized by large incident solar radiation fluxes to the snow covered surface. At well below zero air temperatures, melting of the upper snow layer and water film formation occur in the afternoon hours resulting in a significant albedo decrease. The results of observations at NP-4 drifting station in May 1955 (Chernigovsky and Marshunova, 1965) are presented in Figure 2.23.

The amplitude of albedo changes during the day comprised 13%. The same character of diurnal albedo variations occurred in May 1957 at NP-7 and at NP-22 in June 1979. At NP-22, the difference between the maximum and minimum albedo was equal to 16% (Radionov *et al.*, 1996).

During overcast weather, the diurnal albedo variations do not have as a rule a pronounced character, while its changes throughout the day are minimal. However, before the start of intensive snowmelt in May, the character of diurnal variations was the same as on clear days, while their amplitude did not exceed 6% (Chernigovsky and Marshunova, 1965). The mechanism of this phenomenon remains the same as on clear days: a thawing and watered snow surface layer.

In some cases, the albedo maximum was recorded in the morning and evening hours. The ice crust forming on the snow surface that thawed in the daytime due to radiation cooling at decreasing solar elevation, smoothes the roughness. It possesses a pronounced mirror effect at relatively large angles of incidence of solar rays, leading to an increased albedo in the morning and evening.

Averaging over long time intervals, these effects are smoothed. As a result it can be concluded that the integral albedo values averaged for a large set of observations change little at a solar elevation height of 10–35°. The observed variability is not greater than the accuracy of standard measurements of ±5%.

In summer months in the central Arctic basin there is a large variety of underlying surfaces: from dry fallen snow to open water. In this regard a question about the representativity of standard albedo observations at one point for sufficiently large areas and the entire Arctic region arises. The albedo values measured above the relatively dry areas at the permanent sites turn out to be overestimated. This is confirmed by the results of albedo measurements made in June–August 1958 at the NP-6 station by albedometers mounted at heights of 1.5 m and 12.0 m. During snow cover melting, the albedo values measured by the albedometer at a height of 12 m were 10–12% less compared with those measured at a standard level. This is explained by the fact that a greater active surface area, including puddles and ice areas, is covered from a large height. Beneath a standard albedometer, there was snow (sometimes with bare ice areas). Before and after the start of snowmelt when fresh snow falls out, the albedo at a height of 12 m is less, by 3–5%, than under a standard albedometer. More correct results in these situations can be obtained using data of traverse measurements of albedo of different surfaces taking into account the area they occupy. The materials of such observations are summarized in Chernigovsky and Marshunova (1965) and Marshunova and Chernigovsky (1971). The results of calculations of the average albedo values by area, taking into account the areas of the puddles, for June–August according to data from (Chernigovsky and Marshunova, 1965) are contained in Table 2.9.

No significant differences in the zonal distribution of albedo values in the Arctic basin were detected.

2.3.3 Spectral albedo

The spectral albedo A_λ is calculated as a ratio of the values of semi-spherical fluxes of the reflected and incident solar radiation in sufficiently narrow wavelength

Table 2.9. Mean albedo values in the central Arctic (%).

Latitude (°N)	June	July	August
90	78	62	67
85	78	60	67
80	78	58	67
75	73	57	67

intervals. A relative error of the spectral albedo measurements in Kondratyev (1981) is estimated as 6–8%.

The only measurements of the spectral albedo proper A_λ at the drifting stations were made at NP-22 in June and August 1979 at different states of the underlying surface, illumination conditions and cloudiness (Radionov *et al.*, 1996). On clear days, the measurements were conducted at the solar elevations of 10–37.5° and at continuous cloud cover at noontime.

The spectral albedo was calculated from the results of measurements of the ascending and descending radiation fluxes by a K-3 spectrometer at the wavelength range of 0.4–0.9 μm with a spectral resolution of 20 nm (Radionov *et al.*, 1996).

A plane burnished opal quartz glass was an integrating element. During measurement of semi-spherical radiation fluxes and correspondingly, the albedo calculation, the integrating properties of the receiving site, and in particular its angular characteristic, are of decisive importance. The laboratory studies of the angular characteristics of the plane opal glass used during the measurements showed a significant deviation from the cosine law in cases of solar elevations of 10° to 20°. The measured angular characteristics were taken into account during processing of the results of measurements of spectral radiation fluxes.

The measurements were made at a specially selected site located at a distance of not less than 500 m from the station structures. This was one of the most clean places at the station – little subjected to the daily activity of personnel and technical vehicles.

Figure 2.24 presents spectral variations of average A_λ values at different types of snow surface.

In the spectral area of 0 5–0.7 μm, the variations of A_λ are practically neutral. In the 'blue' and the IR regions of the spectrum, the albedo decreases were observed for melting snow and firn. For dry snow, the decrease of A_λ is noticeable only in the IR wavelength range.

Table 2.10 presents the values of A_λ of freshly blown snow measured on a clear day at different solar elevations. The spectral albedo increased with increasing solar elevation and comprised the maximum value at midday at all wavelengths. This increase had, however, a monotonic character only in the red and IR regions of the spectrum. At wavelengths shorter than 0.6 μm, the increase was not monotonic. The two bottom lines of Table 2.10 demonstrate values measured after noontime. The afternoon albedo decrease was connected not only with solar elevation decreasing but with snow surface melting and water film formation.

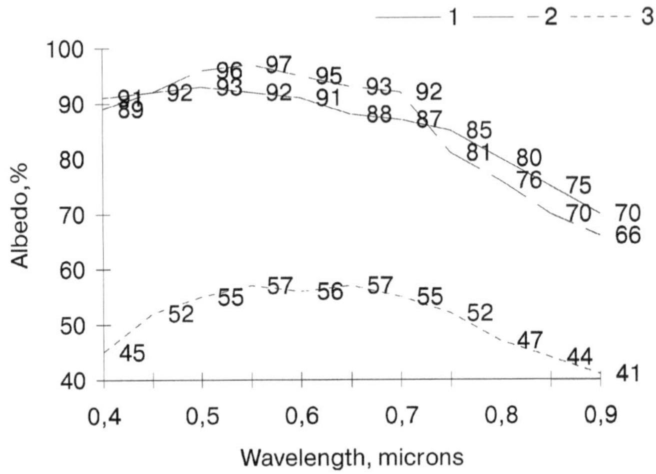

Figure 2.24. Spectral albedo of different types of the snow cover. 1 – coarse grained snow; 2 – coarse grained melting snow; 3 – wet firn.

Table 2.10. Diurnal variations of spectral albedo (%).

	Wavelength (μm)									
Sun elevation (°)	0.4	0.45	0.5	0.55	0.6	0.65	0.7	0.75	0.8	0.9
13.6	75	81	79	78	75	74	70	66	64	60
16.8	82	84	77	73	73	72	71	69	66	–
20.7	61	69	80	65	80	76	80	79	76	–
23.6	77	88	86	86	84	85	85	85	83	77
30.8	74	90	91	91	90	88	89	88	83	80
37.4	94	96	95	96	95	94	92	89	86	74
32.1	–	94	90	92	89	88	86	83	80	71
25.7	85	86	86	85	83	80	83	82	81	76

The largest changes of spectral albedo, and also of the observations of integral albedo, are connected with changes in the physical characteristics of the snow cover under the influence of the meteorological and solar radiation factors.

2.4 ATMOSPHERIC PROCESSES IN THE ARCTIC REGION AND LONG-TERM METEOROLOGICAL FORECASTING

The most complicated and present-day problem addressed by meteorological science is to develop and improve scientifically sound methods of meteorological forecasting – from short- to super long-range forecasts. The development of forecasts and their introduction into practice has the most important scientific and practical significance.

The difficulty of further improving the quality and reliability of meteorological forecasting is attributed to the complexity of the natural atmospheric processes and the phenomena developing as a result of interactions and interdependences of many factors. These difficulties increase to a great extent when addressing the objective of investigating the regularities of development of synoptic processes and weather phenomena in the polar regions with sparse observations. For many years the Arctic Ocean presented the least investigated area of the polar region of the northern hemisphere.

The drift of F. Nansen on board the *Fram* and the voyages and single flights to the North Pole provided valuable, although only general, evidence on the geography of the near-Pole area. Understanding of the atmospheric circulation and the ice cover was insufficiently known. The main area of the Arctic Ocean – the Arctic basin remained blank on the synoptic charts. During the analysis of synoptic charts, one could only use imagination in drawing isobars above the Arctic basin – guessing what occurs with cyclones and anticyclones in the vicinity of the North Pole.

Implementation of a multiyear program for the acquisition of primary meteorological information in the Arctic had several important stages and allowed the generation of a unique (at that time) archive of meteorological observation data and synoptic charts for the purpose of diagnostics and forecasting of atmospheric processes and weather phenomena in the polar region (Treshnikov, 1985; Vangengeim, 1952; Ivanov and Vinogradov, 1999; Ragozin and Chukanin, 1959).

The first stage was connected with creation of a stationary network of polar stations at the coast and islands of the Arctic Seas. Regular observations allowed the investigation of the regime of atmospheric circulation in the Arctic Seas and its peculiarities and relations to the atmospheric processes of other regions. This gave an opportunity to issue short-range weather forecasts based on the analysis of synoptic processes for shipping, aviation and economic activity. Further increase in the quality of these forecasts and their advance period was difficult due to an absence of knowledge of the character of development of the atmospheric processes at higher Arctic latitudes.

The second stage was distinguished by significant progress in the collection of meteorological observation data at high Arctic latitudes at the 'North Pole' drifting stations. The meteorological and upper air observations made at these stations allowed us to gain a completely new understanding of the atmospheric processes and the meteorological regime of the Arctic basin.

With time, data of meteorological observations at the 'NP' stations were supplemented with data obtained in the high-latitudinal airborne 'Sever' expeditions (sever is the Russian word for north). Most important in the program of 'Sever' expeditions was the deployment of 10–25 drifting automatic radio-meteorological stations (DARMS) in the Arctic basin on a yearly basis from 1957.

Regular weather messages transmitted daily by radio gave the possibility to forecasters to follow the advance and changes of air masses in the near-Pole area.

Analysis of a set of meteorological observations allowed the conclusion that in the Arctic basin there is no isolated circulation of the atmosphere and cold 'cap' – as was supposed by many investigators before (Orvig, 1970; Dorsey, 1951; Hare and Boville, 1965). It turned out that the atmospheric circulation in the Arctic center presents one uniform system with the circulation of air masses of the entire northern hemisphere.

Studies at drifting stations showed that the Arctic basin received heat from temperate latitudes. Thus, the Arctic basin influences the circulation above the continents and the oceans, being simultaneously the main area of energy consumption in the large-scale atmospheric circulation.

It turned out that the Arctic basin is an area of frequent and sometimes sufficiently intense cyclonic activity (Vize, 1940; Dzerdzeyevsky, 1954; Leontyeva, 1947; Prik, 1965; Ragozin and Chukanin, 1959). The winter cyclones exit to the Arctic basin predominantly from the North Atlantic both across the Baffin Bay and the Norwegian and the Barents Seas. As a result, well developed low-pressure troughs are observed in winter in these regions. In summer, a low-pressure area forms near the pole.

The cyclones and anticyclones above the Arctic Ocean have almost the same regime as at mid-latitudes. The characteristics of the upper troposphere also remind us of similar perturbations at temperate latitudes. The differences appear in the frequency of occurrence and motion of the air masses rather than in the main characteristics (Orvig, 1970).

The maps of the winter period show several cyclones above the Arctic Ocean, on average, four to six a month above the Norwegian, Barents and Kara Seas. Farther eastward, above the East Siberian Sea, approximately four cyclones are observed in one of the winter months. In the central parts of the basin, winter cyclones intrude most often from the North Atlantic region. Here, the January maps show four cyclones a month, on average. The January cyclones do not practically exit to the Arctic basin from the Bering Strait. However, the cyclones moving to the Arctic from temperate latitudes usually present cyclones at the stage of attenuation, which are gradually filled and disappear at the surface. These cyclones are frequently observed at higher levels (3–5 km), where they can be preserved for a long time as thermally cold cyclones. They are usually characterized by small sizes and low intensity. Insignificant weather deteriorations with winter precipitation fallout in the form of ice crystals are connected with them. An individual cyclone influences a small area of the Arctic basin, and the existence of surface cyclones at high latitudes is usually short – comprising typically several days (Treshnikov, 1985; Orvig, 1973; Ragozin and Chukanin, 1959).

During summer months, a relatively low-pressure area is located above the central polar regions and cyclones exit here from different directions, but most frequently from the North Atlantic and Bering Strait. In July, the cyclones converge in the Arctic basin, up to four on average above the eastern area (from the side of Alaska) and up to six above the Chukchi Sea (Leontyeva, 1947). Similar to winter, their lifetime in the Arctic is brief.

Data on the drift of the NP stations revealed an interaction between the atmospheric processes at mid-latitudes and the processes above the Arctic basin. At temperate latitudes, the synoptic systems determine the weather and climate by means of a constantly changing succession of passing of different air masses. Above the Arctic basin, the types of distributions of cyclones does not influence the climate at the surface, except for the fact that they probably determine turbulence in the inversion layer. The warm and cold weather periods at temperate latitudes more frequently present the phenomena related to advection of air masses. Above the Arctic basin, the periods of cooling are a result of action of the radiation factors, rather than of advection, while more temperate weather conditions near the surface are regulated by the descending vertical transport of warmer air from the areas located higher than the inversion layer.

Large and sharp temperature fluctuations are not a typical feature of the Arctic basin. Most of the year, this region is enveloped by a relatively thin cold air layer, which is isolated to a great extent from the higher lying atmospheric layer. The air temperature at the surface depends predominantly on the ice surface temperature. Sometimes, the advection of warm Atlantic air may cause dramatic air temperature changes of up to 30° in the wintertime near the North Pole. After warm air advection stops, the radiation regime again produces its prevailing influence on the surface temperature, which sharply decreases. Strong winds can also cause an increase of surface temperature as they break the inversion and mix warmer air masses with the masses that are near the cold surface. New cooling follows each wind decrease (Gaygerov, 1962; Drogaitsev, 1949; Orvig, 1973; Prik, 1965).

Advection of warm air masses so strong and pronounced, presents an unusual phenomenon. A more typical condition for the Arctic basin is a stable surface inversion of different intensity depending on the wind and clouds, caused by weak cyclonic perturbations.

Based on data of upper-air measurements conducted at the drifting stations (Gaygerov, 1962; Drogaitsev, 1949), numerous vertical sections of the fields of winds and temperature of the troposphere were analyzed. It was determined that tropospheric jet currents above the Arctic Basin are connected with significant temperature contrasts in the upper troposphere, with fronts reaching the tropopause and warm ridges of high pressure. The high tropopause is observed at the anticyclonic side of the jet current and the low – at the cyclonic.

Data of observations at the drifting stations in the central Arctic basin (Gaygerov, 1962) indicate that the mean monthly wind speed at the surface is low, approximately 4–5 m/s. Nevertheless, according to some observation data, the wind speed in the developed cyclones can reach 25–27 m/s. The mean wind speed rapidly increases with height in the surface inversion layer. This is especially typical of strong winter inversions. The wind speed in the troposphere increases to a maximum, on average, directly beneath the tropopause. At a height of 7–8 km, the highest mean monthly wind speed can comprise 20–25 m/s. The horizontal temperature gradients over the inversion layer can in some cases be significant (about 3°C per 100 km), but on average, they are within 0.5°–1.0°C per 100 km.

The thermal structure of the troposphere is rather variable. One observes significant horizontal temperature gradients in the polar air masses, while mean monthly heights of some isotherms indicate noticeable yearly variations. In summer, the location of the 0°C isotherm above the central Arctic basin can reach heights of more than 4 km. Advection of warm and cold air masses causes sharp changes in the heights of isotherms.

The upper air observations have significantly expanded our understanding of the state of the Arctic anticyclones. Before collecting data at the 'NP' drifting stations, it was supposed that anticyclones play an important role in the Arctic lower troposphere. There was a rather widespread opinion that such systems were always cold anticyclones forming due to low temperatures in a 2–4-km layer of the troposphere. It was however revealed that the highest and most stable of low-mobility Arctic and sub-Arctic anticyclones have a well-developed descending air motion and a warm central part, on average, in the entire tropospheric strata, as compared with the ambient atmosphere (Pogosyan, 1972).

The vertical structure of Arctic anticyclones is non-uniform compared with the structure of anticyclones of mid-latitudes. Within them their frontal surfaces separate the Arctic air from warm air transported from lower latitudes.

In January, the slowly moving anticyclones, coming from their pronounced direction of displacement, originate in the Greenland high pressure area and also in the region of quasi-permanent anticyclonic conditions above the Arctic basin, concentrating at 78°N, 165°W. The exit of migrating anticyclones occurring at the periphery of the winter Siberian High contributes to the latter location of anticyclones. On average, throughout January only one–three anticyclones pass across the basin.

In summer, the Siberian High disappears with only part of the quasi-permanent anticyclone remaining above the eastern Arctic basin. In July, only one–two anticyclones pass across the basin.

It is difficult to make a precise description of the anticyclonic activity in the Arctic. Many of the most stable systems present the extended weakly pronounced high-pressure belts where any pronounced centre is absent (Treshnikov, 1985; Bagrov, 1975; La Seur, 1954). Moreover, some systems move in many cases very slowly and hence, the trajectory of their centres is often of small interest compared with the trajectories of cyclones.

Meteorological forecasts

Many-sided comprehensive studies carried out in the central Arctic, on the basis of multiyear observations at the drifting stations, contributed to the creation and improvement of the service of meteorological forecasts (Ivanov and Vinogradov, 1999). Beginning with the first observations at the NP-1 station in 1937, short-range weather forecasts began to be prepared in the Arctic on a regular basis for servicing the Northern Sea Route.

In the pre-war years, an objective was set to develop a method not only for the short-range, but also for the long-range meteorological forecasts for the Arctic polar area. Specialists of the AARI Long-Range Meteorological Forecasting Department dealt with the problem from 1945 under the direction of G. Ya. Vangengeim.

The diversity of approaches to addressing the problem of long-range forecasting resulted in many forecasting methods based on different principles – taking into account some or other (not always the main) peculiarities of the atmospheric processes in some regions of the northern hemisphere.

During the development of the long-range meteorological forecasting method for the Arctic, the scientific principles of the macro-circulation method, proposed by G. Ya. Vangengeim were chosen (Vangengeim, 1952; Girs, 1974). The main principle among them was that the atmospheric processes observed in the restricted regions of the hemisphere should be investigated and predicted in the background of the processes of at least the entire northern hemisphere – considering them as interrelated and interdependent processes. In other words, the processes observed in some regions of the Arctic are necessary to consider as a result of a manifestation of the general circulation of the atmosphere under the specific physical–geographical conditions of the study area.

The author of the macro-circulation method of study and forecasting of the atmospheric processes introduced two fundamental notions: forms (types) of atmospheric circulation; and natural stages of development of the macro-processes of the general circulation of the atmosphere. These stages are: an elementary synoptic process (3–7 days); a uniform circulation period (7–14 days); a uniform intra-annual period (3–6 months); a stage of circulation epochs (3–5 years); and a circulation epoch (after decades) (Vangengeim, 1952; Girs, 1974). Investigation of the atmospheric processes in time should be made regarding their continuous development with a separation into the real natural stages of development and establishment of the regularities of their changes (Bolotinskaya and Ivanov, 1999; Krutskikh, 1991; Ivanov and Vinogradov, 1996). During the diagnosis of continuous development of the atmospheric processes, temporary boundaries between these stages are identified and the forms and types of the atmospheric circulation and their varieties in the northern hemisphere and the Arctic are determined. All this allows us to record the peculiarities in the character of the modification of the atmospheric processes of different spatial–temporal scales. The results of the diagnosis of the atmospheric processes are presented in corresponding catalogues, bulletins, maps and on magnetic media based on data for the northern hemisphere since 1891 and for the Arctic since 1935 through to the present-day.

For implementing the program of studies and creating a long-range weather forecasting method for such a complicated and little investigated area as the northern hemisphere, complex upper air meteorological, hydrological and ice information was required as well as detailed studies of the spatial–temporal variability in the structure of the general circulation of the atmosphere (GCA) and the processes in the Arctic.

Special emphasis in these studies was given to the main GCA components, which include:

- planetary circumpolar vortices;
- planetary upper air frontal zones;
- long waves;

- extensive stationary cyclones and anticyclones; and
- frontal cyclones and anticyclones.

It was established by these studies that all components of the macro-structure of the troposphere and stratosphere are closely connected. Smaller scale components develop not only in the background of greater scale processes but also influence them. As a result, larger scale objects deform and change. Thus, under the influence of obstruction processes and long waves, the structure of the planetary circumpolar vortex and the system of tropospheric planetary upper air frontal zones change. This leads in turn to the local intensification or attenuations of the processes of cyclo-anticyclogenesis, processes of regeneration of pressure features, transformation of upper air deformation fields and hence to the evolution of the system of planetary upper air frontal zones and the circumpolar vortex. Thus, a closed process of evolution of the general circulation of the atmosphere occurs where all macro-structure components are interrelated and interdependent.

Each macrostructure component of the circulation can be, at a given moment in time, at some or other stage of natural development, therefore, there are a variety of the main forms of macro-synoptic processes as complexes of macrostructure components. In spite of the complexity of the relations between the macrostructure components and plurality of their possible states and combinations, the evolution of macro-synoptic processes always has some specific regularity determining the tendency of their further development. This provided in turn the possibility to develop the classification of atmospheric macro-processes depending on the state of long thermobaric waves (Vangengeim, 1952; Girs, 1974; Ragozin and Chukanin, 1959; Ivanov and Vinogradov, 1995; Ivanov and Vinogradov, 1996).

For more than 60 years, the AARI Long-Range Meteorological Forecasting Department has conducted planned and dedicated studies of the atmospheric processes and issued weather forecasts with different advance periods. Presently, a number of scientific–practical tasks were solved with respect to acquisition, processing, classification of meteorological observation data, development, improvement and introduction into practice of the macro-circulation method of weather forecasting with different advance periods. Among these results, the following can be noted in particular:

(1) Establishment of the dependencies of the different characteristics of the types of atmospheric processes in the Arctic on the circulation character of the northern hemisphere and peculiarities of their short and long-term transformations.
(2) Resolution of the problem of extending the background meteorological forecast up to 12 months so that it would be possible at the end of December of the current year to develop a background forecast in the form of mean monthly characteristics for January–December of the next year.
(3) Using the improved database, a method for running updates of a background forecast for the next three months was developed.
(4) The conditions for formation of some extreme value of meteorological elements and dangerous phenomena (icing, snow storms, etc.) were revealed. A method

for forecasting the large anomalies in the main meteorological elements in the Arctic was developed and practically introduced.

(5) Development and testing, in multiyear practice, of a method of medium-range forecasts of dramatic modifications of the atmospheric circulation processes in the Arctic using elementary synoptic processes.

(6) Typical features of multiyear fluctuations of the atmospheric circulation forms, and related climate changes, were investigated for the purpose of foreseeing prolonged tendencies in the change of atmospheric characteristics and predicting the climatic variability for a super long-range period (up to 30–50 years) (Vinogradov *et al.*, 1999).

The scientific work to further investigate the atmospheric processes and develop and improve the macro-circulation method of long-range meteorological forecasts for the polar area continues. One can hope that resuming the operation of the drifting stations will give a significant impetus to their development.

2.5 CHANGES IN METEOROLOGICAL CONDITIONS IN THE ARCTIC BASIN THROUGHOUT THE LAST CENTURY

A complex of meteorological and oceanographic observations in the Arctic Ocean was carried out for the first time by the Norwegian Polar Expedition headed by Fritjof Nansen on board the ship *Fram* during the voyage along the northern shores of Siberia and then in the ice drift from the New Siberian Islands to the northern Greenland Sea in 1893–1896. In conclusion of the book summarizing the results of his expedition, F. Nansen wrote, in particular: 'In general, it can probably be said that although our expedition was far from solving all problems connected with polar expanses, it did a lot for destroying the mystery surrounding until now these areas and we had an opportunity to obtain a rather clear and correct understanding of that part of the Globe that was before veiled in obscurity permeable only for imagination... However, they still have much in them for investigation requiring years of observations, and a new expedition like the *Fram* would be rather necessary' (Nansen, 1901).

The work performed at the NP-1 drifting station in 1937–1938 initiated planned investigations of nature in the most inaccessible part of the Earth's northern polar area. They were resumed in April 1950 at the NP-2 drifting station that operated until April 1951 and then were continuously carried out at the NP-3–NP-31 drifting stations from April 1954–July 1991.

In spring of 2003, observations in the Arctic basin were resumed at the 'North Pole-32' drifting station (NP-32). The NP-32 drift route from June–December 2003 passed by the vicinity of 85°N. In the same region before, observations at the NP-1 drifting stations in 1937/1938 and then in different years from 1954–1988 at the drifting stations NP-3, 7, 10, 13, 15-17, 19, 23–25, 27, 28 were made (Aleksandrov *et al.*, 1999a). This route was also very close to the drift trajectory of the Norwegian ship *Fram* in June–December 1895 (Chernigovsky and Marshunova, 1965). The

Table 2.11. Mean multiyear air temperature (T) and pressure (P), their RMSD and extremes in the area of the NP-32 drift.

Parameter	June	July	August	September	October	November	December
T (°C)	−2.8 ± 0.8	−0.3 ± 0.2	−1.9 ± 0.8	−9.5 ± 1.8	−19.9 ± 2.7	−28.3 ± 2.9	−30.9 ± 3.5
T_{max} (°C)	3.8	4.3	1.9	1.0	−0.2	−2.9	−1.3
T_{min} (°C)	−14.4	−5.1	−13.1	−32.5	−43.3	−47.3	−48.9
P (hPa)	1,011 ± 5.7	1,010 ± 3.7	1,010 ± 4.6	1,010 ± 3.7	1,010 ± 4.4	1,018 ± 5.9	1,016 ± 6.5
P_{max} (hPa)	1,034	1,030	1,035	1,039	1,036	1,047	1,052
P_{min} (hPa)	981	976	980	983	975	974	961
No. of years	13	13	16	19	22	19	19

indicated circumstances allow us to compare the meteorological conditions of the current drift both with regime conditions typical of the last 50 years and with those observed more than 100 years ago.

Table 2.11 presents the mean multiyear air temperature and pressure values (multiyear average) and their RMSD from observation data of the drifting stations over the period 1937 and 1954–1988, made in the drift area of NP-32. The values of the absolute air temperature and pressure extremes are presented on the basis of observations at all preceding drifting stations that drifted to the north of 85°N (Aleksandrov *et al.*, 1999a). The lower line of the table presents the number of observation years for which the average air temperature and pressure values for each month were calculated.

Table 2.12 presents mean monthly air temperature, atmospheric pressure, wind speed, monthly precipitation totals and their RMSDs in June – December 2003 at NP-32, June–December 1895 on board the ship *Fram* and the mean multiyear values

Table 2.12. Average air temperature (T, °C), atmospheric pressure (P, hPa), wind speed (v, m/s), monthly precipitation totals (p, mm) and their RMSDs.

Characteristic	June	July	August	September	October	November	December
T (NP-32)	−1.8 ± 2.0	−0.1 ± 0.6	−0.9 ± 1.6	−6.1 ± 3.1	−18.3 ± 4.6	−22.9 ± 6.4	−33.2 ± 4.0
T (*Fram*)	−1.9 ± 2.3	−0.1 ± 0.7	−2.3 ± 1.7	−9.7 ± 6.8	−21.4 ± 5.0	−31.3 ± 6.3	−32.8 ± 5.6
T (multiyear average)	−2.8 ± 0.8	−0.3 ± 0.2	−1.9 ± 0.8	−9.5 ± 1.8	−19.9 ± 2.7	−28.3 ± 2.9	−30.9 ± 3.5
P (NP-32)	1,011.9 ± 7.3	1,008.7 ± 7.1	1,007.9 ± 8.2	1,003.1 ± 10.3	1,019.6 ± 8.9	1,014.8 ± 10.0	1,006.1 + 12.2
P (*Fram*)	1,006.8 ± 4.9	1,005.4 + 6.7	1,015.0 ± 11.4	1,005.0 ± 9.2	1,020.6 ± 11.7	1,007.9 ± 12.6	1,015.1 ± 8.5
P (multiyear average)	1,011 ± 5.7	1,010 ± 3.7	1,010 ± 4.6	1,010 ± 3.7	1,010 ± 4.4	1,018 ± 5.9	1,016 ± 6.5
v (NP-32)	4.9 ± 2.5	4.8 ± 2.2	6.0 ± 2.2	5.2 ± 4.7	3.7 ± 2.6	3.2 ± 2.5	4.5 ± 1.8
v (*Fram*)	5.7 ± 2.7	4.9 ± 1.9	5.0 ± 2.2	4.1 ± 2.1	4.4 ± 2.3	4.2 ± 2.7	3.8 ± 2.0
v (multiyear average)	4.5 ± 0.9	4.5 ± 0.7	4.9 ± 1.1	5.0 ± 0.9	5.2 ± 1.2	4.7 ± 0.8	4.6 ± 0.9
p (NP-32)	48.8	36.4	40.3	41.8	21.4	31.9	2.1
p (multiyear average)	9.3 ± 6.2	14.6 ± 7.0	18.4 ± 5.8	19.6 ± 9.5	16.0 ± 5.2	9.8 ± 6.7	8.7 ± 4.5

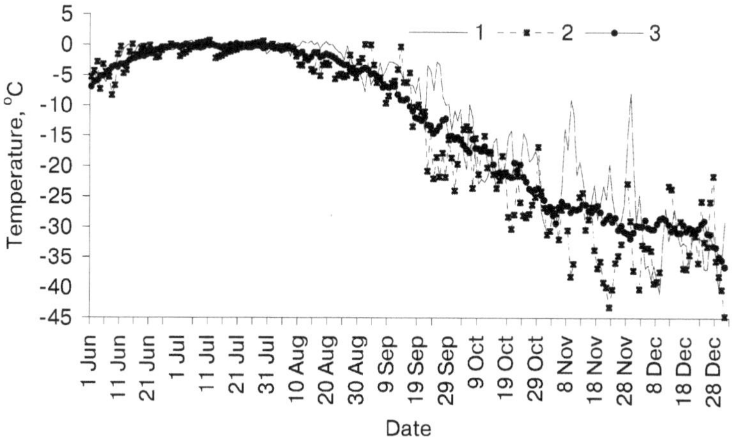

Figure 2.25. Mean daily temperature from June–December at NP-32 in 2003 (1), on board the *Fram* in 1895 (2) and mean daily temperature averaged over the period 1954–1988 (3) for the near-Pole area.

(multiyear averages) over the observation period at the drifting stations in 1937, 1954–1988 for the same months. The statistical characteristics of meteorological parameters at the 'North Pole' drifting stations were calculated from observation data published in Aleksandrov *et al.* (1999a), and for the *Fram* – from data published in the *Arctic Climatology Project* (2000).

The main conclusion that can be made on the basis of the analysis of air temperature data is that the air temperatures observed in August–November at NP-32 were much higher than the temperatures measured a century ago on board the *Fram*. In general from June–November, they were also higher than the multiyear average temperatures calculated for the preceding decades of measurements at the drifting stations.

In all three cases, the minimum temperature variability (RMSD value) was observed in July. In subsequent months, the RMSD values were several times as high as their July values. The dynamics of changes of mean daily air temperatures in the process of the drift of NP-32 and the *Fram* is shown in Figure 2.25, which also presents mean daily air temperatures averaged for 1954–1988 calculated from measurement data from the drifting stations in the study area. As can be clearly seen from the figures, there are quite smooth variations and a good coincidence between the three curves from June to late August. During the following days, large fluctuations of mean daily air temperatures were observed both at NP-32, and on board the *Fram*.

From the second half of September to early December, there is a tendency for the mean daily air temperatures at NP-32 to exceed the multiyear average, while from data of measurements on board the *Fram* in 1895, mean daily air temperatures were mainly below the multiyear average values typical of the second-half of the 20th century in the central Arctic basin.

Data of observations at NP-32 also indicate the increased frequency of manifestation of extreme changes of meteorological parameters – the increased extreme character of weather phenomena. Table 2.13 presents the values characterizing the extreme rates of changes of air temperature and pressure at NP-32 and on board the *Fram*, observed in autumn months.

According to data of multiyear observations at the drifting stations in 1954–1988, the most significant pressure decrease for 24 h in the study area was about 30.3 hPa, and the most significant increase was about 23.4 hPa. These are recorded at the NP-6 drifting station, however, not in the autumn period but in February 1959 (the decrease occurring from 14–15 February, and the increase – from 25–26 February).

On board the *Fram* there were only 3 days throughout the entire summer period with mean daily air temperatures higher than 0°C – reaching 0.1°C on these days. According to observation data from the drifting stations in the 1950s–1990s, the number of days with a positive mean daily air temperatures on average for a year was not more than six. At NP-32, which drifted slightly more to the north than the *Fram*, the number of days was 26. They were not successive, but as a rule in series of three–four days.

The highest mean daily air temperature at NP-32 of +0.9°C was recorded on 10 July, 2003. The positive temperature total for the summer season was 8.6°C. The multiyear average for the area of the NP-32 drift comprises 7.4 ± 4.5°C. The largest positive temperature total for the summer season of 16.6°C was noted at NP-16 in 1971.

In the station drift area in June–November, a large amount of precipitation was recorded. A comparison with multiyear data showed the monthly precipitation totals from June–November to exceed multiyear averages (Table 2.12). In June, the amount of precipitation was more than 5 times greater than a multiyear average; from July–September the monthly multiyear averages were exceeded more than twice and in November by 3.2 times. The NP-32 was characterized by frequent precipitation, from June–September more than 20 days having precipitation were observed in each of the months. In July–August, there was predominantly liquid precipitation, while beginning in September, solid precipitation prevailed. The maximum amount of precipitation fallout for 24 h was observed on 18 June and comprised 12 mm. Before this, the maximum daily liquid precipitation totals in the drift area of NP-32 in June were equal to 6.6 mm – noted in 1988 at NP-28 (Aleksandrov *et al.*, 1999a).

The data of observations at NP-32 supplemented our understanding about the climatic tendencies in the near-Pole area in the present-day. To what extent the warming was manifested in the near-Pole area can be represented by the anomalies of mean daily temperature during the drift of NP-32 relative to their mean multiyear values. It can be seen from Figure 2.25 that mainly positive anomalies predominate with most significant observed in autumn months. This is in full compliance with the temperature field characteristics in the entire area located to the north of 60°N.

In 2003, a positive anomaly of mean annual air temperature in the zone to the north of 60°N, calculated relative to the multiyear average for 1961–1990, was the

Table 2.13. Extreme changes of air temperature and pressure during a 24-h period.

Interval: hour and date of changes	Change of temperature (°C)	Change of pressure (hPa)
NP-32		
12 h, 27 November, 2003–12 h, 28 November, 2003	From −23.9 to −11.8; $\Delta T = +12.1$°C	From 1,017.4 to 1,002.0 hPa; $\Delta P = -15.4$ hPa
06 h, 29 November, 2003–06 h, 30 November, 2003	From −5.5 to −19.0; $\Delta T = -14.5$°C	From 994.1 to 985.7 hPa; $\Delta P = +8.4$ hPa
12 h, 24 September, 2003–12 h, 25 September, 2003	From −7.9 to −0.1; $\Delta T = +7.8$°C	From 1,005.2 to 977.4 hPa; $\Delta P = -27.8$ hPa
18 h, 26 September, 2003–18 h, 27 September, 2003	From −6.0 to −8.4; $\Delta T = -2.4$°C	From 977.8 to 995.6 hPa; $\Delta P = +17.8$ hPa
Fram		
08 h, 10 November, 1895–08 h, 11 November, 1895	From −38.2 to −19.7; $\Delta T = +1\ 8.5$°C	From 1,021.4 to 1,006.6 hPa; $\Delta P = -14.8$ hPa
04 h, 26 November, 1895–04 h, 27 November, 1895	From −35.6 to −19.0; $\Delta T = +16.6$°C	From 1,038.0 to 1,010.4 hPa; $\Delta P = -27.6$ hPa

same as in 1938. Its value was 3 RMSD, which corresponds to 1.4°C. The summer season of 2003 was the warmest for the last 60 years. An anomaly of the average temperature for the summer was 3.5 RMSD (1.2°C), and in autumn – 3.2 RMSD (1.7°C). The warmer autumn seasons were in 1938 comprising 3.7 RMSD (1.9°C) and in 1943 – 3.5 RMSD (1.8°C).

The results of systematic meteorological observations in one and the same area carried out on board the *Fram* and at the NP-1 drifting station allowed us to make a conclusion in the late 1930s that the climatic conditions of the *Fram* drift differed significantly from the conditions typical of the drift period of NP-1 (Vize, 1940). In the 40 years after the Norwegian expedition on board the ship *Fram* was completed, there has been a significant warming in the Arctic. As we now know, a rapid air temperature increase in the northern Polar area throughout the 1920–1930s achieved a maximum in the second-half of the 1930s and was replaced after 1943 by a rapid decrease until the mid-1960s. Beginning in the mid-1960s, the mean annual surface air temperature in the northern Polar area began increasing again. The greatest air temperature increase rate refers to the second-half of the 1990s. After 1936, in the area located to the north of 60°N, more than half the cases with anomalies of mean annual air temperature exceeding double the RMSD from a multiyear average calculated for the period 1961–1990, as in the second-half of the 1990s. The new results of systematic meteorological observations at NP-32 in the Arctic basin are the first, after a 12-year break of instrumental evidence, to show the continued warming of the atmosphere at high latitudes in the Pole area that began in the mid-1960s.

3

Sea ice cover

3.1 ICE OBSERVATIONS IN THE ARCTIC BASIN

A complex of ice observations conducted at the 'North Pole' drifting stations (Romanov *et al.*, 1997) and by the 'North' expeditions (Konstantinov and Grachev, 2000) commonly included the following characteristics:

- geographic coordinates with subsequent calculation of ice drift velocity vectors;
- ice cover deformations of different extent and ice floe rotation;
- ice cover structure and texture;
- distribution of ice temperature, salinity and density at different horizons;
- strength and deformational characteristics of ice;
- dependences of electric parameters of ice on its physical characteristics;
- conditions of electromagnetic radiation of different varieties of sea ice;
- optical and acoustical properties of ice cover;
- natural oscillations and wave processes in the ice cover;
- ice cover morphometry and factors resulting in its changes;
- shearing stresses on the upper and lower ice cover surfaces depending on morphological characteristics and external factors;
- break-ups and shears in the ice cover; and
- distribution of hydrocarbons and other pollutants in the ice cover.

A brief review of ice drift observation techniques applied by the 'North Pole' drifting stations, as well as observations over the movement of drifting automatic radio-meteorolofical stations (DARMS) is given in the work by Gudkovich and Doronin (2001). The same paper also contains the assessments of errors of the procedures and lists their main advantages and disadvantages.

Morphometric study methods included stationary rods, profile and area levelling, ice drilling, sonar and diving observations, aerophotogrammetric surveying and radar mapping and satellite ice information. For measuring temperature, thermistors and resistance thermometers were frozen into the ice. For hydrochemical studies, determination of crystalline structure and the texture of the ice (including measurements of its density), samples were taken by means of a tubular drill – these samples were also used to determine the strength and deformation characteristics of the ice. Larger samples were sawn off the 'cores' cut out of the thick ice cover. Special equipment was used for studying the electromagnetic and electric properties of ice – necessary for developing the remote instruments for measuring the thickness and other properties of the ice cover as well as studying its acoustic processes. Seismic, dip metering and other devices were applied for the study of oscillations and waves. For shearing stress measurements sets of meteorological and oceanological devices were used in the boundary layers of the atmosphere and ocean. Studies on the dynamics, thermodynamics and optical properties of the ice cover would have been impossible without simultaneous actinometric, meteorological and oceanological observations.

Results of the above observations and studies are presented in expedition reports, numerous scientific papers and monographs. Most of these are given in the references of this book.

3.2 ICE COVER STRUCTURE

Changes in the phase state of seawater during sea ice formation are accompanied by the generation of ice crystals separated by layers of water, salts, air bubbles and organic and mineral inclusions initially occurring in water in a suspended state. Accumulations of ice crystals under calm weather form peculiar patches – grease ice – on the water surface; under high waves, grease ice forms whitish ice piles – shuga. The next stage of ice formation is thin opaque ice with its surface wetted from brine – nilas – its thickness usually does not exceed 10 cm. In the case of an intensely freshened surface water layer, a thin shining crust, called an ice rind, forms instead of nilas (Krutskikh, 1984).

With further growth, ice successively passes through the stages of young ice (grey ice 10–15 cm thick, grey-white ice 15–30 cm thick), thin first-year ice (thin first-year ice of the first stage – 30–50 cm and second stage – 50–70 cm), first-year ice of medium thickness (70–120 cm) and thick first-year ice (120 cm or more). Ice which has endured one period of summer thawing becomes second-year ice (up to 250 cm); and after the second period, multiyear ice (about 3 m or more) (Krutskikh, 1984).

Depending on the temperature, dynamic conditions, water salinity and other factors, crystalline *structure* and *texture* of the ice cover form determining its *microstructure*. However, as early as the nilas ice stage, due to the breaking-up, ridging, accumulation of layers and chafing, spatial inhomogeneities occur forming an ice cover *mesostructure*. Formation of cake ice should, probably, be regarded as an initial element of the mesostructure; growing and breaking up, it forms spatial elements of different sizes: from finely broken up (up to 20 m) to giant ice floes (over 10 km). In addition to the horizontal dimensions of ice formations (their fragmentation), the ice cover mesostructure is determined by its compaction, ridge concentration and the structure of its ridge formations, degree of destruction, area and structure of puddles, snow concentration and other characteristics of snow cover on ice (Krutskikh, 1984).

Specific features of ice cover formation and destruction in different parts of the ocean result in the ice cover *macrostructure*. It is determined by: zones of fixed ice connected with the coast – fast ice; a zone where the accumulation of compacted ice prevails – ice massifs, multiyear ice and flaw polynyas; as well as intermediate zones, in which there is a certain system of thickness gradients and other characteristics of ice cover (ridge concentration in particular) (Krutskikh, 1984).

Therefore, depending on the spatial extent, when examining the ice cover make-up its micro-, meso- and macrostructure should be distinguished. The extent of spatial non-uniformity determined by the microstructure ranges from 10^{-1} to 10^{2} cm. For the mesostructure, this extent can be estimated between 10^{2}–10^{5} cm; and for the macrostructure, 10^{5}–10^{8} cm or more. Some of the issues related to this subject are described in numerous scientific publications. Therefore, in the subsequent sections these questions are discussed only in the most general manner.

3.2.1 Ice cover microstructure

Ice microstructure is determined by the shape, size and orientation of crystals. The size of crystals varies extensively: from 0.05–0.07 cm to 100–150 cm. An important structural characteristic is the optical orientation of ice crystals determined by the position normal to the freezing surface. Several types of crystal development are distinguished under static ice formation conditions depending on the value of the vertical water temperature gradient in the surface layer.

The crystalline structure of the ice cover can be ordered and non-oriented. A clear example of a spatially oriented ice structure is that forming under the impact of a constant current. The discovery of ice having such a structure enables a more precise definition of the scheme of currents in a certain sea or part thereof.

Zubov (1945) proposed, as a first approximation, to classify sea ice on the basis of its structure: needle, spongy ice and firn; and as regards its formation depth, into surface, deep sea and bottom ice. In accordance with the structural genetic classification developed by Cherepanov (1976), depending on the hydrometeorological conditions, ice in natural water bodies can be classified into four main groups (A, B, C, D) – each of them, depending on the dynamic nature of ice formation conditions, determines the type of ice. The following ice formation types are distinguished: congelation, congelation–frazil, frazil–shuga, infiltration and secondary recrystallization. Formation of shuga from frazil ice and snow often results in a significant ice volume growth.

Sea ice texture is characterized by the shape and specific arrangement of air and salt inclusions as well as the mineral and organic particles within the ice. A relative volume of inclusions ranges from 5–10% to 40–50%. Based on the character of air inclusions, 8 main ice types can be distinguished; from the outward appearance and shape of salt inclusions, 6 main types are distinguishable (Tyshko *et al.*, 2000). The total number of air and salt inclusions in sea ice is estimated using a 5-score scale.

Under the influence of external heat inflow, seasonal transformations of textural–structural ice characteristics occur (thermometamorphism). Dynamic processes (compression, ridging) also result in recrystallization (dynamometamorphism). Precipitation, snow transportation caused by snow storms and the arrival of water freshened by river flow, have a significant effect on the spatial inhomogeneity of the sea ice microstructure and its seasonal variability.

3.2.2 Ice cover mesostructure

Primary elements of the ice cover mesostructure are ice floes, their fragments, small floes and finely crushed ice (Krutskikh, 1984). Some of these elements represent a conglomerate of ice of different ages. Separated by cracks, leads and fractures (the so-called disturbances of continuity), the primary elements make up larger structural formations: blocks and patches (Borodachev *et al.*, 1981). The average size of homogeneous ice blocks depends on the ice cover age, varying from dozens of metres for nilas ice to several thousands of metres for thin first-year ice. The mean size of conglomerate ice blocks can be even larger.

Ice cover mesostructures, related with systems of continuity disturbances, are subject to seasonal changes under the impact of a changing environment. Its most stable forms in the Arctic basin are: polygonal (in winter), net (in spring), mottled (in summer) and branched–dendritic (in autumn).

Among the characteristic elements of ice cover mesostructure are also thickness irregularities of 'even' ice and snow, puddles appearing on the ice during the thawing period and hummocky formations, which can take the form of standing floes, ice ridges, hummocky ridges and ridge barriers; and on multiyear ice, the form of hillocks (smoothed by melting of ice ridges).

The standard deviation σ_H can be an indicator of ice thickness irregularity resulting from mesostructural elements. The analysis of observation data (Buzuev, 1975) indicates that for the accumulating non-ridged ice over 30 cm thick, standard deviation of its thickness is proportional to its average thickness (H_0) (Buzuev and Dubovtsev, 1978):

$$\sigma_H = 0.11 H_0 \tag{3.1}$$

However, the main contribution to the spatial variability of ice thickness is made by ridgy formations. In the paper by Appell and Gudkovich (1992), on the basis of empirical data on the shape of ice ridges (triangular section of the above-water and underwater part of the ice ridge, which are called the sail and the keel, respectively), as well as on the average ratio of the width of the above-water and underwater part of the ice ridge to its height (H_T), the expressions for the dependence of standard deviation (σ_H) on the values of ridge concentration, average ice thickness H_0 and sail height H_T were obtained. It was assumed that the keel draft equals a four-fold ice ridge height, values of space factors are 0.60 and 0.65, and the ratio between the area of ice ridges and ridge concentration (in points) is adopted in accordance with the paper by Bushuev and Loshchilov (1967). From the expressions received it follows that, for instance, under a sail height equalling the smooth ice thickness, the standard deviation σ_H increases by more than two orders as compared with smooth (non-deformed) ice of the same thickness. The calculations performed show that the average increase of ice thickness is 10% per a one-tenth increase in ridge concentration. It corresponds to the observational data (Gordienko *et al.*, 1967).

Regarding the snow cover thickness, as shown in a number of papers (Nazintsev, 1971a; Buzuev *et al.*, 1979a; Appell and Gudkovich, 1992), the amount of snow on ice of different thicknesses mainly depends on the volume of solid precipitation falling within the existence period of the ice cover of a given age. A significant role in its redistribution over the ice surface is played by snow storms and snow accumulation related to the irregularity of this surface. Therefore, snow cover height is approximately proportional to the ice thickness, the proportionality factor differing in different seas and being strongly dependent on the ice ridge concentration.

Of utmost importance in the formation of mesoscale irregularities of ice cover (particularly, its thickness) are meltwater lakelets – puddles forming on the sea ice surface in the summer season (Spichkin, 1963;). Low reflectivity of puddles and a peculiar convection process under freshwater heating (Appell and Gudkovich, 1979; Makshtas and Podgorny, 1995) intensifies melting, resulting in a quick deepening of puddles and a transformation of a part thereof into openings – thawed patches. This reduces the ice cover strength and promotes ice cover crushing and diverging. Under the impact of certain factors, the evolution of puddles in the process of thawing is accompanied not only by changes of depth, but also of area. The largest area is commonly recorded in mid-July (to 45–60% of ice area) decreasing by the beginning of freezing months to 14–20% (Nazintsev, 1964; Buzuev and Spichkin, 1977; Buzuev *et al.*, 1979a). The maximum size of puddles decreases with the diminishing ice cover concentration and the horizontal dimensions of ice floes as well as with the growing ridge concentration. Puddles have a noticeable impact on ice accumulation in the winter season, increasing the ice thickness irregularity (Buzuev and Spichkin, 1977).

3.2.3 Ice cover macrostructure

Thermal and dynamic processes, which control the main characteristics of sea ice cover, have geographic peculiarities. These specific features result in a large-scale inhomogeneity of different ice cover properties determining its macrostructure. Steady ice formation in the ocean begins at different time. Earliest of all (on the average around the 15–20 August), as is shown by observations, young ice appears in puddles, leads and fractures at high latitudes (near the North Pole). By the beginning of September, these processes extend to the northern boundaries of the marginal Artic seas – gradually covering most of their water area.

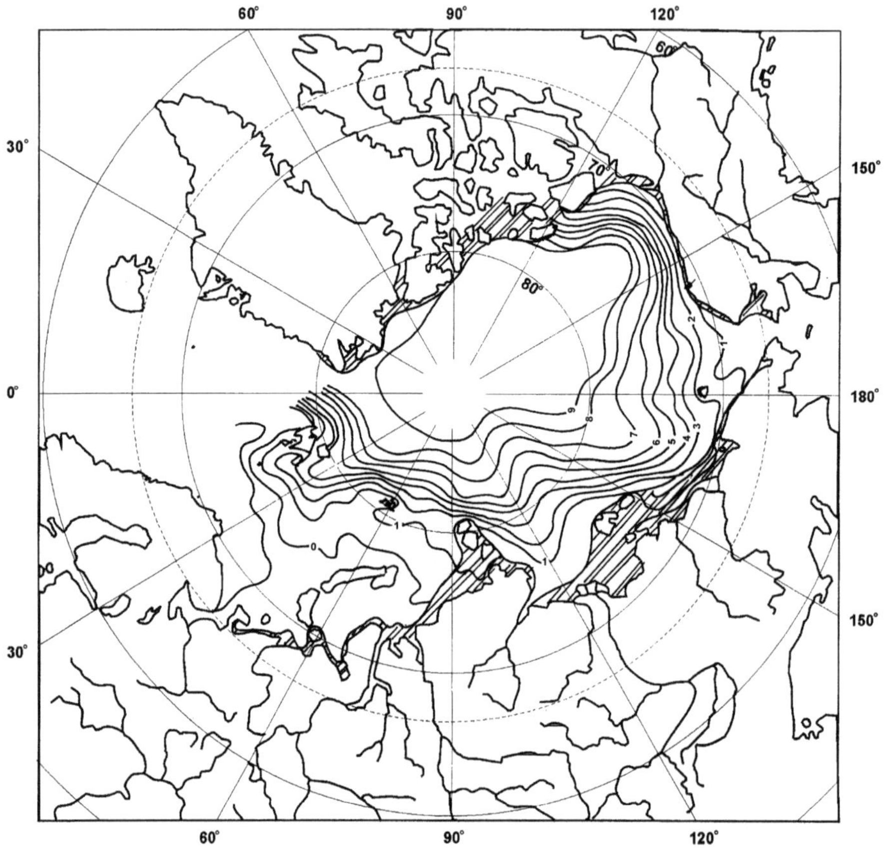

Figure 3.1. The structure of an average field of old ice concentration and the location of landfast ice in February.

Coastline shape, bottom topography, presence or lack of islands on the continental shelf, wind regime, intensity of tidal currents and amplitude of sea level fluctuations – all these factors determine the conditions dividing sea ice into drifting and fixed (fast ice). Differences between these types of ice cover relate to many characteristics of its macrostructure: thickness and concentration, ridge concentration, season, interannual and climatic changes of ice cover conditions, and intensity of its interaction with the atmosphere and ocean. As can be seen from Figure 3.1, fast ice is most developed in the north-eastern Kara Sea and in the regions of the New Siberian Islands and Canadian Arctic Archipelago. Fast ice thickness varies extensively in different regions. At the end of winter (May), the maximum fast ice thickness is recorded in the western Laptev Sea and eastern East Siberian Sea (200–215 cm). West and east of the Novosibirskie Islands, average ice thickness is gradually decreasing. In the western Laptev Sea and eastern East Siberian Sea, fast ice thickness at this time is averaging at about 190 cm; in the north-eastern Kara Sea and the south-western Chukchi Sea, about

170 cm; in the south-western Kara Sea, about 130 cm; in the Bering Strait, about 150 cm; and in the southern Novozemelsky straits, about 100 cm. The above large-scale irregularity of ice thickness distribution is mainly due to climatic features of temperature regime – which controls the ice accumulation period and intensity and snow cover height, which is also strongly affected by ice surface irregularities (Aleksandrov *et al.*, 2004).

Systems of atmospheric and ocean circulation determine the duration of ice cover subsistence, and, thus, the prevailing age of ice formations in different macro-structural zones. As a result, a region forms in the ocean dominated by the multiyear ice cover (Borodachev *et al.*, 1997) (Figure 3.1) having a higher thickness, a specific mesostructure and other properties. On its periphery there is a zone where the amount of multiyear ice is gradually retreating from the boundary of its relative predominance.

Macroscale distribution of multiyear ice corresponds to the location of the high-concentration zone, which forms an oceanic ice massif; its spurs are the main massifs of marginal seas: Spitsbergen, Kara, Taimyr, Ionsky, Wrangelevsky and Alaska (Romanov, 1992).

Higher frequencies of offshore winds in many seas of the Eurasian shelf and onshore directed winds north of Greenland and the Canadian Arctic archipelago, favour flaw polynya formation and ice cover rejuvenation (Zakharov, 1996) and intense ridging and ice accumulation, respectively (Bourke, Garret, 1987). Along with specific thermodynamic conditions this results in large-scale spatial changes in the average ice cover thickness – increasing from the Siberian coasts to the Canadian Arctic archipelago and Greenland coasts. The maximum spatial gradients of ice thickness at the end of winter, caused by ice cover rejuvenation, are recorded in most of the marginal seas of the Siberian offshore (Gudkovich *et al.*, 1972; Gudkovich and Doronin, 2001).

Generalization of the direct ice thickness measurements on drifting stations and calculation results, taking into account the age composition of the ice cover and its ridge concentration (Mironov, 1985; Hibler, 1980), as well as the data obtained on the basis of sonar observations from submarines (Bourke and Garret, 1987), enabled the determination of the average ice thickness distribution in the Arctic basin at the end of winter (Figure 3.2, see colour section). As is seen, the maximum ice thickness exceeding 7 m is recorded at the northern coasts of the Canadian Arctic archipelago – gradually decreasing to about 2 m in the seas offshore Siberia. A significant share of macroscale spatial irregularity is due to the distribution of ice ridge concentration. Thus, according to Romanov (1992), the average height of ice ridges near Ellesmere Island is 300–345 cm, whereas on the fast ice in Novosibirsk region is does not exceed 75–100 cm. It should be noted that in accordance with the research conducted by the Arctic and Antarctic Research Institution (AARI) under the Northern Sea Route International Programme (INSROP) (Brestkin *et al.*, 1998), there is a rather close non-linear link between the average height of the ice ridges (H_T, cm) and the average thickness of the surrounding smooth ice (H_0, cm) expressed by the formula:

$$H_T = nH_0^m \tag{3.2}$$

In this formula, m and n are parameters determined by the least-squares technique to have the following values:

- for drifting ice $m = 0.62$, $n = 6.59$; and
- for fast ice $m = 0.94$, $n = 0.88$.

Since the thickness of non-deformed ice is on average decreasing from the northern coasts of the Canadian Arctic archipelago toward the Siberian offshore seas, the average height and draft of ice ridges are also diminishing in the same direction. However, the maximum draft of sea ice (keel dimension) is well in excess of the smooth ice thickness. North of Greenland and the Canadian Arctic archipelago, keels over 14–16 m were recorded with a maximum exceeding 40 m (Zakharov, 1996).

With growing solar radiation and air temperature in the spring season, thawing of snow and then of ice begins. Average initial thawing time in the area of the Artic seas south of 72°N occurs in the last ten-day period in May; in the first ten-day period of June it commonly reaches 75°N; by the end of the second ten-day period of this month it reaches the northern boundaries of the marginal seas (Gudkovich *et al.*, 1972); and by the end of June, reaches the near-Pole area (Figure 3.3, see colour section). The standard deviation, characterizing the interannual variability of the time period when thawing starts, is about 10 days. The rate of ice thickness decrease, caused by thawing in the Arctic seas, varies extensively. In the Arctic basin, the average value of ice thickness decrease, within a season, ranges from 75–100 cm near the northern margins of the seas to 0–20 cm in the near-Pole area (Figure 3.4, see colour section).

In marginal seas, ice thawing results in the gradual removal of the ice cover from a larger or smaller part of their water area, and sometimes from the entire sea area. As a result, the ice cover extent changes (i.e., the share of sea taken up by the ice cover). Ice cover destruction points are commonly flaw polynyas, river mouths, advection areas of relatively warm water of oceanic origin and clean water areas (which do not freeze in winter due to a significant heat inflow from water). Ice cover destruction is accompanied by divergence. The pattern of this process, which proceeds differently in different seas, is considered in detail in a number of publications (Gudkovich *et al.*, 1972; Zakharov, 1996; Gudkovich and Zakharov, 1998).

In the Arctic basin, outside the marginal seas, under current climatic conditions, the ice cover is retained all the year round. In summer, its slight divergence is recorded, which, to a major extent, depends on dynamic processes. The so-called *equilibrium ice thickness* H_p (i.e., thickness of multiyear ice), which under the given climatic conditions does not change from year to year, can serve as an indicator of the thermal interaction between the atmosphere and the ocean under the multiyear ice conditions. Zubov (1945) estimated the H_p value, having equated the winter ice thickness increase with summer thawing. A detailed analysis of the factors controlling both the equilibrium ice thickness and its formation time was given in the paper by Doronin and Kubyshkin (2001). Using the example of the calculations for two Arctic regions (1, Wrangel Island; 2, Franz Josef Land), the authors of this paper demonstrated that under the current climatic conditions and with a heat flow from the water of $F_w = 7.2\,\mathrm{kJ/cm^2}$ per year, the equilibrium ice thickness in the first region is 258 cm (a formation time 12 years); and in the second one, 285 cm and 26 years, correspondingly. At $F_w = 0$, the value of these indicators would change to 340 cm (18 years) and 475 cm (50 years), respectively. If in the latter case there is no snow cover, the formation time for the equilibrium ice thickness would change to 11 and 38 years, correspondingly. Therefore, the factors increasing the amplitude of seasonal ice thickness fluctuations diminish the H_p value and its formation time. Taking into account the fact that, as a result of drifting, the ice cover moves from one climatic zone to another, and also that the time period during which there is multiyear ice at high latitudes of the Arctic basin is limited, its thickness commonly does not reach the equilibrium value.

3.3 ICE DRIFT OBSERVATION RESULTS

The results of ice drift observations during drifting expeditions, supplemented by the data on the drifting of automatic spar buoys and buoys, allowed investigation of the basic characteristics of ice drift velocity and its patterns of space–time variability.

3.3.1 Statistical characteristics of drift velocity in the Arctic basin

For many scientific and practical tasks it is necessary to know the climatic (numerical and probabilistic) characteristics of ice drift velocity. These characteristics can be obtained on the basis of observational data received applying different techniques. The largest amount of information on ice drift velocity and direction in the Arctic Ocean (AO) was obtained from observations of drifting stations and automatic radio buoys. As to the observation 'density' (number of measurements per unit area), this Ocean surpasses other World Ocean water areas of a similar size.

The analysis shows that ice drift velocity vectors, as a rule, follow the law of a normal elliptical distribution. The simplest numerical characteristics of this distribution are the following parameters (Marchenko, 1964):

- V_s and σ_s – the average value of module and scalar mean square deviation of the drift velocity vector;
- V_r and σ – the module of the average resultant vector and its vector mean square deviation;
- l – the degree of dispersion ellipticity; and
- q and λ – the climatic stability parameters of ice drift velocity.

Formulas, from which the above numerical characteristics were calculated, are given in the paper by Gudkovich and Doronin (2001). All the characteristics change noticeably in space and time. Taking this into account, the initial observation materials were grouped on the basis of eight areas of the Arctic basin (shown in the above publication) and four seasons: January–March, April–June, July–September and October–December.

An important feature of the statistical characteristics of ice drift velocity is their dependence on the time interval, for which the resultant vector is determined. The

Table 3.1. Statistical ice drift characteristics for different time periods averaged over a year.

Periods	L	Q	λ	V_s	σ_S	σ	V_r	n
24 hours	–	0.30	0.28	6.5/7.5	3.5/4.0	7.0/8.1	–	9,816
Ten days	0.28	0.48	0.47	42/4.9	25/2.9	44/5.1	20/2.3	1,682
Month	0.41	0.65	0.74	90/3.5	50/1.9	84/3.2	59/2.3	582
Season	0.53	0.80	0.11	215/2.8	88/1.1	152/2.0	168/2.2	269
Half a year	0.56	0.88	0.48	388/2.5	151/1.0	232/1.5	345/2.2	159

Note: V_s, σ_S, σ and V_r values in the numerator are in km for the corresponding period while in the denominator they are in cm/s. n represents the number of observations.

following intervals were adopted: (a) a conditional 24 hours – time interval between astronomic determinations of coordinates on drifting stations (this interval averaged at about 30 hours); (b) a ten-day calendar period (the coordinates which are closest in time to the 1st, 11th and 21st day of each month were used; (c) a calendar month; (d) three-month intervals, corresponding to the above seasons; (e) half-year intervals for the summer (April–September) and winter (October–March) months; and (f) yearly intervals (October–September and April–March).

In addition to the observation materials of the 'North Pole' drifting stations and the Russian automatic radio devices (until 1975), information on the location of expeditions on the *Fram* (1893–1896), *Maud* (1922–1924) and *G. Sedov* (1937–1940) vessels were used in calculations as well as information from the American drifting stations 'T-3' (1952–1970) and 'Arlis-2' (1961–1965). For half-year periods, account was also taken of observations over the drift of expeditions on *Tegethoff* (1872–1873), *Jeannette* (1879–1881), *Dymphna* (1882–1883), *Svyataya Anna* (1912–1914), *Karluk* (1913–1914), *Chelyuskin* (1933–1934) and *Lenin* (1937–1938) vessels and also Sturkerson's drifting camp (1918). This information was used to calculate the statistical characteristics of ice drift velocity for the six time intervals, both areally and seasonally. Table 3.1 presents their values for all the regions for different periods averaged for a year – enabling us to reveal the pattern of seasonal changes of these characteristics and their dependence on scale.

Mean values of ice drift velocity module, obtained on the basis of all the data, point to their well expressed dependence on timescale. With increasing averaging period (T), the average ice drift velocity decreases from 7.5 cm/s for 24-hour periods to 2.2 cm/s for the period of one year, that is almost a 3.5-fold decrease. And, contrariwise, the values of ellipticity (l) and stability (q and λ) parameters increase significantly with increase in the period. Dependences of V_s, σ and σ_s on T are adequately approximated by the power functions obtained applying the least-squares technique:

$$V_s = 7.5T^{-0.217} \tag{3.3}$$

$$\sigma = 8.7T^{-0.333} \tag{3.4}$$

$$\sigma_s = 4.4T^{-0.290} \tag{3.5}$$

Here, V_s, σ and σ_s values are expressed in cm/s and T is in days.

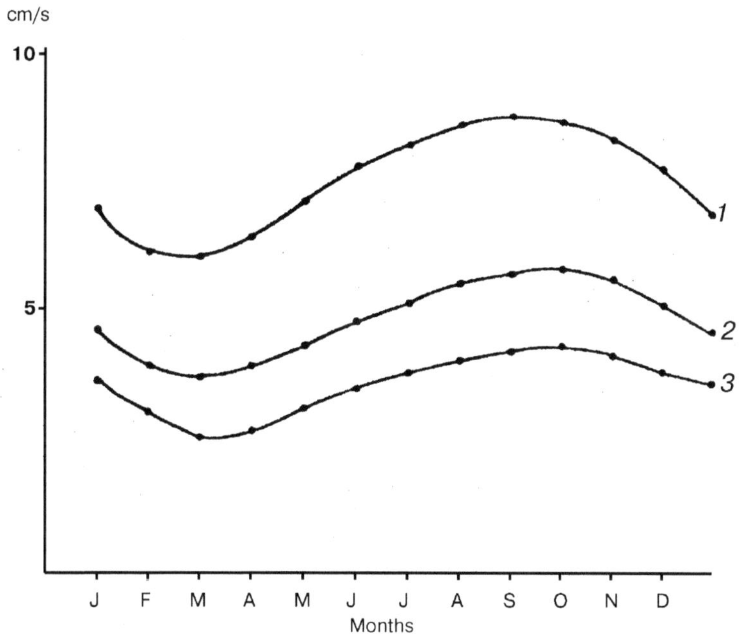

Figure 3.5. Annual variations in the mean ice drift velocity module for daily (1), 10-day (2) and monthly (3) periods.

Statistical characteristics of ice drift velocity display annual variations. As is seen in Figure 3.5, the average modules of drift velocity for the 24-hour, ten-day and monthly periods change in a coordinated way within a year: the lowest average drift velocities are recorded in March; the highest ones, at the end of summer and the beginning of autumn (August–October). Similar to the average velocity, the amplitude of its seasonal variations decreases with growing extent. Let us note that even within a three-month period the average velocity modules change slightly within a year and become similar for the six-month periods, since a seasonal velocity increase by the end of summer is compensated for by its diminishing stability.

Statistical characteristics of ice drift velocity vectors enable the calculation of such important parameters as the main mean square deviations; and on their basis, the parameters of the dispersion ellipse (direction and value of its main axes) which are associated with the expectancy for a random vector of hitting the corresponding ellipse. Using Figure 3.6(a,b) as an example, vectors of the average resultant ice drift velocity for the month and dispersion ellipses for the Arctic basin areas for two seasons are shown. Their analysis allows us to reveal the features of space–time changes: an increase of interannual velocity variability from the winter season to the summer one and the growth of the dispersion ellipticity degree in the areas adjoining the Fram Strait and the Chukchi and Beaufort Seas, particularly in winter.

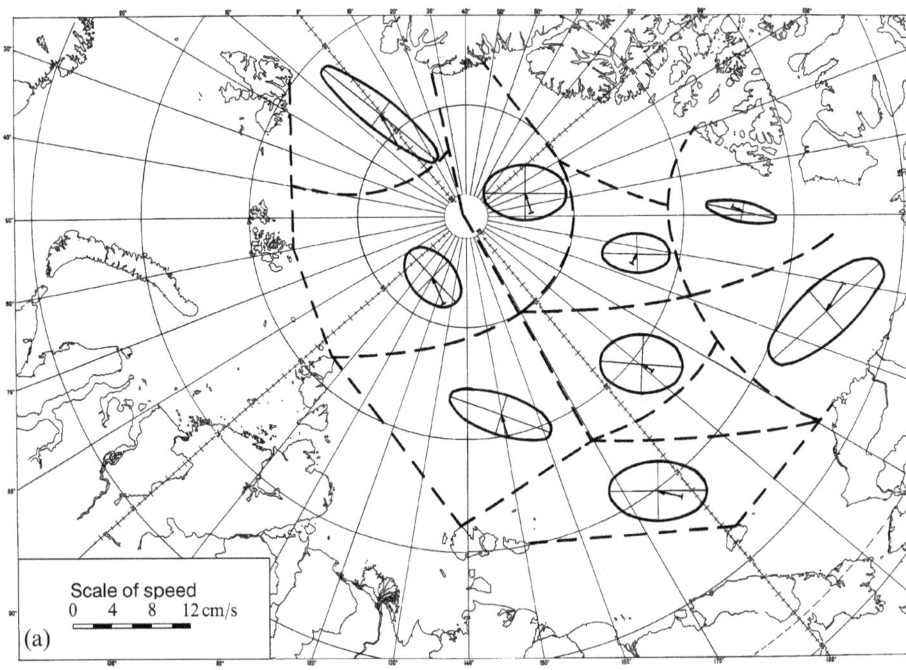

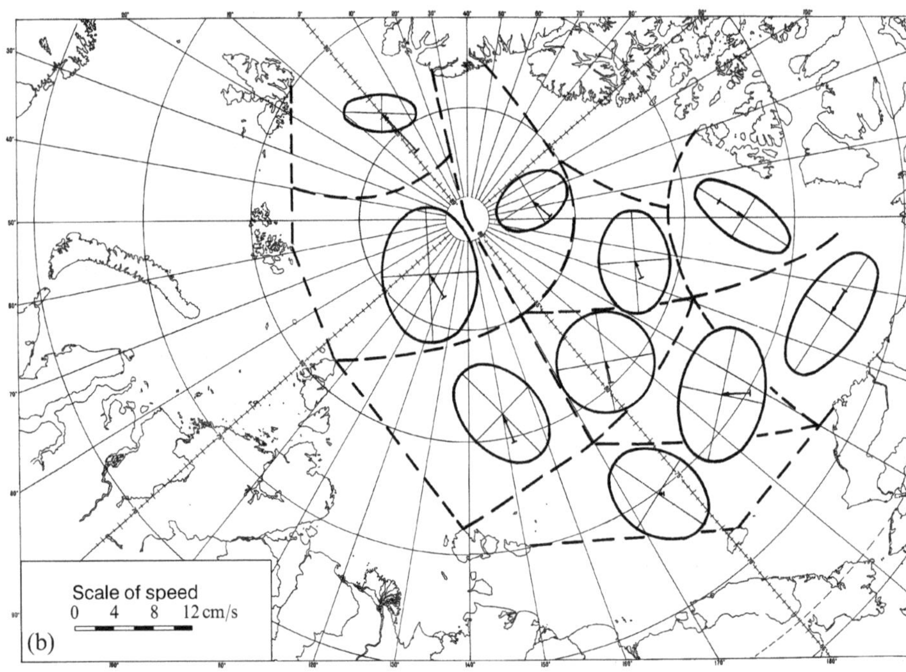

Figure 3.6. Vectors of the resulting mean ice drift velocity for one month and dispersion ellipses ($P = 80\%$) in (a) January–March and in (b) July–September, by region.

Table 3.2. Values of the parameters of the module distribution function of ice drift velocity depending on the resultant velocity averaging period.

	Averaging period			
Parameters	1–30 days	3 months	6 months	12 months
γ	1.58	1.96	1.85	2.6
δ	0.80	0.85	1.27	1.17

Values of statistical characteristics allow, proceeding from the law of the normal circular and elliptical distribution, the calculation of simple ice drift roses, which characterize drift reiteration and average drift velocity for each compass point when the number of observations is not sufficient for reliable calculations based on the empirical evidence (Figure 3.7).

For resolving a number of hydrometeorological and engineering tasks, it is necessary to have an expression of the density function of the distribution probability of the drift velocity vector. This function is well approximated by the formula proposed in the paper by Anapolskaya and Gandin (1958):

$$P(V > v) = \exp\left[-\delta\left(\frac{V}{V_s}\right)^{\gamma}\right] \tag{3.6}$$

where γ and δ are certain parameters depending on the variation coefficient σ_s/V_s.

Table 3.2 gives the values of these parameters depending on the vector averaging scale.

Adequate results in the case of describing the distribution of the ice drift velocity module are also given by the function expressed by the Pearson curve of the first type (Gudkovich *et al.*, 1989). Thus, knowing the average value of the ice drift velocity module in the target area and season, the probability of certain velocity values can be calculated from equation (3.6) as well as the value of the maximum velocity of the specified probability (P).

3.3.2 Main constituents of ice drift velocity

Being located at the boundary of such a mobile media as the atmosphere and the ocean, sea ice cover is subject to a continuous mechanical impact of these media resulting in the movement of the constituent ice formations in space (drift). An exception is made in the case of ice cover sections of limited area, connected with the coast by the sea floor (fast ice, stamukhas), as well as relatively short-term stops of compacted ice drift under conditions of onshore directed winds near the coasts and fast ice.

The first systematic observations on ice drift conducted by Nansen (1902) during the Transarctic expedition on the *Fram* (1893–1896) concluded that ice drift in the

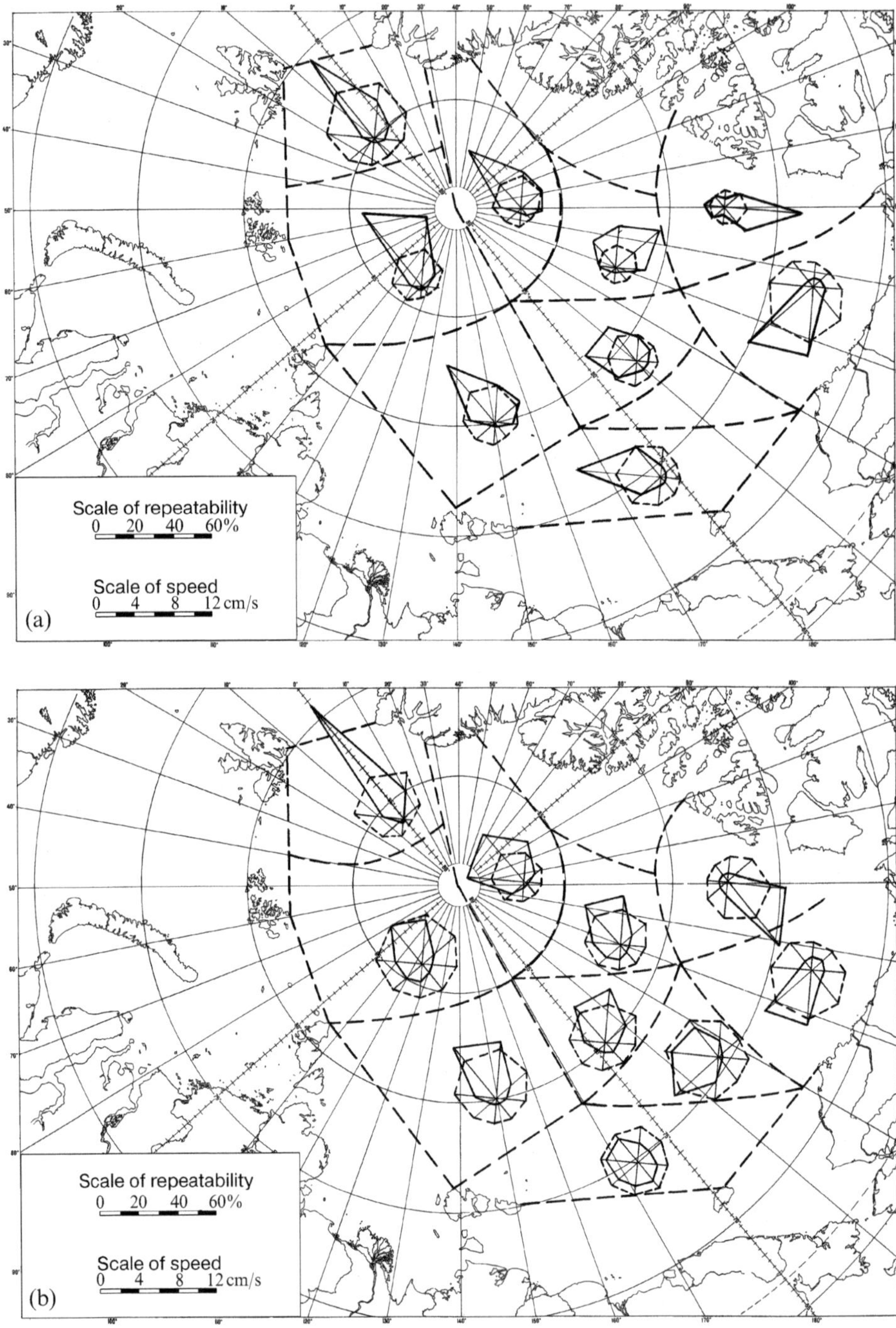

Figure 3.7. Roses of monthly ice drift velocities in (a) January–March and (b) July–September, by region.

Arctic basin had two basic components: a purely wind component closely related to local wind, and a 'non-wind' component, which was essentially independent of local wind at the time of observation. The first component is caused by tensions on the ice cover upper surface resulting from relative air movement (wind); the second one is determined by different interrelated factors, such as gradient currents, sea level slopes, tidal phenomena, etc.

For separating the above components both direct observations of the currents and water masses, and different indirect procedures (based on comparing the data on the resultant ice drift and wind or atmospheric pressure gradient for certain time intervals) can be applied (methods used by Nansen (1902), Sverdrup (1928), Vize (1933), Buynitsky (1951), Watanabe (1967), Gudkovich (1965), Thorndike and Colony (1982), Thorndike and Colony (1983)).

Indirect methods were extensively used for studying ice drift in the Arctic basin. The pattern of changes in total drift velocity and its components in space and time was revealed, particularly, their seasonal variations (Buynitsky, 1951; Vorobyev and Gudkovich, 1976). Assessments of the role of the 'non-wind' component in the general ice drift in the Arctic basin show that it averages at about 60%. The role of the current is highest in the middle of summer and lowest in transitional seasons. It increases with increasing time interval (averaging scale), which is accounted for by a much higher stability in the gradient current (mostly of baroclinic nature) as compared with the wind. The role of the current also increases toward the strait between Greenland and Spitsbergen, where it averages at 70%, reaching 90% in the Greenland Sea at latitudes of 74–78° in the summer (Gudkovich and Pozdnyshev, 1995). According to the above pattern, on the one hand, general ice drift is to a large extent determined by the system of gradient currents, and on the other hand, drift variability, both within the short and long-term periods primarily depends on the wind flow anomalies.

A characteristic feature of the wind ice drift in the Arctic basin is the practically linear dependence of its velocity on the speed of the wind causing it; the ratio of these velocities (the so-called wind coefficient) is on average close to 0.02. Wind drift direction deviates from the wind direction (in northern hemisphere) by an angle of about 30° to the right.

Expressions for wind drift velocity projections (V_x, V_y) on the axis of Cartesian coordinates (x, y), connecting them with the corresponding wind velocity projections (W_x, W_y), have the following form:

$$\left.\begin{aligned} V_x &= k(W_x \cos\alpha + W_y \sin\alpha) \\ V_y &= k(W_y \cos\alpha - W_x \sin\alpha) \end{aligned}\right\} \tag{3.7}$$

where k and α are the wind coefficient and angle of ice drift deviation from the wind.

Observational data indicate that the wind component of the drift rather quickly follows changes in the wind: commonly, in 3–6 hours after the onset of a stable wind, ice drift acquires a similar fixed character (Buynitsky, 1951). Quite often, as a result of transferring internal stresses through the ice cover, the ice drift changes precede those of wind (Bushuev *et al.*, 1967; Gudkovich, 1957). The consequence of inertia, ice cover inhomogeneity, impact of internal interaction forces and some other factors is a more or less significant spread of empirical values of the wind coefficient and drift deviation angle from the wind; however, it might be presumed, to a rather great approximation, that for the compacted multiyear ice the wind coefficient as well as the angle between drift and wind directions are constant for the given season.

Wind coefficient k and angle α vary noticeably within a year: they increase from the end of winter to the mid-summer. The value of seasonal k variations amounts to 50% of its mean value, and α variations, to 40%. Causes of annual variations of k and α values are due to seasonal fluctuations of certain forces determining ice drift. The most significant role in this process is played by stratification of the atmospheric and oceanic boundary layers and the irregularities of horizontal ice surfaces, which control shearing stresses on these surfaces as well as changes in ice cover thickness, concentration and strength affecting the Coriolis forces and internal resistance. Obstacles (coast, fast ice) have a significant effect on the wind drift value and direction.

Since the velocity of surface ('ground') wind is related to the atmospheric pressure gradient at sea level, the dependence of the wind component of the ice drift velocity (V_B) on the atmospheric pressure gradient ($\partial P/\partial n$) can be expressed by the formula first proposed by Zubov (1945):

$$V_B = K(\partial P/\partial n) \tag{3.8}$$

where K is the so-called isobaric coefficient, its numerical value being obtained empirically or calculated from the wind coefficient k on the basis of empirical relations between the speeds of geostrophic and 'ground' wind. The fact that the deviation angle of the latter from the geostrophic wind direction is on average close to that of the wind drift from the wind (the above angles being of contrary signs) enabled Zubov to formulate the law of isobaric drift: ice drifts along instantaneous isobars with a velocity proportional to the atmospheric pressure gradient.

Actually, absolute values of angles between ice drift and wind (α), as well as between geostrophic and 'ground' wind directions (γ) are not equal to each other – therefore, wind ice drift is not strictly directed along the isobar. Using the known formulas for the projections of geostrophic wind speed on x and y axes, it is possible to express the projections of wind ice drift velocity by means of atmospheric pressure gradient components:

$$\left.\begin{aligned} V_x &= K\cos\beta\left(\frac{\partial P}{\partial y} + tg\beta\frac{\partial P}{\partial x}\right) \\ V_y &= K\cos\beta\left(\frac{\partial P}{\partial x} + tg\beta\frac{\partial P}{\partial y}\right) \end{aligned}\right\} \tag{3.9}$$

Here, $K = kk_1/\rho_a f$, where ρ_a and f are the air density and the Coriolis parameter; k_1 is the ratio of ground and geostrophic wind speeds $\beta = \alpha + \gamma$.

In the ice drift calculations the second terms of equations (3.9) expressing ageostrophic corrections can be neglected. However, as will be shown below, these terms play an important role in the investigations and calculations of ice cover compaction, divergence and deformation. Equations (3.9) represent a generalization of the above-mentioned Zubov's law of isobaric drift, which holds true provided $\beta = 0$.

Annual variation of the mean value of the isobaric coefficient is similar to changes in the wind coefficient. Deviation angle of wind ice drift velocity from the isobar in the middle of summer, according to the observations in the Arctic basin, averages at $+20^\circ$ decreasing to -5° by the end of winter. Analysis of the observation results over the drift of automatic radio devices in the Arctic seas of the Russian offshore has demonstrated that the mean values of isobaric coefficients are 25–40% higher, and the angles of deviation drift from the isobar 5–10° lower than in the Arctic basin. Such differences can be accounted for by a smaller ice thickness in the Arctic seas as well as by the influence of barotropic gradient currents, the role of which should increase with decreasing sea depth and increasing proximity to the coasts. It should be noted that in case of ice thickness below 100–120 cm as well as under non-stationary conditions with a changing wind, β angle can assume negative values not only in winter but also in summer, which is of fundamental significance for the processes of ice cover divergence, compaction and compression.

As noted above, surface currents have a major effect on the ice cover drift. The action of a drift current in the Ekman layer, caused by wind-induced movement of ice relative to the water, forms the resistance force, which is one constituent of the balance of forces determining the wind ice drift. Reverse ('drag') action of currents on the ice cover, together with a direct impact exerted on it by the gravity projection on the inclined sea surface, are mainly due to the gradient currents caused by a sloping of the free sea surface and the corresponding horizontal pressure gradients in water. Ice cover movement determined by these factors is sometimes called gradient drift (Felsenbaum, 1959).

The system of gradient currents in the (AO) mainly forms under the impact of the following factors:

- large-scale wind fields determined by atmospheric circulation patterns;
- freshwater balance of the AO, the main components of which are river runoff and atmospheric precipitation in excess of evaporation;
- inhomogeneous distribution of water density resulting from the impact of mechanical and non-mechanical factors;
- global tilting of the sea level surface between the Pacific and Atlantic;
- conditions of water exchange with neighbouring oceans; and
- morphological features of the AO (configuration of coasts, bottom topography).

The actions of these factors are closely interrelated. Wind fields influence the system of gradient currents as a result of adaptation of the mass field (water density) to the resultant wind circulation. Therefore, the system of currents in the deep sea part of the AO is mainly baroclinic. This circumstance is associated with a high stability of the system of currents, since a restructuring of the water density field mainly determined in the Arctic by the distribution of salinity, is associated with movement of water particles over large distances. Under these conditions, the formation time of such currents for the scope of the Arctic basin is measured in years

(Gudkovich and Nikiforov, 1965), which provides grounds for calling these currents 'constant', 'non-wind', etc.

Analysis of the results of observation of ice drift and wind as well as the materials of oceanological surveying accomplished in different years allowed the construction of a rather detailed scheme of surface water circulation in the AO (*Atlas of the Oceans*, 1980; Gudkovich, 1957, 1961a; Treshnikov and Baranov, 1972). In accordance with these schemes, in the Amerasian sub-basin, above which an anticyclonic wind field prevails, an extensive anticyclonic water circulation occurs, around the periphery of which a broad Transarctic current passes originating from the Pacific current of the Chukchi Sea and coming out of the Arctic basin through the Fram Strait. Its extension in the Greenland Sea is the East Greenland current. Towards the Greenland Sea, the Transarctic current is replenished by the surface runoff from the Russian Artic seas. A characteristic feature of the system of currents in these seas are cyclonic circulations also covering the adjacent areas of the Arctic basin. In the straits of the Canadian Arctic archipelago, surface water movement is directed from the Arctic basin to the Baffin Sea, from where the Labrador current originates. Average velocity of the Transarctic current increases towards the Fram Strait from 1–2 cm/s to 5–7 cm/s (in the Strait proper) and to 18 cm/s (at a latitude of 72° in the Greenland Sea) (Gudkovich and Pozdnyshev, 1995). Current velocity varies within the year: in the Arctic basin, it is the highest in summer and the lowest in winter; in addition to the annual component, these variations also display a half-year constituent – in accordance with a seasonal variation of water inflow through the Bering Strait. Seasonal variations in the Fram Strait display a winter maximum of current velocity resulting from the corresponding variations of the prevailing wind.

Oceanological surveying of the Arctic basin accomplished in different years by high-latitude air expeditions 'North' enabled the revealing of interannual variations in the system of surface currents – primarily concerned with the ratio between areas with anticyclonic and cyclonic circulation systems as well as the position of the Transarctic current core. These variations display a cyclicity with a period of about 8–10 years (Gudkovich, 1961a; Gudkovich and Nikolayeva, 1963). A similar fluctuation rhythm is displayed by ice conditions in the Artic seas, ice export to the Greenland Sea, inflow of Pacific water and a whole series of other indicators of the hydrological regime in the AO (Gudkovich and Kovalev 1967; Shpaykher, 1969).

The main factor determining interannual variability of the system of surface currents in the Arctic basin is the wind field above the AO, which is closely related to the atmospheric pressure distribution. Intensification of the Arctic anticyclone and its displacement towards Eurasia results in the expansion of anticyclonic circulation and reduction of cyclonic, weakening of anticyclone and its displacement towards the American continent – to the opposite consequences. It is characteristic that similar variations of atmospheric circulation and the corresponding system of barotropic currents in the Arctic basin were confirmed in a number of recent studies (Proshutinsky and Johnson,1997; Polyakov *et al.*, 1999), where they were called the 'Arctic fluctuation'. It should be noted that, unlike the design fields of barotropic currents, in which the circulation direction changes completely from one epoch to another, in the Arctic basin system, only water areas covered by circulation systems of a certain sign change. This points to the above-mentioned high stability of baroclinic currents in the study area.

In shallow Arctic seas, particularly near the coasts, water density distribution does not fully reflect the structure of gradient currents and the corresponding sea surface slopes. Here, gradient currents, to a major extent, have a barotropic character. Since the formation time of such currents is comparatively short, the gradient component of the current and, hence, of the ice drift, correlates well with the wind, changing together with changes in the latter, hence the empirically determined higher mean values of wind coefficients and their dependence on wind direction relative to the coast. The latter factor also has a marked effect on the angle between the ice drift and the wind direction. River runoff, which strongly increases in the spring–summer season, also has a certain impact on the system of currents in the Arctic seas.

Empirical evidence, in general, testifies to the applicability of the Ekman theory of gradient currents to ice drift in coastal areas (Glagoleva, 1950; Gorbunov, 1956). According to this theory, the highest velocity of the gradient current directed along the coast is 1.4-fold as compared with the wind current velocity on the surface. Consequently, the wind coefficient for the total ice drift should be at most 2.4-fold as compared with its value for the wind drift. The observational data show that the wind coefficient value is even higher in certain straits, where due to the narrowing of the stream, gradient currents can be much stronger than those found near a straight coast.

Currents near the ice edge have a specific character resulting from the action of both mechanical and non-mechanical factors. As was shown by Vize (1944b), horizontal gradients of hydrological and meteorological elements are commonly more acute here. According to hydrological observations (Gudkovich and Nikolayeva, 1963; Nikolayev *et al.*, 1986; Frolov, 1987), near the ice edge, peculiar lenses of water freshened by thawing were recorded, which are associated with anticyclonic eddies in water and ice cover. In the paper by Proshutinsky (1993), the generation model of such eddies is considered; and the paper by Gudkovich and Zakharov (1998) presents the results of numerical experiments confirming the impact of ice edge dynamic processes on the ice cover distribution in the Arctic seas.

In some areas of the Arctic basin, particularly in the marginal seas, tidal currents have a marked effect on the ice drift resulting in the appearance of a periodical tidal component to the ice drift. Short-period (diurnal, semi-diurnal) and long-period (from several days to several years) tidal current (ice drift) components are distinguished.

Proceeding from the results of several investigations, the direction and velocity of short-period tidal ice drift and tidal current components practically coincide. It is only for compacted ice cover that the tidal drift velocity is 10–20% lower than the tidal current velocity in the upper water layers. There is evidence indicating that the onset time of the maximum velocities as well as tidal ice drift change to the ebb-tidal ice drift and vice versa, is slightly (1–3 hours) ahead of the corresponding times for the tidal currents. This can be due to the direct impact of sea level slope on the tidal wave.

According to Dvorkin (1970), the maximum velocity of semi-diurnal tidal currents over most of the Artic seawater area is 10–30 cm/s. Only within certain limited areas does it exceed 1 m/s. Maximum tidal ice drift velocity in deepwater parts of the Arctic basin, as a rule, does not exceed 5 cm/s. In quadrature, it is 2–2.5 times less than in syzygy.

Tidal ice drift trajectories in most of the regions are close to circular. It is only near the coasts that they commonly become reversible. The estimates based on the long waves theory show that in most parts of the Artic seas the maximum ice movement on the tidal wave is 2–4 km, rarely reaching 20 km, whereas in the Arctic basin it, as a rule, does not exceed 1 km. However, practical significance of tidal drift is due to an essential irregularity of its velocity in space, which is associated with the phenomena of ice cover concentration, divergence and compression. An important role is played by the tidal currents in the process of fast ice formation. In some papers (Proshutinsky, 1983; Kowalik, 1981) the impact of non-linear effects generated in the process of tidal level fluctuations, currents and interaction of the latter with the ice cover, on the excitation of residual (Stokes) currents is discussed. However, the role of these currents in ice drift is, apparently, insignificant for most of the AO regions.

Of doubtless interest are the attempts to investigate the impact of long-period tidal components on the ice drift. The analysis of results of ice drift observation in the Arctic basin enabled Maksimov to conclude that tidal ('astronomic') drift in the central Arctic is one of the constituents of total ice drift resulting in a loop-like character, particularly during periods of wind weakening (Maksimov, 1959). Local significant semi-monthly and monthly variations in currents and ice drift were recorded by analysing the results of observations over the currents at buoy-based stations as well as the data from the drift of radio spar buoys in the Arctic seas (Vorobyev, 1966).

Due to the fact that the oscillation amplitude of the current and ice drift velocity resulting from the impact of long-period tides is rather low, when studying these phenomena it is necessary to leave out the components caused by the impact of the wind. A quasi-periodicity can be recorded in its changes; its periods can (accidentally or regularly) appear to be close to the periods of long-term tides. Application of a special procedure, excluding wind drift, to a 140-day series of daily measurements of ice drift velocity at one of the drifting stations (Gudkovich and Evdokimov, 1971) enabled the receipt of the real characteristics of semi-monthly and monthly tidal components of the ice drift. The maximum velocity of the above components appeared to be equal to 1.5 and 0.9 cm/s, correspondingly, which is commensurable with the gradient ice drift velocity; however, it adds up to barely 1/3 of the mean wind drift velocity for a 24-hours period. In this case, a unidirectional ice floe displacement under the impact of tidal drift does not exceed 10–12 km. Thus, the influence of long-period tides results in noticeable variations of gradient ice drift in the Arctic basin; however, their role in the total drift is insignificant. This conclusion generally agrees with the results of analysing the considered phenomena in the Arctic seas (Voynov, 1999).

In addition to the tidal movement, drift ice is also subject to inertial oscillations in a horizontal plane leading to the appearance of loop-like trajectories, their radius depending on the initial drift velocity and width of the area. In the Arctic seas, where the period of inertial oscillations is close to that of principal tidal semi-diurnal waves, the radius of inertial orbit is commonly around 1 km. It is rather difficult to separate the inertial and tidal ice movements. For this, long-term and precise observations are required (Hunkins, 1967).

3.3.3 General ice drift in the Arctic basin

Maps of ice drift roses and dispersion ellipses discussed in the previous section contain a lot of useful information on the space–time variability of ice drift velocity. However, they give no idea of the gradual change, within the space of the region, of such characteristics as the module and direction of the average resultant drift vector, which expresses the most probable ice movement for a certain period. Each projection of this vector signifies a statistical field, which, according to Koplan-Dix (1968), can be represented as a certain relief surface assigned to this period and region, near which all the observed values are found deviating from it upwards and downwards at different random distances restricted by the limiting values.

In this case, each of these surfaces can be approximated by the analytical function, for instance, a bidimensional polynomial of the form:

$$\tilde{V}_{x,y} = \sum_{q=0}^{t} \sum_{p=0}^{m} (a_{pq})_i x^p y^q \tag{3.10}$$

where x, y are coordinates of the drift vectors $\tilde{V}_{x,y}$ origin; m is the degree of the polynomial based on the p argument; t is the degree of the polynomial based on the q argument; and a_{pq} is the polynomial coefficient.

From the system of equations of the type (3.10), generated from the observation materials, an extended matrix of normal equations is formed. The a_{pq} coefficients can be found by least-squares methodology. A root-mean-square value of the difference of actual and predetermined values of (σ_δ) functions can serve as a measure of approximation accuracy; it is associated with values of mean square deviations of the corresponding sample σ_i by the ratio:

$$\sigma_\delta = \sqrt{1 - r_i^2} \tag{3.11}$$

where r_i is the correlation coefficient between the actual and computed values of the functions. Hence:

$$r_i = \sqrt{1 - \left(\frac{\sigma_\delta}{\sigma_i}\right)^2} \tag{3.12}$$

From this formula, r_i values were calculated for each set of polynomial coefficients (equation 3.10), approximating the fields of the average resultant ice drift for four three-month seasons, two half years and one full year. Calculations were performed for two versions of exponents: $t = m = 2$ and $t = m = 3$. The summary of σ_i, σ_δ and r_i values obtained on the basis of the second version, which has a slight advantage over the first (the number of polynomial coefficients in this case equals 16), shows that with an increasing length of period the closeness of the associations increases. This is due to the fact that with an extension of the period the relative role of drift variability in time decreases compared with its variability in space. The approximation error is higher in the summer season, when the time variability of the drift is particularly high.

From equation (3.10), on the basis of the assigned initial coordinates in the Arctic basin, it is easy to give a climatic forecast of the drift of stations and other objects for 3, 6 or 12 months. Probability of a certain error in each case can be determined on the basis of a corresponding dispersion ellipse. These equations were used for constructing the schemes of the most probable ice drift in the Arctic basin within certain time periods. Some of these schemes are given in the *Atlas of the Oceans* (1980). Essentially, these were the first quantitative schemes of drift, which were not obtained by calculations based on the dynamic models, but were entirely based on the statistical analysis of observed results. An advantage of such a scheme is the level of detail, which is typical of design diagrams, along with a reliability commonly characterizing the schemes based on the field data.

Figure 3.8(a,b) presents schemes of the average resultant ice drift in the Arctic basin for the winter and summer half years calculated from the relevant approximation formulas. These schemes agree with the earlier published qualitative schemes of the general ice drift (Gordienko, 1958; Gudkovich, 1957). They reflect the existence of the Transarctic ice flow between the near-Pole area and the northern margins of the seas of the Eurasian offshore directed to the Greenland Sea. From the left, a great amount of ice from the Russian Artic seas joins this flow; from the right, it is adjoined by an area of anticyclonic circulation, the centre of which is at about 78°N, 150°W.

Ice getting into the Transarctic flow from the Kara Sea is carried to the Greenland Sea in 1–2 years; from the Laptev Sea, in 2–3 years; from the East Siberian Sea, in 3–4 years; and from the Chukchi Sea, in 4–5 years. The ice circulation period within the anticyclonic circulation ranges between 4 to 10–12 years: the lower limit is related to the area of quasicircular ice drift in the Beaufort Sea and north of it; the higher one, to the trajectories passing through the stagnant area adjoining the northern coasts of Greenland and the Canadian Arctic archipelago. In summer, ice from this region often gets into the straits of the Canadian Arctic archipelago and further into the Baffin Sea (Belov and Vedernikov, 1968; Winchester, 1962).

Ice drift observation results enable the determination of the probability that the drifting objects become involved in anticyclonic circulation. The paper by Gudkovich (1965b) determines the position of two borders, one of them delineating the Arctic basin, wherefrom ice does not get directly into the Greenland Sea. In

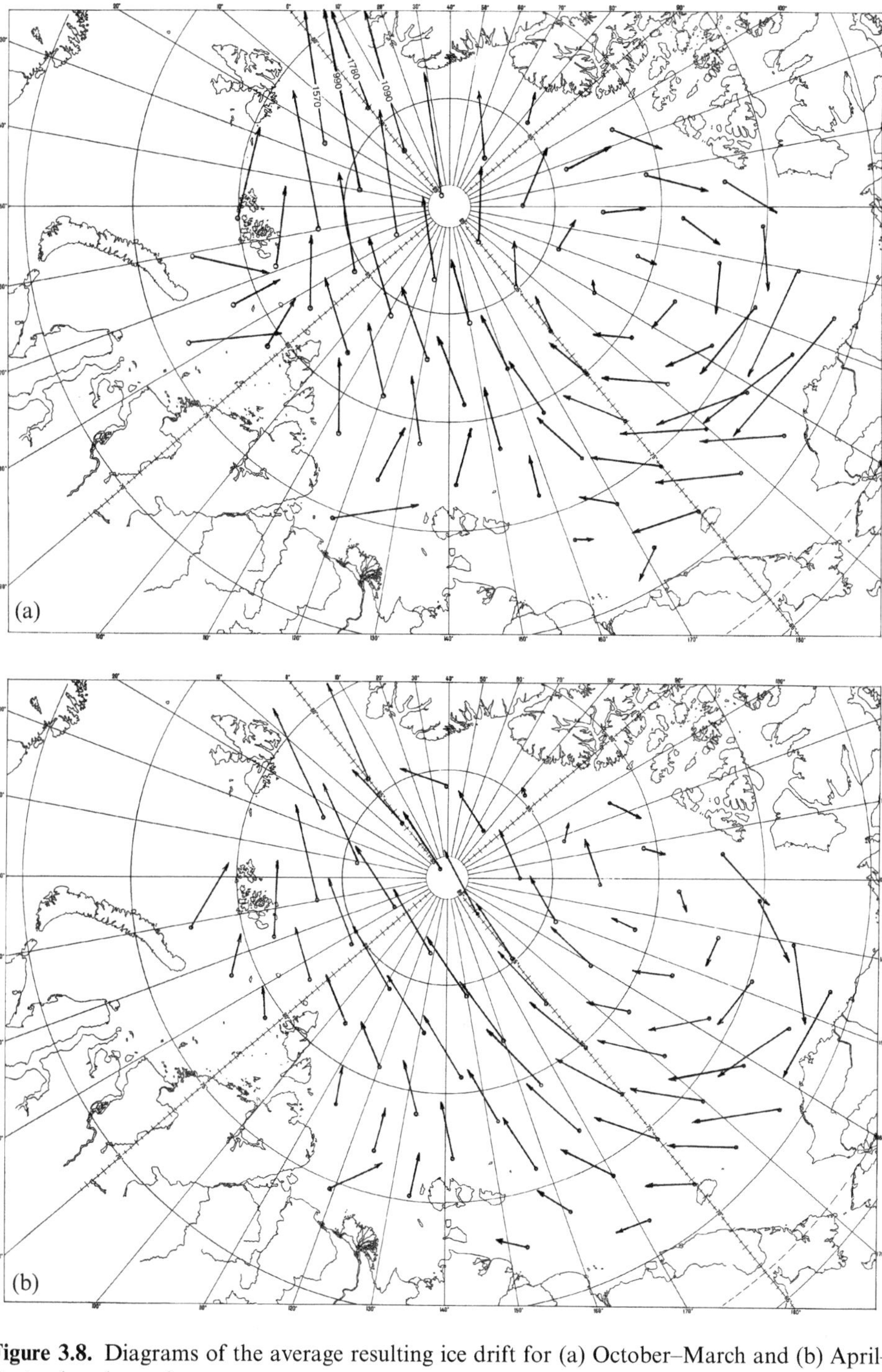

Figure 3.8. Diagrams of the average resulting ice drift for (a) October–March and (b) April–September, by regions.

summer, a part of ice from this area moves into the straits of the Canadian Arctic archipelago; however, most of it either remains within the circulation, or on its completion, is incorporated into the effluent Transarctic flow. The second boundary outlines an area, wherefrom ice is either carried to the Greenland Sea, or thaws within the marginal seas, never becoming involved in the anticyclonic circulation. Between the above boundaries, there is a transitional zone, wherefrom ice in certain cases is carried to the Greenland Sea; and in other cases, gets into the circulation area. Due to the transitional zone, the area of ice involved in circulation can vary from 2.5–3.5 million km^2 (approximately 40–60% from the Arctic basin).

Variability of the general ice drift scheme also shows up in a marked deviation of certain branches of the Transarctic flow from their common position. As a result, in certain years or seasons, closed cyclonic ice circulation occurs in the Laptev Sea and in the area of Wrangel Island; and a commonly existing effluent flow from the Kara Sea to the Artic Basin reverses its direction. Most frequently, such changes occur during the summer season.

From the comparison of the ice drift schemes for the winter and summer half years given in Figure 3.8 certain differences between them are revealed. Most apparent are the differences in ice drift velocity at the approaches to the Fram Strait, where in winter the velocity grows toward the strait. Another feature is the displacement of the Transarctic flow core from Eurasia to America and the corresponding decrease of anticyclonic circulation area from winter to summer.

A more detailed analysis of the ice drift schemes for three-month seasons calculated from equation (3.10) demonstrates that for two winter seasons (October–December and January–March) a scheme approximately corresponding to the above-described average annual character of the general drift is typical. In this case, the position of the Transarctic flow axis, which commonly extends along the northern margins of the Artic seas, can change significantly from year to year, affecting the ice export from these seas. In spring (April–June), the drift scheme acquires the simplest of patterns: in most areas of the Arctic basin and marginal seas, except for a comparatively small stagnant area north of the Canadian Arctic archipelago, ice drifts toward the Greenland Sea. In summer (July–September), the Transarctic flow core is displaced towards the line joining the Bering Strait and the Fram Strait. However, interannual variability of the ice drift in this period is most pronounced. In certain years, the drift character, typical of winter seasons, is retained, while in others the cyclonic system with its centre commonly located north of the Laptev Sea develops abnormally. These changes have a strong effect on the ice exchange value between the marginal seas and the Arctic basin (Gudkovich and Nikolayeva, 1963).

3.4 PATTERN OF IRREGULAR ICE COVER MOVEMENT

3.4.1 Types of irregular movement

In addition to the forward movement of ice – drift proper, of great scientific and practical interest in the studies of ice cover movement are the effects arising due to the changing velocity of movement in space and during the seasons, years. These changes are determined by the ice drift velocity field. At any moment, V_x and V_y components of the velocity vector V in this field are functions of x and y coordinates, i.e.:

$$V_x = V_x(x, y) \qquad V_y = V_y(x, y)$$

Having selected the point of origin in a random point x_0 and y_0, let us break down these expressions into Taylor series:

$$\left.\begin{aligned} V_x &= V_{xo} + \left(\frac{\partial V_x}{\partial x}\right)x + \left(\frac{\partial V_x}{\partial y}\right)y \\ V_y &= V_{yo} + \left(\frac{\partial V_y}{\partial x}\right)x + \left(\frac{\partial V_y}{\partial y}\right)y \end{aligned}\right\} \tag{3.13}$$

Here, the higher order terms are discarded, which are insignificant if rather small areas are considered where the velocity might be regarded as changing linearly.

From the four space derivatives comprised into equation (3.13) the following four characteristics of plane motion irregularity can be obtained:

$$\operatorname{div} \vec{V} = \frac{\partial V_x}{\partial x} + \frac{\partial V_y}{\partial y} \tag{3.14}$$

$$\operatorname{rot} \vec{V} = \frac{\partial V_y}{\partial x} - \frac{\partial V_x}{\partial y} \tag{3.15}$$

$$\operatorname{def}_1 \vec{V} = \frac{\partial V_y}{\partial x} + \frac{\partial V_x}{\partial y} \tag{3.16}$$

$$\operatorname{def}_2 \vec{V} = \frac{\partial V_x}{\partial x} - \frac{\partial V_y}{\partial y} \tag{3.17}$$

Equations (3.14) and (3.15) express, correspondingly, the velocity field divergence and rotor (vorticity); equations (3.16) and (3.17), angular and linear deformations. The first two characteristics are invariant in respect of the coordinate axes choice (i.e., they retain their value and sign in case of changes in the coordinate system). The two latter characteristics depend on the coordinate system, and when it is turned by 45°, pass into each other (similar to the changes in the Cartesian coordinates of the point in case of axes rotation by 90°).

It is not difficult to prove that expression:

$$\operatorname{def} \vec{V} = \sqrt{(\operatorname{def}_1 \vec{V})^2 + (\operatorname{def}_2 \vec{V})^2} \tag{3.18}$$

is invariant with respect to the coordinate system. Therefore, equation (3.13) can be written in such a form:

$$\left.\begin{aligned} V_x &= V_{x0} + \frac{x}{2}\operatorname{div} \vec{V} + \frac{x}{2}\operatorname{def} \vec{V} - \frac{y}{2}\operatorname{rot} \vec{V} \\ V_y &= V_{y0} + \frac{y}{2}\operatorname{div} \vec{V} - \frac{y}{2}\operatorname{def} \vec{V} + \frac{x}{2}\operatorname{rot} \vec{V} \end{aligned}\right\} \tag{3.19}$$

The kinematic essence of equations (3.19) is as follows: ice cover can be transferred from one position to another one by forward transfer (drift proper), divergence (i.e., expansion or compression), deformation (in the narrow sense, this term means a change in ice body shape without changes in its dimensions) and, finally, rotation. A common feature of the movement described by the last three terms is the spatial irregularity of the velocity field (which is typical of it). Therefore, equations (3.14)–(3.17) can serve as kinematic indicators of the irregularity of the ice cover movement, each of them expressing quite definite features of this irregularity which are independent from each other.

It is common that drift and rotation can occur without the displacement of ice floes, whereas divergence and deformation involve displacement. Normally, all types of ice movement are simultaneously involved. However, each type of movement reflects different processes and phenomena occurring in the ice cover and, thus, can be studied separately. In the subsequent sections the main results of these investigations are presented.

3.4.2 Ice cover rotation

Studies of the patterns of rotational ice cover movement are not only of scientific interest, but also of great practical interest. Data on ice floe rotation allow us to make judgments of the certain transverse irregularity of the wind and gradient ice drift.

Ice floe rotation in the process of complicated ice cover movement has been recorded by many researchers (Buynitsky 1951; Hackel and Khmyznikov, 1938; Gudkovich, 1955a; Lappo, 1958; Timokhov, 1970.). The statistical data given below are based on extensive observations of 'North Pole' drifting stations. These data indicate that the rotation speed changes with time, its average value depending on the timescale and season. In summer months (from June–September), rotation speeds increase significantly. It is in this period that in certain cases the rotation speed approaches a value of 90° per day. This is commonly associated with intense ice cover divergence – when ice floes appears to be isolated from the surrounding ice.

Table 3.3 gives the statistical data on ice floe rotation in the Arctic basin for the 24-hour, 10-day, month and 3-month periods. From the data in this table it follows that the mean values of clockwise (ω_+) turns, counter-clockwise (ω_-) turns, mean values for the entire set (ω), mean absolute values ($|\omega|$), as well as the main deviations (σ) increase noticeably from winter to summer. Besides, these values grow with extension of the period, whereas the asymmetry (Λ) and excess (ε) diminish. The latter circumstance indicates that with increasing averaging scale the distribution of rotation speeds approaches a normal one.

In calculations of the characteristics given in Table 3.3 the above-mentioned cases of values falling out abruptly from the corresponding data set were excluded in accordance with the extremes truncation criteria adopted in the statistics. Their number does not exceed 3–4% of the number of observations, and in most cases is associated with the summer period. Use of Kolmogorov's concordance criteria at a

Table 3.3. Statistical data on ice floe rotation (in degrees) for different periods in winter (numerator) and summer (denominator).

Period	ω_+	ω_-	ω	$\lvert\omega\rvert$	σ	Λ	ε	Number of cases	Notes
One day	$\frac{0.9}{2.2}$	$\frac{-1.0}{-2.2}$	$\frac{0.12}{0.33}$	$\frac{0.6}{1.7}$	$\frac{1.0}{3.3}$	$\frac{-0.56}{-0.07}$	$\frac{15.9}{5.5}$	$\frac{7{,}254}{4{,}166}$	Excl. 3 cases Excl. 44 cases
10 days	$\frac{3.0}{5.2}$	$\frac{-2.9}{-4.9}$	$\frac{1.35}{1.50}$	$\frac{3.0}{5.0}$	$\frac{3.9}{6.8}$	$\frac{-0.26}{-0.25}$	$\frac{2.2}{1.5}$	$\frac{721}{387}$	Excl. 1 case Excl. 32 cases
1 month	$\frac{6.4}{12}$	$\frac{5.2}{10}$	$\frac{3.9}{4.6}$	$\frac{6.0}{12}$	$\frac{7.0}{14}$	$\frac{-0.19}{-0.13}$	$\frac{0.7}{1.2}$	$\frac{244}{128}$	Excl. 12 cases
3 months	$\frac{19}{24}$	$\frac{-8.0}{-18}$	$\frac{12}{16}$	$\frac{17}{23}$	$\frac{15}{23}$	$\frac{0.05}{0.16}$	$\frac{1.1}{-0.2}$	$\frac{93}{28}$	Excl. 1 case Excl. 7 cases

5% significance level allows us to assume the distribution of ice floe rotation values for the ten-day, month and three-month periods as following the normal law.

Different opinions were proposed concerning the causes of ice floe rotation: non-simultaneous wind and current changes at opposite sides of the ice floe (Hackel, Khmyznikov, 1938), inhibitory influence of coasts and compacted ice massifs (Zubov, 1945), levelling of the main ice floe axis in the eastward–westward direction under the influence of the Earth's rotation (Crairy, 1958), curvilinear ice drift trajectories in the baric systems and transverse irregularity of constant currents (Buynitsky, 1951), inertial forces (Lappo, 1958), etc. Timokhov (1970), solving the balance equation of the moments of external forces applied to ice, derived an equation, where angular acceleration of the ice floe is a function of wind speed rotor, currents as well as ice floe concentration and size.

It is known that the rotation speed in continuous media is associated with the characteristic of transverse speed irregularity – rotor (vorticity) – by the ratio:

$$\frac{\partial \omega}{\partial t} = -\frac{1}{2}\operatorname{rot}\vec{V} \tag{3.20}$$

A minus sign indicates that under a positive rotor value, the rotation proceeds counterclockwise. Using the above ratios of drift velocity to atmospheric pressure gradient, we obtain:

$$\operatorname{rot} V \approx k \operatorname{rot} W \approx K\nabla^2 P \tag{3.21}$$

where ∇^2 is a bidimensional Laplace operator.

Hence, the wind component of ice movement velocity rotor (as well as ice cover rotation speed) is proportional to the wind speed rotor or Laplacian of atmospheric pressure. The validity of equation (3.21) is qualitatively confirmed by the data given in the paper by Buynitsky (1951): during the drift of the *G. Sedov* ice-breaking vessel, in 94% of all cases, clockwise ice floe rotations occurred under an anticyclone above the drift area; and counter-clockwise rotations in 76% of cases were recorded during the passage of cyclones. Quantitatively, this expression was checked by observation at the 'NP-2' drifting station (Gudkovich, 1955a), and in greater detail – on the basis of observations on ice testing areas in the Arctic basin, in 1961, 1962 and 1966 (Bushuev *et al.*, 1967). Values of correlation coefficients of such dependences for ten-day averaging scales are 0.75–0.85. Empirical linear dependences, as a rule, include free terms, which express the influence of transverse irregularity of gradient currents independently of the wind.

Two schemes are proposed in the paper by Volkov *et al.* (1971), accounting for the discovered coordination of separate ice floe rotation with rotation of more or less extensive ice cover areas. The first scheme suggests that the entire massif is turning around a common rotation centre as a single solid plate. In accordance with the second scheme, rotation of ice floes is determined by the friction of their edges under conditions of rectilinear drift in the presence of shear deformation. Under real conditions, ice floe rotation can occur in accordance both with the first, and with the second scheme. Within certain short time periods, the character of rotation can change; however, within long time periods, in certain regions, one scheme would commonly dominate, whereas in other regions the other would prevail. Thus, in the area of anticyclonic ice circulation in the American sector of the Arctic basin, clockwise ice cover rotation will mainly follow the first scheme. In the zone of the Transarctic current the drift velocity decreases from the current core to its margins; therefore, here the rotation according to the second scheme prevails; left of the core, ice floes commonly rotate counter-clockwise; and right of the core, clockwise.

Ice floe rotation in the open pack and the discontinuous ice of the marginal Artic seas, where the role of turbulent processes is significant, has its own specificity. As a result of repeated aerial surveying of the ice cover in the Chukchi Sea the following was ascertained (Gorbunov and Timokhov, 1968):

- Randomly alternating ice floe groups with a unidirectional rotation are distinguished in the ice cover.
- Rotation speed varies rather extensively in space – from 0 to ±100 degrees/hour; for the majority of ice floes this speed lies within ±30 degrees/hour and has a distribution close to normal.
- Mean-square rotation speed is inversely proportional to the average ice drift velocity; it decreases with increasing ice cover concentration and increases with diminishing ice floe size.

It should be noted that the kinetic energy of ice rotational movement is several orders less than that of the forward movement.

3.4.3 Velocity divergence and changes in ice cover concentration

As follows from the expression of plane motion rate divergence (equation 3.14), div V characterizes the divergence or convergence velocity of the ice floes making up the ice cover. Its numerical expression conveys changes in the value of the ice cover elementary area assigned to a unit area, for a unit of time. Therefore, the value and sign of div V control the intensity of the ice cover compaction or divergence.

According to Nikiforov (1957), the mechanical change of ice concentration N ($0 < N \leq 1$) within a time unit is described by the following equation:

$$\frac{\partial N}{\partial t} = N \operatorname{div} \vec{V} - \vec{V} \operatorname{grad} N \tag{3.22}$$

The first term in the right-hand side of this equation characterizes the change in concentration due to the spatial non-uniformity of drift velocity; and the second term, a change in concentration due to advection. The influence of the latter depends on non-uniformity of concentration, drift velocity as well as the angle between the direction of drift and concentration gradient. Should the observer move together with the ice, the intensity of concentration changes around him would be determined only by the first term of equation (3.22). At $\operatorname{div} V > 0$, a decrease in concentration (diverging) occurs; at $\operatorname{div} V < 0$, an increase in concentration occurs, which at $N = 1$ (10 tenths) will pass into compression and ridging.

Divergence of ice movement velocity can be caused by the following factors:

- spatial non-uniformity of current velocity and sea level;
- spatial non-uniformity of wind fields;
- influence of coasts (or fast ice);
- ice cover inhomogeneity (its thickness, concentration, ridge concentration, which determines the roughness of the upper and lower ice surfaces as well as the weight per surface unit);
- horizontal diffusion processes; and
- non-stationary character of external impact.

Due to low velocities of 'constant' currents over most of the Arctic basin and its marginal seas, the velocity divergence caused by these currents is, as a rule, low. Nevertheless, in the areas of convergence and divergence of such currents associated with specific features of bottom topography and coast configuration, the corresponding anomalies in reiteration of concentrations or divergences of the ice cover can be recorded, their value also depending on the action time of the processes.

Divergence of velocity resulting from tidal currents depends on the type, height and length of a tidal wave. In the case of a progressive wave, the maximum divergence (i.e., the highest intensity of ice cover divergence) is recorded at the time when the tidal current gives way to the ebb tidal current, whereas the highest convergence (maximum intensity of compaction, and at $N = 1$, the highest compression) should occur at the time when the ebb tidal current gives way to the tidal current. Ice cover concentration reaches its maximum on the tidal wave crest (high tide) and minimum at the time of ebb tide (low tide) (Legenkov, 1958, 1963). The amplitude of tidal changes of divergence, all other things being equal, increases with decreasing wavelength depending on the wave period and sea depth. Therefore, tidal phenomena in the ice cover, which are commonly observed in offshore areas appear to be insignificant in the deep sea. However, even in the case of a shallow depth and a relatively high velocity of tidal currents the resulting changes in concentration – due to a comparatively short process duration – are minor and rarely exceed 1 tenth.

Nevertheless, these phenomena play an important role for navigation conditions – strongly depending on the presence or lack of compressions.

It should be noted that tidal phenomena under real conditions are more complicated than the pattern above – obtained on the basis of the simplest relations of the long waves theory. These phenomena are not adequately studied under continental slope conditions, where the tidal range changes abruptly, and a partial reflection of tidal waves occurs, which can, moreover, have a progressive standing character.

Unfortunately, the question of drift velocity divergence caused by inertial phenomena has been inadequately studied. As is known, in the case of ice drift velocity changes with time related to wind strengthening or weakening, inertial oscillations arise in the horizontal plane both in the ice cover, and water: the trajectory of ice floes assumes a loop-like character. The radius of inertial orbits depends on the drift velocity and latitude of the area. In the Arctic seas, it is about 1 km. The period of inertial oscillations here is rather close to that of the principal semi-diurnal tidal waves – as a result, it might be difficult to separate the inertial and tidal movements of ice floes. Long-term and precise observations would be necessary for that (Hunkins, 1967).

Inertial ice cover oscillations can be of interest in relation to this question, since changes in the velocity in space, related with these phenomena, should be determined by a non-simultaneous character of wind changes. This depends on the movement speed of baric systems, which is, on average, several times lower than that of tidal wave propagation under conditions of deepwater. Hence, it might be expected that the velocity divergence resulting from the inertial ice cover movement can under certain conditions be several times stronger than that under tidal drift.

Numerous studies (Buynitsky, 1951; Gudkovich, 1957; Drogaytsev, 1956; Zubov, 1945; Felsenbaum, 1957; Khanaychenko, 1946) show that the greatest role in the mechanical alteration of ice cover concentration away from the coasts and ice edge is played by ice movement velocity divergence resulting from non-uniformity of the wind field. Both the non-uniformity proper of the wind fields and the degree of wind drift deviation from isobaric (geostrophic) relations are significant, since under isobaric ice drift, as follows from expressions (2.8) and (3.14), $\operatorname{div} V = 0$. As was mentioned in Section 3.2.3, deviation of the real ice drift from isobaric drift shows in the angle between the ice drift direction and instantaneous isobar $\beta \neq 0$, and a non-linear dependence of drift velocity on atmospheric pressure gradient.

In order to reveal the association of ice movement velocity divergence with wind and baric fields let us use the ratios of the components of drift velocity, wind and baric gradients considered in Section 3.2.3. Then:

$$\operatorname{div} \vec{V} = \frac{k \sin \beta}{\sin \gamma} \operatorname{div} \vec{W} = K \sin \beta \nabla^2 P \tag{3.23}$$

Since γ angle always has a negative value, from expression (3.23) it follows that at $\beta > 0$ the divergence of ice movement velocity is opposite in its sign to the wind speed divergence. Only at $\beta < 0$ (thin ice, winter) do the signs of divergence of drift velocity field and wind coincide.

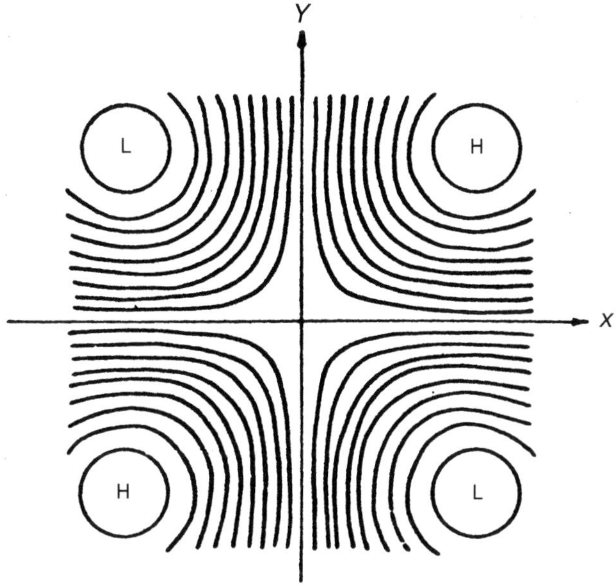

Figure 3.9. Diagram of the deformation pressure field at $L_x = L_y$ (isobars are drawn across 1 hPa).

In order to reveal the dependence of drift velocity divergence and redistribution of ice cover concentration on non-uniformity of the wind fields let us use the model of a stationary baric field proposed by Drogaytsev (1956) and later applied by other authors. Such a field is described by the expression:

$$P = P_o \sin\frac{2\pi}{L_x}x \sin\frac{2\pi}{L_y}y \tag{3.24}$$

where P is deviation of atmospheric pressure from a certain average value; P_o is the amplitude of pressure variations; and L_x and L_y are distances parallel to coordinate axes between atmospheric pressure maxima and minima.

The appearance of this field at $P_o = 10$ millibar and $L_x = L_y = 4,000$ km is shown in Figure 3.9.

Drift velocity components were calculated from equations (3.9) at $\beta = 20°$, taking into account the empirical dependence of the value of isobaric coefficient K on the atmospheric pressure gradient. Ice concentration was calculated from equation (3.22).

Figure 3.10 presents the results of calculating the concentration redistribution in the baric field within 10 days under a uniform initial concentration equalling 8 tenths. As is seen in the figure, the greatest changes in concentration occur near the centres of cyclones (diverging) and anticyclones (compaction). On the deformation field axes these changes are slight; thus, the role of non-linearity of drift velocity

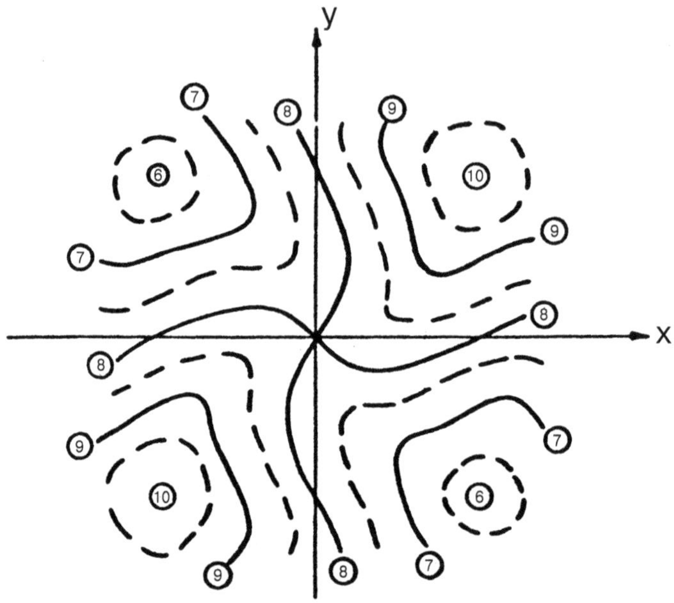

Figure 3.10. Calculated distribution of ice cover concentration in the stationary pressure field (Figure 3.9) at $N_0 = 0.8$ (8 tenths) and $T = 10$ days.

dependence on the baric gradient is of secondary importance. In the hyperbolic point of the field there is no change in concentration.

The character of ice concentration changes revealed on the basis of the listed calculation results is confirmed by numerous observations of aerial ice patrols. During the years when cyclonic fields prevail in summer above the Arctic basin, observations record a significant ice cover divergence sometimes reaching 4–6 tenths. Contrariwise, in the years when the reiteration of the Arctic anticyclone is higher, ice cover remains concentrated. This pattern is also confirmed by the results of calculating areas of triangles and quadrangles constructed on the basis of simultaneous coordinates of the Russian and American drifting stations and from observations at specially organized testing grounds. Correspondence between changes in the area and ice cover rotation was revealed: an increasing area (ice diverging) was accompanied by a counterclockwise ice cover rotation; a diminishing area (ice compaction), a clockwise rotation (Figure 3.11). This pattern agrees with the conclusions based on the joint consideration of the above formulas.

In winter, the association between drift velocity divergence and rotation direction changes in accordance with variations in the angle of drift deviation from the isobar (β), though it might be difficult to grasp these changes, since the mean value of divergence in this period is much lower. Nevertheless, the available data do not contradict such a conclusion (Bushuev *et al.*, 1967; Volkov *et al.*, 1971; Hibler *et al.*, 1972).

Above, the distribution of ice drift velocity divergence, and compacting and

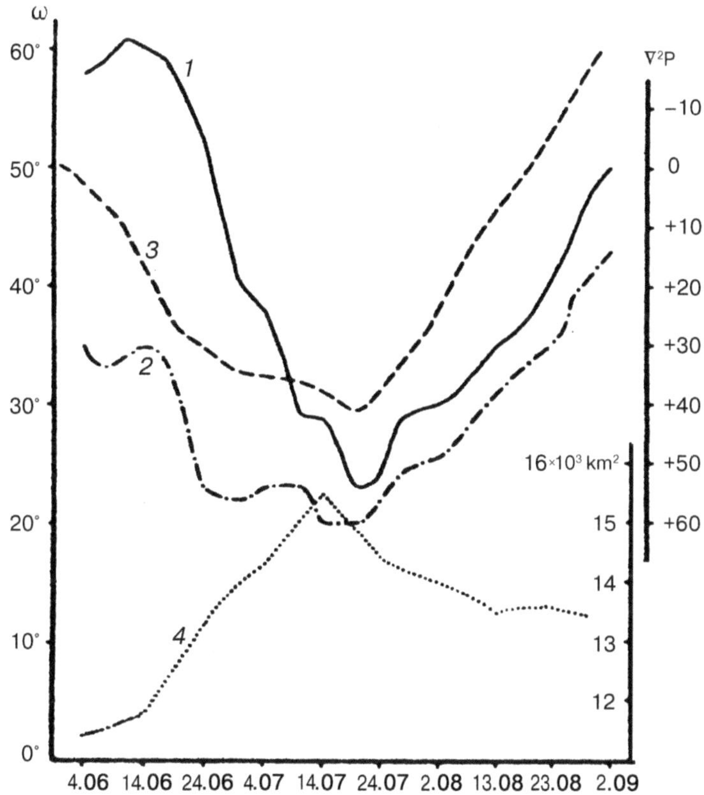

Figure 3.11. Change of the mean orientation of ice floes (1), ice polygons (2), , $\nabla^2 P$ value (3) and polygon area (4) in the summer of 1966.

diverging zones in stationary baric systems were considered. However, quite a number of dynamic effects are associated with non-stationarity of baric fields proper – their movement in space and changes with time. Under these conditions, into equations (3.9) (connecting the wind speed components with baric gradients), additional terms should be introduced, taking into account the 'isallobaric' wind directed from the area of atmospheric pressure increase to that showing a decrease (Hibler *et al.*, 1972). After this, a correction should be included into equation (3.23) which is proportional to the Laplacian change rate of atmospheric pressure with time.

The paper by Gudkovich (1979a) features an example of calculating div V in a baric system moving at constant velocity and comprising a cyclone and an anticyclone – along the axis passing through their centres. Calculations show that in the case under consideration, in front of the cyclone, processes should arise resulting in ice compaction; and in front of the anticyclone, in its divergence. This task is resolved more strictly in the paper by Pozdnyshev (1990). The above pattern is also confirmed by observations over ice compression in the autumn season: the most intense compressions are often recorded under the stronger winds connected with an approaching cyclone. In certain cases, zones of the highest ice compression coincide with the position of peaks of isallobaric atmospheric pressure drops, which confirms the adequacy of the considered model. However, special field experiments, a more complete and detailed analysis of observational results and their correlation with theory are still to be performed.

It is known that the inhomogeneity of ice cover thickness, as well as the roughness of its upper and lower surfaces mainly depending on ridge concentration, cause spatial non-uniformity of the ice drift velocity, resulting in internal stresses and deformations. Large-scale inhomogeneity of thickness can be particularly pronounced in ice export seas, where a gradual ice cover 'rejuvenation' occurs, and a well-expressed ice thickness gradient originates directed from the thick first-year ice towards the flaw polynyas, which in winter are quickly covered with young ice. High horizontal gradients of ice thickness are also recorded near the boundary of multiyear ice. A significant non-uniformity of ridge concentration can be recorded near the outer boundary of fast ice under conditions of prevailing onshore directed drift as well as in the zone of intensely crushed residual ice.

The influence of non-uniform ice thickness and ridge concentration on the formation of divergence and deformation velocity fields was investigated in a paper by Arkhipov *et al.* (2001) by means of a simple stationary free-drift model. Non-uniform distribution of thickness and ridge concentration was assigned by matrices with constant gradients: thicknesses ranging within 0.5–3.0 m and ridge concentration within 1–5 points. The target fields of characteristics of ice drift velocity non-uniformity were calculated for 8 cases of uniform wind, the direction of which changed by 45° relative to the thickness and ridge concentration gradients. Values of strain coefficients on the upper and lower surfaces of smooth ice, and their dependence on ridge concentration, were preliminarily determined on the basis of aerodynamic investigations of ice floe models (Gudkovich *et al.*, 1963).

Figure 3.12 shows the dependences of ice cover divergence and deformation rate on wind direction relative to the thickness and ice ridge concentration gradients obtained by numerical experiments. The thickness (Figure 3.12(a)) and ridge concentration (Figure 3.12(b)) values are shown; on the left-hand side of Figure 3.12(a,b) can be seen the conditional wind directions relative to the ordinate axis. From Figure 3.12(a) it follows that div V attains the maximum positive value under the wind directed along isolines of equal thicknesses if the ice thickness increases from left to right relative to the wind direction. An opposite wind direction causes maximum convergence.

From Figure 3.12(b) it follows that the divergence and convergence maxima caused by non-uniform ice ridge concentration are confined to the zone of elevated ridge concentration (3–4 points). The value of div V is the highest under winds directed toward higher ridge concentration, whereas under winds of opposite directions the maximum convergence is recorded. In the zone of lower ridge concentration, the vergence values are low. The above pattern is of great significance in the process of ice cover deformation associated with formation of linear continuity breaks – cracks and leads. This is considered further below.

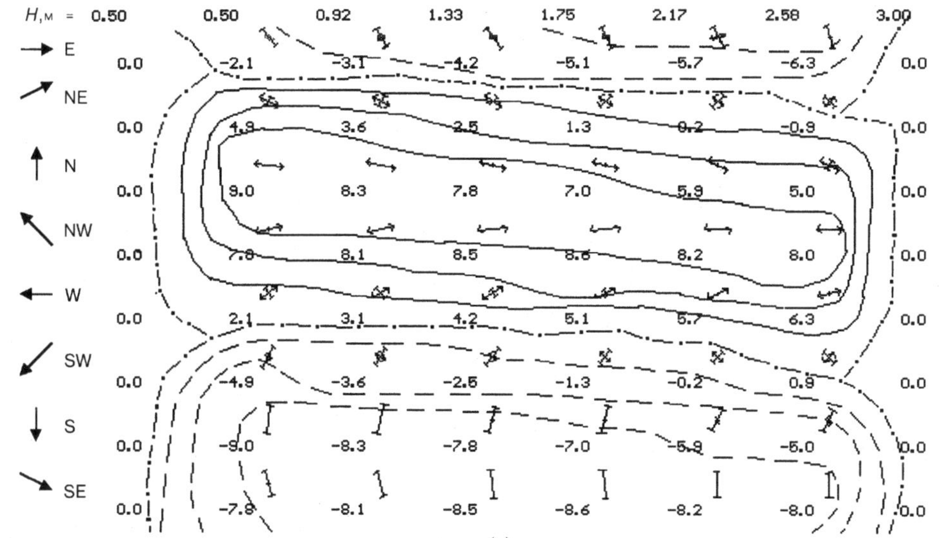

(a)

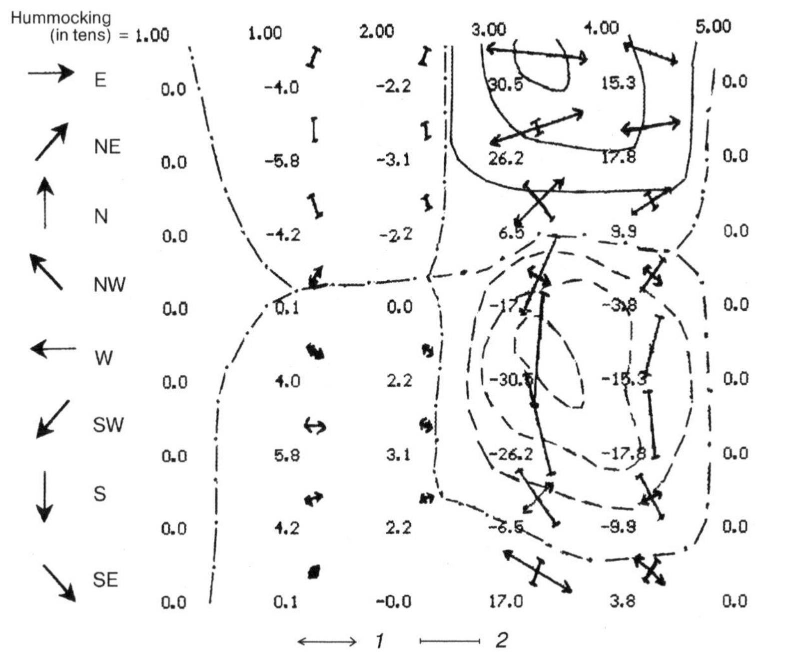

(b)

Figure 3.12. Isopleths of the calculated values of drift velocity divergence and characteristics of deformation ellipses at the non-uniform thickness of the ice cover (a) and its hummock and ridge concentration (b) depending on the wind direction (arrows on the left) at V = 5 m/s (1 – stretching axis, 2 – compression axis).

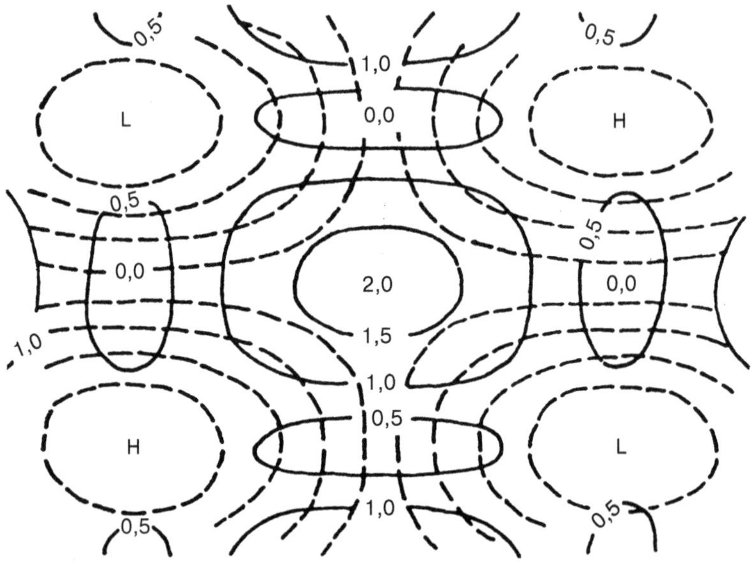

Figure 3.13. Distribution of the calculated deformation rate values (in $10^{-7}\,s^{-1}$) in the deformation pressure field at $L_y/L_x = 0.7$.

3.4.4 Ice cover deformation rate

Ice cover deformation, similar to divergence, occurs under a spatial non-uniformity of the ice cover movement velocity. Dependence of the deformation rate on non-uniformity of the wind (baric field) can be easily found by substituting expressions (3.9) into equations (3.16), (3.17) and (3.18) arriving at:

$$\operatorname{def} \vec{V} = K\sqrt{\left(\frac{\partial^2 P}{\partial x^2} - \frac{\partial^2 P}{\partial y^2}\right)^2 + \left(2\frac{\partial^2 P}{\partial x \partial y}\right)^2} \tag{3.25}$$

From this formula, it follows that the maximum deformation rate ('maximum shear rate') does not depend on the angle between the ice drift and geostrophic wind direction (β), but is only determined by the value of the isobaric coefficient and specific features of the baric field. Using equation (3.24) to describe the latter we can get the distribution of def $\vec{V}$ values (Figure 3.13) in a deformation baric field similar to that of Figure 3.9. Unlike in the latter, $L_y/L_x = 0.7$ was adopted (i.e., it is slightly elongated along the x axis).

The analysis of equation (3.25) shows that the deformation value taken into account by the first term of the square root is the highest in the baric centres and equal to zero on neutral lines of the baric field. At $L_y = L_x$, this term turns to zero, wherefrom it follows that in the baric systems close to circular ones this component of the deformation rate is practically lacking. The component described by the second term of the square root of equation (3.25) reaches its maxima at hyperbolic

Table 3.4. Types of baric formations during the *G. Sedov* ice-breaking drift in 1939 on days with and without shears.
From Buynitsky (1951).

Shears	Saddle number of cases (%)	Cyclonic number of cases (%)	Anticyclonic number of cases (%)
Recorded	73/68.5	20/18.5	14/13.0
Not recorded	57/23.5	74/31.0	109/45.5

points of the baric field and equals zero in the centres of cyclones and anticyclones and along the lines joining neighbouring centres.

From Figure 3.13 it follows that under a moderate 'elongation' of the baric systems the area of maximum deformation lies near the baric saddle, whereas in the centres of cyclones and anticyclones the deformation is insignificant. The result obtained accounts for the pattern known from observations: the strongest ice shears – break-ups and ridging – are reported from areas of baric saddles characterized by weak winds (Buynitsky, 1951; Gudkovich, 1957). This is confirmed by the data in Table 3.4 produced by Buynitsky by analyzing the observations of the *G. Sedov* ice-breaking expedition (Buynitsky, 1951).

Data from Table 3.4 confirm the above pattern. As we see, the distribution of deformation value in the baric fields differs from that of the divergence value distribution in them. Kinematics also shows the above indicators in the ice cover dynamics differ, allowing consideration of their movements separately. Nevertheless, another approach is also of interest (i.e., when the ice cover divergence and deformation are considered as a single deformation process – in the broad sense of the word – including not only changes in shape, but also in size (in our case – ice cover area)) (Timokhov and Kheysin, 1987; Hibler *et al.*, 1972). Using tensor description, this process can be expressed through a bidimensional linear tensor of deformation rate:

$$\dot{\varepsilon}_{i,j} = \frac{1}{2}\left(\frac{\partial V_i}{\partial x_j} + \frac{\partial V_j}{\partial x_i}\right) \tag{3.26}$$

having the following four components expressed through the rate derivatives along coordinate axes x, y.

$$\dot{\varepsilon}_{xx} = \frac{\partial V_x}{\partial x} \tag{3.27}$$

$$\dot{\varepsilon}_{yy} = \frac{\partial V_y}{\partial y} \tag{3.28}$$

$$\dot{\varepsilon}_{xy} = \dot{\varepsilon}_{yx} = \frac{1}{2}\left(\frac{\partial V_x}{\partial y} + \frac{\partial V_y}{\partial x}\right) \tag{3.29}$$

ε_{xx} and ε_{yy} values characterize linear deformation and the ε_{xy} value, angular deformation. As is seen, the sum of expressions (3.27) and (3.28) represents div V; their difference, $\mathrm{def}_2 V$; and the third component coincides with expression $\mathrm{def}_1 V$.

It is known that through any point on the plane two mutually perpendicular straight lines can be drawn, along which there will be no angular deformations; and of two of the components of linear deformation one will be the maximum, and the other one – the minimum. Such axes (ξ, η) are called the principal axes of strain, and the components of linear deformation rate along the principal axes – the principal deformation rates. The value of the principal deformation rates is determined from the formula:

$$\dot{\varepsilon}_{\xi\eta} = \frac{1}{2}\left[\mathrm{div}\,\vec{V} \pm \sqrt{\left(\mathrm{def}_2\vec{V}\right)^2 + \left(2\mathrm{def}_1\vec{V}\right)^2}\right] \tag{3.30}$$

and the direction of the principal axes (relative to the x axis) can be found from the expression:

$$\theta = \frac{1}{2}\mathrm{arctg}\frac{\mathrm{def}_1\vec{V}}{\mathrm{def}_2\vec{V}} \tag{3.31}$$

For a visual demonstration of deformation over a certain time interval (or deformation rate) it is convenient to draw a deformation (deformation rate) ellipse. The direction of the major axis of the ellipse corresponds to the maximum extension (or minimum compression); while the minor axis corresponds to the minimum extension (maximum compression). On the basis of the deformation ellipse one can approximately determine the direction of cracks in the ice cover (Hibler *et al.*, 1972). If a crack appears, ice deformation is determined by a shift along the crack and expansion (compression) in the perpendicular direction. Therefore, under the stretching deformation the major axis of the ellipse should be at an angle of 45–90° to the crack (90° – in the case of pure extension; 45°, in the case of simple shear). Under compressive deformation the major axis of the ellipse should deviate from the directions of cracks by an angle of 0–45° (0, for a pure compression; 45°, for a simple shear). According to Gorbunov *et al.* (1998), the directions of ruptures in the ice cover are commonly perpendicular to the elongation axis of the deformation ellipses. This pattern is accounted for by characteristic systems of disturbances in the sea ice cover continuity, among them arch-like systems discovered on approaches to the straits (Gorbunov *et al.*, 1986).

Turning back to the results from the analysis of divergence and deformation rate dependence on the non-uniformity of thickness and ice ridge concentration, let us consider the parameters of the deformation ellipses schematically represented in Figure 3.12. Taking into account the above remarks, ruptures in the ice cover directed approximately along the isolines of equal thicknesses should be most developed – if the wind blows along these isolines leaving thicker ice to the right (Figure 3.12(a)). This pattern is confirmed by aerial ice patrol results: in seas, where ice thickness increases from south to north, extensive leads often form in the ice cover under winds from the eastern quarter. Ruptures forming in the case of a non-uniform ridge concentration, as follows from Figure 3.12(b), should also be directed along the corresponding isolines; however, their greatest width (or frequency) will be recorded under the winds directed toward higher ridge concentration.

An ice cover deformation model appearing under the influence of onshore directed winds near coasts or stationary ice (uniaxial compression) is considered in the paper by Gudkovich and Romanov (1970). The results of calculations based on this model confirmed the known fact that there is a growing deformation intensity toward the coast, which is corroborated by the data on ice ridge concentration distribution near coasts and external boundaries of fast ice in the areas where onshore directed winds prevail. The model also allowed the determination of the inverse relationship of deformation rate and ice thickness.

The paper by Appell *et al.* (1976) considers a characteristic case of ice cover deformation, when there is an obstacle of limited size in the path of drifting ice. Internal interaction forces before the obstacle result in a decreasing ice drift velocity and angle of deviation from the wind. The zone of substantial impact to the obstacle, as was shown in Section 3.2.3, can spread over a great distance depending on the ice concentration and thickness and also the size of the obstacle. Due to the fact that outside this zone the resistance forces are much lower, deformation occurs in ice near its boundaries resulting in ruptures in the ice cover near the right margin of the zone affected by the obstacle (looking along the wind direction), ice pressure and ridging at its left margin. Such phenomena, which can be of practical value for navigation, are confirmed by observation.

3.4.5 Horizontal turbulence and diffusion phenomenon in the ice cover

One of the causes of spatial non-uniformity of ice drift velocity is horizontal turbulence. Turbulent motion should result in diffusion of ice cover properties, such as concentration, thickness, etc. (i.e., in levelling of their spatial non-uniformity and even displacement of ice edges toward the open sea (Appell and Gudkovich, 1992)). These processes are also associated with changes in concentrations of ice cover contamination of natural and anthropogenic origin, as well as with a gradual change in distance between the drifting objects mounted on ice. Such processes are also confirmed by the discovery of certain objects from abandoned drifting stations at great distances (to 50 km) from each other (Gudkovich, 1979b). The turbulent motion of ice floes is also related to their internal forces; interaction in dynamic models of the ice cover (Section 3.5.2), which by analogy with models of oceanic currents are sometimes identified with horizontal turbulent friction, though the momentum transfer mechanisms differ markedly in the ice cover and water.

For studying these processes in the ice cover it is probably possible to use the known provisions of A. N. Kolmogorov's local isotropic turbulence theory (Kolmogorov, 1941). According to this theory, horizontal turbulence is regarded to be resulting from a superposition of a great number of vortices of different scales, the energy of which not only comes from the averaged motion and is transferral through the hierarchy of scales from larger to smaller, but to a major extent – from the corresponding vortices in the atmosphere and sea and certain energy-supplying frequency zones (Ozmidov, 1968). It should be noted that the turbulent motion of sea ice is not affected by vortices, the size of which are much smaller than the characteristic size of ice floes.

The most important parameter of horizontal turbulence theory is the coefficient of horizontal turbulent exchange A_l, or diffusion coefficient, the study of which in the sea is discussed in numerous publications. For determining this parameter it is convenient to use the formula proposed by Obukhov (1941) and first applied for the studies of horizontal turbulence in the ocean by Richardson and Stommel:

$$A_l = \frac{(\overline{\Delta l})^2}{2\Delta T} \tag{3.32}$$

where Δl is a change in the distance between two ice floes within a ΔT time interval (the line above means averaging for a series of observations).

Synchronous observations of the movement of two or several ice floes at different distances from each other serve as the material for studying ice cover motion. For the first time, such a study was conducted on the basis of observation results at a special testing site in the area of the 'NP-10' drifting station in 1962 (Volkov *et al.*, 1971); and in 1966, on a broader scale, in the area of 'NP-15'. One year earlier, the results of aerial ice surveying in the Chukchi Sea were used for this purpose (Gorbunov and Timokhov, 1968).

Table 3.5 gives the mean values of the A_l coefficient calculated from results of half-year observations at the testing ground in 1966 for four spatial scales $l = 10$–780 km.

From the data in Table 3.5 it follows that the value of the horizontal turbulent exchange coefficient for concentrated ice in the Arctic basin, within the scale range 10^6–10^8 cm varies from 7×10^4 to 2.6×10^6 cm^2 s^{-1}.

According to the theory of local isotropic turbulence, the dependence of the horizontal turbulent exchange coefficient on the scale l should follow the 'law of 4/3' empirically discovered by Richardson and theoretically grounded by Obukhov (1941):

$$A_l = c\varepsilon^{1/3} l^{4/3} \tag{3.33}$$

where c is a universal dimensionless constant close to 0.1 and ε is the turbulent energy dissipation rate (energy flow along the above-mentioned cascade of vortices).

A_l and l dependence (equation 3.33) viewed on a logarithmic scale is expressed as a straight line, the slope tangent of which equals 4/3. Using this coordinate system, a diagram was constructed based on A_l values obtained from the above-mentioned observational results (Figure 3.14). As is seen from this figure, changes in the turbulent exchange coefficient within the scale interval of 100–1,000 km agree well

Table 3.5. Dependence of the A_l coefficient on l distance.

l (km)	10	110	300	780
$A_l \times 10^{-6}$ (cm^2 s^{-1})	0.07	0.14	0.71	2.56

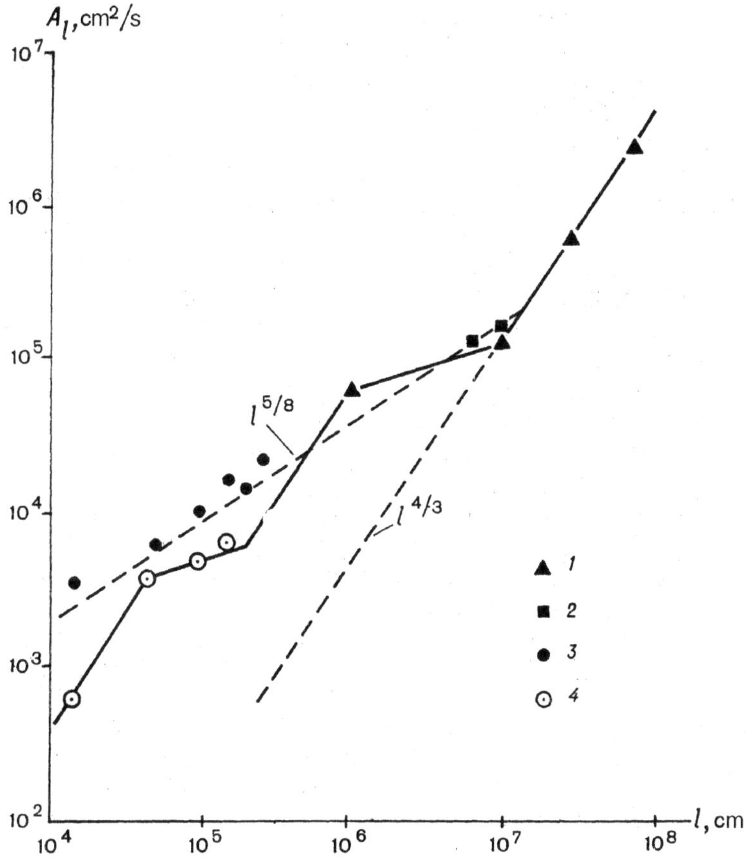

Figure 3.14. Dependence of the values of coefficient A_l on l from observational data at the ice polygons in the Arctic basin in 1966 (1) and 1962 (2), in the Chukchi Sea in 1965 for all cases (3), and for close ice conditions (4).

with the law of Richardson and Obukhov. Turbulent energy flow for this interval (at $c = 0.1$) equals $\varepsilon \cong 5 \times 10^{10}\,\mathrm{cm}^2\,\mathrm{s}^{-3}$, which is four to five orders lower than its typical values for the open ocean.

For the scale intervals less than 10^6 cm (10 km) the empirical evidence contradicts the 'law of 4/3'. According to the investigative results (Gorbunov and Timokhov, 1968), the exponent at l in this range appeared to be equal to 5/8. It should be noted that the data obtained for the Arctic basin and the Chukchi Sea do not quite agree, since the latter mostly characterize open pack and discontinuous ice with a predominance of broken and fragmented ice floes, whereas the former are assigned to compacted ice dominated by ice floes. The values of time intervals ΔT differ strongly (0.5–5.0 days and 66 minutes, correspondingly). The influence of concentration is particularly great: those points assigned to the ice cover with a concentration of 7–9 tenths in the diagram are significantly below the average values (see Figure 3.14).

Table 3.6. Values of turbulent energy flow for inertial intervals of horizontal turbulence scales in a compacted ice cover.

l (cm)	$c\varepsilon^{1/3}$ ($cm^{2/3}\,s^{-1}$)	ε ($cm^2\,s^{-3}$)
10^7 to 10^8	8×10^{-5}	5×10^{-10}
5×10^5 to 10^6	8×10^{-3}	5×10^{-7}
10^4 to 8×10^4	2×10^{-2}	8×10^{-6}

If only these data are taken into account the $A_l = f(l)$ dependence can be interpreted in full compliance with R. V. Ozmidov's statements (Ozmidov, 1968) concerning oceanic turbulence. There are several (in this case, three) inertial intervals in the turbulence spectrum, where the inflow of turbulent energy from the outside is practically non-existent, and, thus the principles of the theory of local isotropic turbulence hold true. The above intervals are separated by zones with a considerable energy inflow. One of these zones is near the 10^5-cm (1-km) scale. It corresponds with the interval within which tidal waves as well as inertial oscillations are the source of additional turbulent energy (as was specified in Section 3.2.3, the orbit radius of both oscillations is about 1 km). The second zone lies within the 10–50-km interval. Possibly, energy inflow from the outside within this range is due to the influence of a mesoscale ice cover inhomogeneity in respect to thickness, ridge concentration, ice floe size etc., since the above factors, to a certain extent, determined the process of wind energy transfer to the ice cover.

The source of energy at the beginning of a low-frequency inertial interval (at $l > 1{,}000$ km) are, probably, synoptic atmospheric vortices (cyclones and anticyclones). When passing from one inertial interval to the other one lying within the zone of smaller scale, the ε value increases. Table 3.5 presents the calculated ε values for each of the three inertial intervals.

As is seen from Table 3.6, due to energy inflow from the outside in two energy-supplying zones the total turbulent energy flow increases by four orders. However, within a high-frequency interval its value is also approximately four orders lower than in the open ocean (Ozmidov, 1968). The cause of such low values of turbulent energy flow in the ice cover as compared with its values for the open ocean can be explained as follows: a considerable part of this energy is spent on the work of the internal interaction forces between ice floes.

Interesting data on A_l coefficient values determined for open pack ice (1–2 points), based on aerial surveying results, are given in a paper by Losev and Gorbunov (1979). The authors revealed an alternation of zones with low or high intensity of horizontal turbulence.

It was mentioned above that for scales l smaller than the horizontal dimensions of the ice floes the law of changes in the A_l coefficient depending on l is no longer valid. Thus, A_l values calculated from results of theodolite observations over the shears of adjacent ice floes (for $l = 800$ m scales) appeared to be quite low: between 40–300 $cm^2\,s^{-1}$ (Legenkov and Romanov, 1972). This distinguishes the horizontal turbulence process in the ice cover from a similar process in water, where turbulent energy eventually transforms into the thermal motion of molecules. In the ice cover it is spent for ice deformation occurring during ice floe interaction.

3.4.6 Mechanical oscillations and waves in the ice cover

Among the dynamic processes which can lead to non-uniformity in the ice drift velocity field are such specific phenomena as oscillations and waves in the ice cover. The main specific features of tidal waves, and those which are of most importance, were considered in Section 3.2.3. Interest in the studies of shorter period waves is due to the influence of these phenomena on ice cover strength, its deformation and destruction. The history of studying oscillations and waves in the ice cover is described in the monograph by Smirnov (1996). These studies, including both the analysis of materials of field observations and experiments, and the development of theoretical models of the wave dynamics of ice, enabled the basic pattern of wave processes in the ice cover of the Arctic and Antarctic to be revealed.

A significant contribution to the study of the considered phenomena was made by Russian scientists: Linkov (1958), Kheysin (1967, 1987), Timokhov (1987), Sytinsky and Tripolnikov (1964), and Smirnov (1972b, 1996). The main results obtained in these papers and many other publications are briefly presented in this section.

Oscillations and waves discovered in the ice cover have different origins and are noted for their diverse spectral compositions, amplitudes and propagation speeds. Different devices are used to record them: seismographs, accelerometers, pendulum inclinometers, tide-gauges with higher sensitivity, rod strain meters, pressure (stress) meters. The processes recorded by these devices are in many respects determined by the interaction between the ocean–ice–atmosphere system, therefore, their study should be accompanied by meteorological and oceanological observations.

The simplest type of ice cover oscillations are the vertical movements of free float oscillation types in water. The period T_n of such oscillations is determined from the formula (Hunkins, 1962):

$$T_n = 2\pi\sqrt{H\rho/g\rho_w} \tag{3.34}$$

where H and ρ is ice thickness and density; ρ_w is the water density; and g is the acceleration due to gravity.

According to this formula, with increasing ice thickness from 0.25–5.0 m the period of float oscillations increases from approximately 1–4 s. Near icebergs, this period can reach 30 s or more. The pulse taking an ice floe out of the state of rest can be created by both the direct impact of the wind and sea waves arriving as ripples from open sea areas or generated under ice cover due to the wind ice drift.

The excitation of float oscillation is connected partly with ripple energy. Due to this, progressive waves with periods less than that of free oscillations of ice floes cannot propagate through such ice. The average size of broken ice forming under the impact of float oscillations depends on the wavelength (hence, the period). Away from the edge, ice cover thickness and the period and length of the arriving waves increases. Thus, the average size of broken ice also increases. This phenomenon is recorded in the course of ice patrols on courses traversing the ice edge in the Barents Sea and other regions. In addition to vertical oscillations, ice floe oscillations relative to the axes in the horizontal plane (similar to vessel rolling or pitching) are also commonly recorded.

Therefore, the penetration of ripples through the belt of broken ice is accompanied by filtration of short-period components. Attenuation of long waves propagating under the drifting ice cover is insignificant, being more noticeable in fast ice. Passing under a continuous ice cover, the gravity waves (ripples) turn into flexural gravity and flexural waves, their parameters depending on the elastic and plastic properties of the ice cover. Studies of the formation processes of such waves were conducted in different years on the 'North Pole' drifting stations, during expeditions on ice-breakers and on the fast ice of the Arctic seas. It follows from observations, that within the range 0.1–1,000 Hz practically all wave types can be recorded: longitudinal and transverse waves, and flexural and flexural gravity waves. The latter have a period ranging between 0.1–10 s and can arise during ice ridging. They commonly propagate as trains moving with a group velocity. The phase velocity of flexural waves with a period of 0.05–1.00 s is 5–500 m/s, and the group velocity is 1,000 m/s. Khejsin (1967) considers the theory of certain phenomena associated with the propagation of different waves in the ice cover: excitation, reflection at the edge, refraction and diffraction, dispersion, etc.

When considering flexural and flexural gravity waves in the ice cover, the phenomenon of frequency (different frequency corresponds to different velocity) and amplitude (acceleration or deceleration of the wave crest relative to its base) dispersions should be recorded. Passing through the entire Artic basin, the waves gradually change their spectral composition at the expense of a quick cancellation of short-period waves. The period of 'filtered' waves is 25–30 s, and their height decreases by two orders. The parameters of such waves indicate that the stresses originating in the ice cover, as a rule, do not attain a destructive effect. However, under significant internal stresses of non-wave nature the '... external wave additive can serve as a trigger mechanism resulting either in an instantaneous ice cover break-up, or relaxation of internal stresses due to formation of multiple micro cracks' (Smirnov, 1996, p. 98).

Observations of oscillations and waves in the ice cover point to collisions of ice floes accompanied by bends, tilting, ice acceleration and generation of self-oscillations (Smirnov, 1987, 1996). The latter characterize the wave motion of ice cover and point to compressions and shear stresses as well as shock interaction between ice floes commonly generated when cyclones approach. It is in the front part of cyclones, as was shown above, that the convergence zone of ice drift velocity appears – caused by its 'isallobaric' component. In addition to the trigger mechanism, these phenomena favour the fatigued breakdown of ice under alternating, multiply repeated loads. We are not considering waves generated when icebergs break away from outlet glaciers in the Arctic and Antarctic. This phenomenon may, under certain conditions, be caused by ripples as well as tidal sea level fluctuations.

In addition to relatively short-period waves of flexural and gravity in nature, ice cover observations from drifting stations revealed waves, which were called 'slow waves' (Smirnov, 1972b). The period of such waves is measured in minutes (from 3–30), the maximum amplitude of vertical ice displacements reaching 5 mm; and phase propagation velocities reaching 0.2–2 m/s. The analysis of simultaneous observations over water temperature and electrical conduction at different horizons enables the proposition of a hypothesis stating that slow waves mainly form under the impact of internal waves in density jump layers. Preconditions for excitation of internal waves are created by wind waves under the ice cover. At the same time, in certain cases, long-period ice oscillations occurred under intense ridging in the summer season (Smirnov, 1996).

3.5 MATHEMATICAL MODELLING OF SEA ICE COVER

Investigation of a complex series of physical processes in the sea ice cover based on results of long-term observations of drifting expeditions enabled the realization of a most important task – constructing mathematical models of the evolution of ice cover conditions. According to Frolov *et al.* (1997), depending on the specific tasks facing the scientists, mathematical models can have a research or applied significance. The former are of a cognitive character and aim at investigating the role of different factors and their interaction in ice cover state changes, including climatic changes. The latter shall respond to the specific practical requests, solving the tasks related with planning, organization and management of different economic (and other operations) in ice covered seas (navigation, mineral production, etc.). Naturally, the improvement of applied models depends on progress in the sphere of modelling of research tasks; and the choice of the trend of the latter is, in many respects, determined by the results of using applied modelling.

Commonly, mathematical models of ice cover are grouped under three main classes: *dynamic models* targeted at reproducing ice movement, its concentration, divergence and deformation under the impact of wind and currents; *thermodynamic models*, taking into account only thermal processes – ice thawing and accumulation; and *dynamic–thermodynamic models* targeted at the most complete reproduction of ice cover evolution under the impact of both dynamic and thermal processes.

Dynamic and thermodynamic models have a limited application: the former are mainly used for short-term calculations and forecasts, since within time periods exceeding 3–5 days, the impact of thermal processes on ice redistribution becomes visible; the latter – for closed water bodies, fixed ice and in certain climatic studies, for which the impact of ice dynamics can be neglected. Therefore, both can be considered (and are often used) as the corresponding blocks of dynamic–thermodynamic models. When developing the latter, a broad range of tasks is commonly solved, which can be classified within three categories:

- modelling of heat destruction and ice cover accumulation;
- description of forces affecting the ice cover and calculation of drift velocity field; and

- determination of kinematic changes in parameters of ice cover conditions during its movement.

3.5.1 Modelling of thermodynamic processes

Thermodynamic processes, along with dynamic processes, determine changes in ice cover conditions in time and space. The role of thermodynamic processes is particularly important due to the fact that they determine the formation (accumulation) and destruction (thawing) of ice. The intensity of ice accumulation or thawing depends on the heat balance on the upper, lower and lateral surfaces of ice formations. The heat balance characterizes the value and direction of total heat flows (Q_Σ) through the corresponding surface. It has four main components: radiation (R_d), turbulent (F_t), convective (F_w) and advective (F_a):

$$Q_\Sigma = R_d + F_t + F_w + F_a \tag{3.35}$$

The last term in the right-hand side of this equation depends on water temperature gradient and sea current velocity. It is taken into account in dynamic–thermodynamic models and is not considered here.

An important stage in the *ice accumulation process* is its initial appearance on the sea surface, free from ice cover. For the beginning of stable ice formation on the open sea surface the following conditions should be fulfilled: (a) the temperature of the homogeneous surface water layer should drop to water freezing point taking into account the salinity of this layer; (b) a negative heat balance for the water surface, as an absolute value, should not be less than the heat inflow to the surface layer due to convection and advection.

The sea cooling process in the autumn season is discussed in detail in papers by Doronin (1969), Gudkovich *et al.* (1972) and Frolov (1981). Modelling of this process involves, in addition to calculation of the heat balance of the open sea surface, the calculation of the surface layer temperature and salinity, increase of heat content of the layer due to advection and turbulent heat flows, as well as the increase of its thickness during convection. The condition of the latter is the inequality $(\rho_w)_i \geq (\rho_w)_{i+1}$, where i and $i+1$ indices refer, correspondingly, to the previous and subsequent steps in time. The model allows calculating and predicting the time period of stable ice formation and further ice accumulation, taking into account the release of the latent heat of water → ice phase transition occurring during the course of it.

Taking into account the fact that radiation heat flow includes a short-wave (Q) and a long-wave (E) part; and the turbulent one, corresponding heat exchange with air (F_a) and heat consumption for evaporation (F_l), and neglecting the presence of heat sources and sinks within ice and while assigning all the flows to a horizontal surface unit, we get:

$$L\rho \frac{\partial H}{\partial t} = -[Q(1 - A) + E + F_a + F_l + F_w] \tag{3.36}$$

where L is heat released during the formation of a unit amount of ice (L dependences on presence of brine in ice and changes in the volumetric heat capacity depending on water temperature are not taken into account here) (Doronin and Kubyshkin, 2001); H and ρ are ice thickness and density; and A is the albedo of the snow–ice surface.

From equation (3.36), we receive the formula for ice thickness determination:

$$H(t) = H_0 - \int_0^t \frac{Q(1 - A) + E + F_a + F_l + F_w}{L\rho} dt \tag{3.37}$$

where H_0 is the ice thickness at the initial point in time. Heat flows are regarded as positive if they are directed towards ice.

Formula (3.37) can be used both under ice accumulation conditions and under thawing conditions. However, in the former case, due to low values of the heat balance constituents from which it is comprised, its determination accuracy is low; therefore, for determining ice thickness another approach is used (Doronin and Kubyshkin, 2001). Since vertical heat flow through ice is proportional to the thermal conductivity of ice λ and temperature gradient, and the weight of forming ice is proportional to crystallization heat L, the heat balance equation at the ice–water interface will have the form:

$$L\rho \frac{\partial H}{\partial t} = \lambda \frac{\partial T}{\partial z} - F_w \tag{3.38}$$

Expression (3.38) does not take into consideration the almost constant presence of the ice surface snow cover, which has a strong impact on ice thickness growth. Taking this impact into account, a vertical temperature profile in the ice and snow mass can be represented as a broken line passing through temperature values T_s, T_0 and Θ. This means that vertical heat flow in both media does not change with ice thickness, i.e.:

$$\lambda \frac{\Theta - T_0}{H} = \lambda_s \frac{T_0 - T_s}{h} \tag{3.39}$$

where h and λ_s are snow thickness and thermal conductivity; T_0 and T_s are ice and snow surface temperature; and Θ is water freezing temperature.

From equation (3.39), we arrive at an expression for the temperature at the snow–ice interface:

$$T_0 = \frac{H\lambda_s T_s + h\lambda\Theta}{h\lambda + H\lambda_s} \tag{3.40}$$

Precise solution of equation (3.38) taking into account equations (3.39) and (3.40) cannot be obtained due to complicated real conditions during ice accumulation (variable vertical temperature gradient in the ice mass, dependence of thermal ice characteristics on temperature and salinity, etc.). In some papers, approximate formulas were used – the calculations made on their basis agree with the observational data. Having accepted that snow thickness as well as thermal ice and snow characteristics remain constant within the time interval, Nikolayeva and

Shesterikov (1970) arrived at the following formula for calculating ice thickness:

$$H = -\left(\frac{\lambda}{\lambda_s}h + \frac{F_w}{L\rho}\tau\right) + \sqrt{\left(\frac{\lambda}{\lambda_s}h + \frac{F_w}{L\rho}\tau\right)^2 + H_0^2 + \frac{2\lambda}{L\rho}(\Theta - T)\tau - \frac{2h\lambda}{\lambda_s}\left(\frac{F_w}{L\rho}\tau - H_0\right)} \tag{3.41}$$

Calculations from formula (3.41) are performed by a successive approximation technique, in which ice thickness calculated at the previous stage with τ duration is introduced into the square root.

If numerical values λ and λ_s equalling, correspondingly, 2.2 and 0.3 J/m s degree are substituted into formula (3.41), and heat flow from water (F_w) is neglected, we get a simple formula:

$$H = -7.0h + \sqrt{(7.0h + H_0)^2 + 0.001\,22(\Theta - T_s)\tau} \tag{3.42}$$

Here, ice and snow thicknesses are expressed in metres, and τ – in days. The influence of heat flow from water on ice thickness can be taken into account with a slight error by subtracting $F_w\tau/L\rho$ from expression (3.42). Heat arriving to the lower ice surface during the accumulation period is mainly due to the development of convection under sea water cooling and salinization. It has a noticeable effect on ice thickness variation. A certain part of the heat flow from water forms at the expense of turbulent heat exchange between the relatively warm deepwater of Atlantic origin and colder and freshened surface water.

For calculating ice thickness from the above formulas, it is necessary to know the snow surface temperature (T_s). This can be obtained from the air temperature using the snow surface heat balance equation. The paper by Shesterikov (1963) describes the procedure of T_s calculations and contains the nomograms testifying that in the case of ice thickness over 50 cm the difference between the snow surface and air temperature at the 2-m level is mainly due to the snow cover height. If there is no snow on thin ice, the influence of ice thickness becomes noticeable.

Formula (3.42) has a form similar to many empirical formulas including the known formula of Zubov (1945). Here, the $(\Theta - T_s)\tau$ value represents the sum of degree–days of frost. Calculations based on this formula show, that with increasing initial ice thickness, its growth rate diminishes, which results in ice thickness levelling with its accumulation.

Accumulation of residual and multiyear ice has its own features. During the period of thawing, the liquid phase content increases in the ice cover, particularly, in its lower layers. Puddles remain on the ice surface during the onset of freezing. As a result, heat sources of phase transitions appear in the ice mass, distorting the quasi-linear character of vertical temperature variations. Under these conditions, equation (3.39) becomes incorrect. The process of puddles freezing and having a liquid phase in the ice mass results in a significant (sometimes 2–3 months) delay for the onset of ice freezing onto its lower surface. This circumstance, as well as additional snow accumulation in the zone of freezing puddles, promotes a growth of mesoscale irregularity in the ice thickness distribution (Buzuev and Spichkin, 1977).

Modelling of the *ice thawing process*, as was mentioned above, is based on the heat balance equation (3.36). A number of factors controlling the components of this equation shall be taken into account.

It is known that the incoming flow of short-wave radiation Q, representing the sum of direct and diffuse solar radiation, depends on astronomic factors (Sun height over the horizon, and, hence, the latitude of the location and the season of the year) as well as transparency of atmosphere and cloudiness. Inter-annual variability of short-wave radiation within rather long time periods (10-day period, month) in the Arctic is not significant (10–15%). Variations of total radiation due to cloudiness is, to a major extent, compensated by the corresponding changes in sky counter-radiation. For our calculations we have used the long-term means of global radiation Q for each of the corresponding 10-day periods.

The share of absorbed solar radiation practically depends on the state of the underlying surface. Its reflectivity (albedo) varies extensively: from 0.08 for water to 0.98 for freshly fallen snow. The albedo of the underlying surface is one of the most important factors determining the variability of the heat balance during the period of thawing, and, hence, of ice conditions in the Arctic seas. Albedo is closely associated with ice cover destruction – observed changes in its surface (snow darkening, appearance and development of puddle areas and depths, formation of thawed patches, etc.).

Experimental determination of the albedo value is discussed in a great number of papers. Thawing of ice cover is characterized by a simultaneous presence of different types of surface: snow cover (A = 0.7–0.9), 'dry' ice areas (A = 0.55–0.75) and puddles (A = 0.10–0.40). Albedo strongly depends on the degree of ice contamination.

The second term on the right-hand side of equation (3.36) characterizes the effective radiation representing the difference between long-wave surface radiation and sky counter-radiation:

$$E = E_n - E_a \tag{3.43}$$

Values incorporated into the right-hand side of this formula depend on air temperature and humidity, temperature of the underlying surface and cloudiness. Due to a low repeatability of major cloudiness and humidity anomalies in the Arctic during the thawing period, while calculating E_n and E_a the mean annual cloudiness and humidity values can be adopted. On these grounds, in the paper by Marshunova and Chernigovsky (1965) simple semi-empirical formulas are proposed for this purpose.

The third and fourth terms of the right-hand side of equation (3.36) characterize turbulent flows of latent and evident heat. Both flows are strongly effected by a difference in the temperature of the air and the underlying surface. Under the conditions of the Arctic seas, in the summer season, this difference is interrelated with moisture deficit, which enables significant simplification of the expression for calculating turbulent flows (Appell and Gudkovich, 1992). Provided that ice surface temperature during the thawing period is equal to a zero, the above simplifications enabled the following expression for the heat balance of the underlying surface to be obtained (Guzenko, 2001):

$$L\rho \frac{\partial H}{\partial t} = Q(1 - A) - 0.08\sigma T_0^4 + (3.68\sigma T_0^3 + K_{tv})T_a \tag{3.44}$$

where σ is the Boltzmann constant; $T_0 = 273°K$ is the surface temperature; T_a is the air temperature (°C); and K_{tv} is the coefficient characterizing turbulent heat exchange with the atmosphere.

Formula (3.44) allows, assuming values of total radiation, albedo of the snow–ice surface and air temperature are known, calculation of ice thickness variation at each step in time.

It should be noted that when considering the integral mass balance of ice on sea water area, the absorbed short-wave radiation plays a primary role. However, an indirect impact of turbulent heat flows, depending on air temperature, on ice thawing is particularly significant during the initial period, when they determine the time of the onset of thawing and the corresponding changes in the ice cover albedo. Under certain conditions (near the coast and ice edge) turbulent heat flows can exceed the amount of absorbed short-wave radiation.

For an adequate reproduction of sea ice evolution it is necessary to obtain reliable air temperatures, to a major extent, determining the time of the onset of thermal ice destruction and the intensity of its thawing. Air temperature distribution above the sea depends on a diversity of factors, most important among them being the state of the underlying surface and, primarily, ice distribution as well as heat advection by air transfer. Statistical association of the 10-day average and monthly average air temperature at polar stations with the above factors is well-grounded, the role of ice cover extent increasing markedly within the period of thawing. Doubtless, in the second half of summer and at the beginning of autumn, the forecast of mean air temperature even on the coast of the Arctic seas is not only a meteorological, but also an oceanological task (Appell and Gudkovich, 1992).

Pronounced inhomogeneity of the underlying surface in the Arctic in summer, when warmed up land often adjoins the ice covered sea surface, raises the question of the necessity of taking into account air temperature transformations above the sea. This problem is analysed in comparatively few papers, both theoretical (Doronin, 1959; Nikolayev, 1963), and empirical (Nikolayeva and Shesterikov, 1970; Appell and Gudkovich, 1992). The authors of the last mentioned paper have developed an effective procedure enabling the calculation of the distribution of mean 10-day air temperature above a seawater area taking into account its transformation above the ice cover and open sea.

The above procedure made it possible to calculate, simultaneously with ice thawing, the changes in the heat content of an active water layer (Q_w) under the impact of radiation and turbulent heat flows on the open sea surface, including fractures. In the latter case, heat absorbed by water (Q_M) is proportional to the area of fractures:

$$Q_M = Q_B(1 - N) \tag{3.45}$$

where Q_B is heat arriving to a unit water surface and N is the function of ice cover concentration.

A part of this heat:

$$Q_L = \varepsilon_1 Q_M \tag{3.46}$$

is used for ice thawing, the other part:

$$Q_w = (1 - \varepsilon_1) Q_M \tag{3.47}$$

for water mass heating (i.e., for increasing the heat content of the active sea layer).

Heat used for thawing ice can be consumed from the side (Q_s) and from below (Q_b) the ice:

$$Q_s = \varepsilon_2 Q_L \tag{3.48}$$

$$Q_b = (1 - \varepsilon_2) Q_L \tag{3.49}$$

Attempts to investigating the redistribution pattern of the heat absorbed by open water areas (Nikolayev, 1963; Doronin, 1969, etc.) show that they depend on many factors: ice thickness, concentration, size of ice floes, intensity of turbulent mixing, etc. Lack of empirical data does not allow a reliable general solution. Therefore, different simplifications are used in the available thermodynamic models. In the paper by Zubov (1945), $\varepsilon_1 = \varepsilon_2 = 1$ is used (all the heat absorbed in fractures is consumed for ice thawing from the side). In the paper by Appell and Gudkovich (1992), $\varepsilon_1 = N$, $\varepsilon_2 = 1$ is used (a part of this heat is spent for heating water; however, thawing from below is not taken into account). Meanwhile, the experiments show that for areas of ice formations deeply submerged into water, thawing (erosion) occurs throughout the year (Karelin, 1943; Zubov, 1945; Grishchenko, 1981; Bogorodsky and Sukhorukov, 1983).

Penetration of heat absorbed in the summer, by open sea areas, into the underlying layers (down to the pycnocline layer), as a result of turbulent and radiation heat exchange, leads to a 'warm interlayer' phenomenon in the Arctic seas. This heat during the next winter arrives at the lower ice surface increasing the F_w component in the heat balance equation (3.36). As a result, the accumulation of ice thickness diminishes by several dozens of centimetres (Gudkovich *et al.*, 1979; Petrov and Frolov, 1984; Gudkovich and Kudryashov, 1985).

The process of ice thawing is significantly complicated by irregular snow cover and ice thickness, which, to a certain extent, control the process of puddle formation and development. This process is described in detail in papers by Appell and Gudkovich (1992) and Guzenko (2001), where the following specific features of the process are considered:

- the beginning of snow thawing under a downward heat flow due to thermal conductivity;
- the accumulation of meltwater under snow on the upper ice surface and its partial freezing;
- the rise of meltwater level to the snow surface and the initiation of puddle formation;

- the quick increase of puddle areas resulting from intense snow and ice thawing resulting in decreasing albedo;
- the influence on this process of the ice surface slope depending on the average thickness and ridge concentration as well as the role of the concentration (ice floes of average size);
- the development of convection in puddles under heating from freshened water and quick deepening of puddles;
- the penetration of radiation heat through puddles and intensification of thawing from the lower surface of the ice cover;
- the meltwater filtration through the ice and its running-off into cracks and thawed patches – accompanied by diminishing puddle area; and
- the partial freezing of meltwater on contact with the underlying cold and salt seawater.

Calculations of thawing illustrating the above features enable us to account for the time gap (about 15 days) revealed by observations between the starting date of thawing and appearance of puddles on ice, trace the evolution of the relative area of puddles and thawed patches, and also calculate the change of meltwater balance and density functions of ice thickness probability during the process of thawing. This increased the informative nature of the calculations used for estimating the pilotage speed and choosing the optimal navigation routes (Buzuyev and Ryvlin, 1976).

3.5.2 Modelling of sea ice dynamics

Modelling of sea ice dynamics is based on the consideration of the balanced equation of forces acting on the ice cover, which is called the motion equation. It is based on the application of Newton's second law: the resultant value of all the forces acting on an ice cover unit area is equal to the product of ice mass assigned to this area by acceleration, i.e.:

$$\rho_i H \frac{d\vec{V}}{dt} = \vec{\tau}_a + \vec{\tau}_w + \vec{C} + \vec{G}_n + \vec{G}_t + \vec{R} \tag{3.50}$$

where ρ_i and H are ice density and thickness; $\vec{V}$ is the drift velocity vector; τ_a and τ_w are the shear stresses on the upper and lower ice surfaces; $\vec{C}$ is the Coriolis force; $\vec{G}_n$ is the gravity projection onto the sea surface; $\vec{G}_t$ is the horizontal component of the tide-generating force; and $\vec{R}$ is the internal interaction force.

Assessments of the relative value of forces comprised within equation (3.50) depend significantly on the average speed, and, hence, the scale of movement. Thus, at an average wind speed of about 1 m/s (characteristic of a large-scale movement within long time periods), the values of shear stresses and Coriolis force are of the same order, whereas at a wind speed of 10 m/s (characteristic of 24-hour intervals) the assessments of the former appear to be at least an order higher than those of the latter. The value of the G_n forces at sea level slopes about 10^{-7}, characteristic of the system of baroclinic currents in the Arctic basin, becomes commensurable with the Coriolis force only in the case of a large-scale ice movement. Therefore, the short-term calculations of these forces can be neglected. Their value becomes significant at level slopes about of 10^{-5}–10^{-6}, which is characteristic of the ebb pile-up processes in shallow areas and tidal waves of diurnal and semi-diurnal range. The inertial terms related with changes in drift velocity with time and comprised within the motion equation, appear to be significant for diurnal velocity variability; therefore, in the medium and long-term calculations it is possible to use stationary models. Inertial terms resulting from the space irregularity of velocity should be taken into account in calculations of mesoscale ice redistribution (spatial scale of about 1 km). Assessments of components of the internal interaction forces show that normal components within them are well in excess of the tangential ones (lateral friction). The influence of the latter is noticeable only within a narrow coastal zone several kilometres wide. Assessments of tide-generating forces acting on ice show that their value is several orders less than other components of the motion equation; therefore, for describing tidal phenomena in ice dynamics account is taken of the forces related to the velocity of tidal currents and the corresponding sea level slopes.

This demonstrates that when solving the motion equation account should be taken of the purpose of the developed ice dynamics model. Hence, the great diversity of dynamic ice cover models proposed up to now. In these models, for solving equation (3.50), the components comprised therein are expressed by dependences based on the laws of physics (mechanics). As a rule, such dependences include different parameters, which are determined during field or laboratory experiments. Numerous experiments were conducted at the Russian and foreign drifting stations as well as in special expeditions on drifting ice – such as the well-known AIDJEX field experiment (Maykut *et al.*, 1972).

Dynamic ice cover models differ both in their degree to which account is taken of the acting forces, and the means by which they are expressed. A review of the studies on ice dynamics modelling can be found in many monographs (Doronin and Khejsin, 1975; Timokhov and Kheysin, 1987; Appell and Gudkovich, 1992; Frolov *et al.*, 1997; Gudkovich and Doronin, 2001). In the majority of the earlier models, ice cover was regarded as an infinite flat plate, to which tangential forces invariable in space (τ_a and τ_w) and Coriolis acceleration were applied (Sverdrup, 1928; Shuleykin, 1938; Gevorkyan, 1941; Shvets, 1946). In subsequent models, which began with the works of Laykhtman (1958) and Campbell (1964), ice cover was considered as a film of viscous fluid occurring between two layers of less viscous fluids – air and water. Thus, in these models, the internal interaction force is associated with the irregularity of the drift velocity field.

The first-term of the right-hand side of equation (3.50), expressing shear wind stress, is determined through the vertical gradient of the averaged wind speed:

$$\tau_a = \rho_a k_a \frac{\partial W}{\partial z} \tag{3.51}$$

In the case of a logarithmic wind profile typical of the near-ice atmospheric layer, this formula is transformed to the following form:

$$\vec{\tau}_a = \rho_a c_a |W| (\vec{W}) \tag{3.52}$$

Here, the c_a coefficient is the function of aerodynamic roughness z_0, the Karman constant and wind observation height. This formula, instead of wind speed module and vector, should contain the corresponding differences of wind speed and ice drift velocity. Since drift velocity is two orders lower than the wind speed, its value can be neglected. The mean value of z_0 determined in many Arctic expeditions is 5×10^{-2} cm. A review of the experimental procedures on determining the value of the roughness parameter and their results can be found in the following publications: Doronin (1980), Timokhov and Kheysin (1987) and Untersteiner and Badgley (1965). However, the value of this parameter strongly depends on ridge concentration in the ice cover, thus, affecting the value of c_a.

Studies of z_0 and c_a dependence on ridge concentration, thickness and size of ice floes were accomplished in a paper by Arkhipov *et al.* (2001). They are based on the results of laboratory experiments (Gudkovich *et al.*, 1963; Gudkovich and Nikiforov, 1963). According to the results of investigation, the impedance coefficient c_a increases 4.1-fold with increasing relative areas of non-ridged ice from 0–45% (ice thickness is 3 m). The value of this coefficient also depends on ice cover fragmentation. If in formula (3.52), instead of wind speed at the height of 10 m the speed of geostrophic wind is used, the value of c_a should be multiplied by 0.36 (Thorndike and Colony, 1982).

Shear stress on the lower ice surface (τ_w) is commonly determined applying the same procedures as τ_a:

$$\tau_w = \rho_w c_w |V_r| \vec{V}_r \tag{3.53}$$

Here, ρ_w is water density and c_w is the dimensionless impedance coefficient.

Calculations are complicated by the circumstance, that V_r is relative velocity equalling the geometric difference of drift and current velocity vectors at the lower boundary of water sublayer, in which shear friction stress displays a rather weak vertical variation. The thickness of this sublayer, according to the results of different investigations, is 0.5–1.0 m. The roughness parameter of the lower surface of ice, according to the observation results of sub-ice currents (Johannessen, 1970) varies extensively from 0.2–10 cm. Correspondingly, the variation range of the resistance coefficient is significant (Belyakov, 1974); it depends on relief of the lower surface (ridge concentration) and is also subject to the influence of seasonal variations in vertical gradients of water density depending on the processes of ice thawing and accumulation. In the above-mentioned paper (Arkhipov *et al.*, 2001), the procedure of c_w evaluation was determined depending on the ratio of ice thickness and horizontal dimensions of ice floes, as well as on ridge concentration. The influence of these parameters on the resistance coefficient on the lower ice surface, as was expected, appeared to be much higher than on its upper surface.

As follows from formulas (3.52) and (3.53), stress values τ_a and τ_w depend on the module to vector ratio of the corresponding velocities. For instantaneous values, the above assessments coincide, pointing to a quadratic form of stress dependence on velocities. With increasing time period, within which the average resultant values of stresses should be obtained, the differences between the average velocity module and its resultant change. At the same time, the average velocity module tends to a certain constant value, which depends on the area and season. Under these conditions, the resultant tangential stress becomes proportional to the first-degree of the corresponding velocity, and friction coefficients, characterizing the intensity of turbulent exchange of momentum, appear to be dependent on the average module of instantaneous velocity. The use of quadratic dependences to the averaged wind speed (ice drift velocity) values leads to distortion of stress calculation results and, hence, to the errors in ice drift velocity calculations.

Standard expressions are used to determine the third and fourth terms in the motion equation. Coriolis force, representing a horizontal projection of the deflecting Earth's rotation force, is expressed by the formula:

$$\vec{C} = \rho_i H f \times \vec{V} \tag{3.54}$$

where ρ_i is the ice density; $f = 2\omega \sin \varphi$ is the Coriolis parameter; ω is angular Earth rotation velocity; and φ is the geographic latitude.

The Coriolis force is directed at 90 degrees to the right (in the northern hemisphere), to the direction of ice drift velocity V. Gravity projection onto the sea surface G_n is determined from the formula:

$$G_n = -\rho_i g H \nabla \zeta \tag{3.55}$$

where $\Delta\zeta$ is the slope of the sea level surface.

In formulas (3.54) and (3.55) ice thickness H should be used as the thickness of deformed ice H_d, which should be determined on the basis of ridge concentration points T and thickness of smooth ice H using the empirical dependence given in the paper by Gudkovich and Romanov (1970) approximated by the expression:

$$H_d = \gamma(1,2T/5 + 1) \tag{3.56}$$

where γ is the average coefficient of ridge filling.

In the majority of papers, when the internal interaction forces in the ice cover are determined, the expressions similar to the ratios used in the continuum mechanics were applied. In this case, the interaction force represents a divergence of the internal stress tensor (σ_{ij}):

$$\frac{1}{H} R_i = \frac{\partial \sigma_{ij}}{\partial x_j} \tag{3.57}$$

For finding force R, it is necessary to assign the governing equation expressing the connection between the internal stress tensor and parameters of ice cover state or drifting field characteristics. Among the numerous publications describing sea ice evolution there are models assigning viscous (Laykhtman, 1958; Campbell, 1964; Hibler, 1977a; Appell and Gudkovich, 1992), elastic (Ovsienko, 1976), plastic as well as combined elastic–viscous (Kheysin and Ivchenko, 1973; Ivchenko and Khejsin, 1974), elastic–plastic (Coon *et al.*, 1974), viscous–plastic (Nikiforov and Timokhov, 1978; Hibler, 1977b; 1979; Gudkovich and Klyachkin, 2001) or elastic–viscous–plastic (Nikiforov and Timokhov, 1974; Kulakov and Timokhov,

1979) properties to the ice cover. These papers only partly reflect the diversity of approaches to determining the interaction force.

When developing the ice dynamics models account should be taken of the fact that ice cover responds differently to shear and normal stresses, and stretching and compressive forces, which can be recorded simultaneously during ice movement. Internal interaction forces depend on the ice cover state – its thickness, concentration, strength, fragmentation as well as the availability and orientation of the ordered systems of continuity disturbances (cracks, leads) in compacted ice. Shows of hydrostatic pressure in the ice cover are not adequately studied. It is difficult to simultaneously take all the above features into account in a single dynamic model; therefore, commonly, each model is adapted to a certain region, season and space–timescale of movement.

3.5.3 Dynamic–thermodynamic models of ice cover evolution

As an example, let us consider certain versions of ice cover evolution modelling. In a paper by Appell and Gudkovich (1992) using viscous rheology (stationary model), normal internal interaction forces are determined in the coordinate system coinciding with thc main axcs of the deformation rate tensor. Direction of the main axes is determined by the successive approximation technique, starting with free-drifting. Internal stresses are proportional to the values of the main deformation rates, and the proportionality coefficients depend on ice thickness, its concentration and sign of the corresponding deformation rate.

From the solution of the motion equation with regard to the boundary conditions, which are assigned by a non-trivial procedure, the ice drift velocity field is determined. The latter allows calculating kinematic changes in the ice cover parameters. The basis of calculations is a differential equation, which is the consequence of the law of mass conservation. Particularly, for the ice cover concentration function it has the form:

$$\frac{\partial N}{\partial t} = -\mathrm{div}(N\vec{V}) + S_N \tag{3.58}$$

Here, $0 \leq N \leq 1$ is the ice concentration function and S_N is the thermodynamic term.

For calculating advective changes in parameters, which do not depend on the velocity divergence (for instance, smooth ice thickness H), the following equation is used:

$$\frac{\partial H}{\partial t} = V_x \frac{\partial H}{\partial x} + V_y \frac{\partial H}{\partial y} + S_H \tag{3.59}$$

where S_H is the thermodynamic term.

Should it turn out in the calculation process that in a certain cell $N > 1$, then the velocity field is corrected by means of iterations, and ice thickness correction is calculated (due to ridging), which is added to the average ice thickness of the cell. An efficient procedure is used to suppress the effect of computational viscosity based on taking into account not only the mean value of the calculated parameter, but also the first and the second moments of its spacial distribution in the cell.

This model has shown good results when comparing the calculated estimates with observational data on the drift of American buoys EGOS both within relatively short time periods (5, 10, 30 days), and within longer periods (Appell *et al.*, 1994). In the latter case, account was taken of the mean flow field, which in the form of a bidimensional polynomial approximates the fields of geometric differences of design wind drift and observed ice drift. Comparison of calculations with observations has shown that the model allows a fairly precise calculation not only of ice drift vectors, but also of the divergence value of its velocity. Combination of the dynamic block of the model with the thermodynamic one enables the performance of ice cover calculations and predictions in the Arctic seas within the navigation period (Appell and Gudkovich, 1992). Another version of the dynamic–thermodynamic model (Frolov, 1981) was used for calculations and forecasts in the winter season.

An example of non-stationary dynamic–thermodynamic models intended for short-term calculations and forecasts on ice-covered sea areas of a limited size can be the model described by Gudkovich and Klyachkin (2001). It uses viscoplastic rheology. The criteria for the onset of plastic deformation is the disturbance of the stability of an ice plate lying on an elastic foundation with respect to buckling (Kheysin and Ivchenko, 1973):

$$\sigma_{\lim} = \sqrt{\frac{\rho_w g E H}{12(1 - \mu^2)}} \tag{3.60}$$

where $\sigma_{\lim}$ is the stress limit, at which the stability of the plate is disturbed; E is the Young modulus; and μ is the Poisson ratio.

The procedure of internal stress calculations does not differ essentially from that accepted in the previous model. Should this stress exceed $\sigma_{\lim}$, ridging will occur (ice thickness increases). Ice cover is represented by a set of markers, each of them being characterized by spatial coordinates, thickness, ridge concentration and velocity. Thermodynamic ice accumulation, similar to the ridging result, is calculated for each marker. Calculation of heat flow from water is performed similarly to the procedure provided for in the model (Frolov, 1981). For this purpose, on the basis of the heat and salt balance equation, changes in temperature, water salinity and thickness of the surface quasi-homogeneous layer are calculated. Calculation of complete flows within the friction layer enables the determination of sea level fluctuations, and on their basis, components of the motion equation resulting from a level slope (barotropic components). The model has demonstrated a good correlation with field evidence; however, it is necessary to conduct tests on the basis of longer term observational results.

Different Russian and foreign dynamic–thermodynamic models of sea ice evolution became widespread not only in the practices of scientific and operational support of economic activity in ice covered seas (Kochetov, 1973; Zuyev, 1983; Ovsiyenko and Efroimson, 1983; Kolesov, 1990; Ashik, 1997; etc.), but also in climatic studies (Hibler, 1979; Semtner, 1976, 1987; Washington and VerPlank, 1986; Makshtas *et al.*, 1988).

3.6 ICE EXCHANGE BETWEEN THE ARCTIC BASIN AND ITS MARGINAL SEAS

Arctic seas through which the Northern Sea Route (NSR) passes are, actually, bays of the Arctic basin; therefore, there is an almost constant ice exchange between these seas and the basin. With respect to marginal seas, the Arctic basin acts as a reservoir capable of receiving large ice masses carried out from the seas and supplying these seas with thicker multiyear ice. Ice exchange direction and intensity depend, to a great extent, on air transfer near the boundaries of the Arctic basin and its marginal seas. However, local winds do not solely fully determine the ice exchange extent. Ice export from marginal seas, to a certain degree, depends on the extent of ice export from the Arctic basin to other regions, mainly to the Greenland Sea. Ice supply from the basin to the seas under onshore directed winds is strongly limited by the influence of coasts and fixed ice (fast ice). Another factor influencing ice exchange, is water circulation in the basin, which represents a single system with currents in the marginal seas. The pattern of ice exchange in the marginal Arctic seas with the Arctic basin is analysed in detail in Gudkovich and Nikolayeva (1963).

Since no systematic observations over ice drift at the borders of the Arctic basin with marginal seas are conducted, the data on ice exchange can be received only from calculations. Most difficult are the estimates of ice exchange between the Arctic basin and the Greenland Sea, because of the necessity of taking into account a poorly studied East Greenland Current, with its velocity varying extensively in space and time, as well as to the variability of ice flow width and some other factors. Therefore, even the evaluations of the average area of ice passing through the Fram Strait within a year differ by more than double between different authors (Gudkovich and Nikolayeva, 1963; Lebedev and Uralov, 1977; Vinje and Finnekasa, 1986; Harder *et al.*, 1998). Since all the empirical calculation procedures are based on comparing the observational data on ice drift and currents in the Fram Strait with components of baric gradients in the direction transverse to the main flow, the design values obtained by different authors appear to be well correlated.

Long series of data on the average monthly ice export velocity to the Greenland Sea through the Fram Strait were listed in the paper by Gudkovich and Nikolayeva (1963). In the proposed calculation procedure major emphasis is placed on the calculation of the gradient component of ice drift, which plays a significant role in ice drift throughout the East Greenland Current. Results of calculations for 1946–2003 based on this procedure have shown that the average annual ice export from the Arctic basin to the Greenland Sea is 650,000 km^2 or about 2,000 km^3 per annum, 2/3 of this value occurring in the cold half-year period (October–March).

For assessing ice exchange between the Arctic basin and the 'export' seas, from which ice export to the Arctic basin is recorded throughout most of the year (Laptev, Kara and Barents Seas), the simplest calculation procedure is based on the ratios of 'isobaric drift' proposed by N. N. Zubov (equation 3.8). If the x axis is directed along the 'entrance' section with length l, approximately coinciding with the northern boundary of the sea, by integrating expression (3.8) along this axis from the western boundary of the sea to the eastern one we get:

$$S = \int_l V dx = K \int_l \frac{dP}{dx} dx = K(P_l - P_0) = K \Delta P \tag{3.61}$$

Here, S is the area of ice cover, which passed through l sections per time unit. The S value is determined by the dimensionality of the K coefficient and the corresponding averaging scale of the baric map. In case of using the monthly average maps of atmospheric pressure at sea level the dimensionality of the isobaric coefficient is: km^2/gPa per month. Values of the latter were obtained from observational data on the total ice drift in the Arctic basin increased by 25% (according to the empirical ratio of the corresponding average annual values).

From formula (3.61) it follows that the resultant area of ice exchange is clearly independent of the section length and is proportional to the difference of atmospheric pressure at its ends. It is presumed that ice cover (irrespective of ice concentration) does not disappear throughout the section. Naturally, for a correct evaluation of ice exchange volume the information on ice concentration and thickness and their variations with time are necessary.

Table 3.7 presents the design data on the average ice exchange between the Arctic basin and the Greenland, Barents, Kara and Laptev seas for three-month periods. Data on ice exchange with the Greenland Sea were obtained applying the procedure discussed in the paper by Gudkovich and Nikolayeva (1963) and were supplemented by Teitelbaum up to 2003. Values referring to the other three seas are calculated from formula (3.61). Calculation results show that ice export from the Barents Sea to the Arctic basin, on average, proceeds from November to March; from the Kara Sea, from October to April; and from the Laptev Sea, from September to May. During the other months of the warm season the supply of ice to these seas from the Arctic basin is commonly recorded. It is characteristic that the duration of the period within which ice cover is exported from the Arctic basin to the above seas diminishes from west to east.

Table 3.7. Average ice exchange between the Arctic basin and Greenland, Barents, Kara and Laptev Seas in three-month periods (in thousand km^2).

	Months			
Sea	October–December	January–March	April–June	July–September
Greenland	190	220	150	90
Barents	−15	−25	20	15
Kara	−75	−90	15	30
Laptev	−140	−155	−25	−40

Positive values represent export from the Arctic basin; negative values represent supply to the basin.

Undoubtedly, the design data can contain errors due to underestimation of the influence of gradient currents and other factors. For evaluating the reliability of design values the position of the boundary of compacted ice (7–10 tenths) in the Russian Arctic seas at the end of September was compared with the boundary of the area dominated by old ice (two-year and multiyear) in March of the next year. Difference of latitude in the position of the indicated boundaries on the meridians (at 50 of longitude) was accepted as the value of ice movement along the corresponding meridians within 6 winter months (October–March). Ice exchange area (in km^2) of the sea with the Arctic basin within the indicated period was determined from the formula:

$$S = 12{,}350\, \Delta\varphi_{av}\, \Delta\lambda \cos\varphi_{av} \tag{3.62}$$

where $\Delta\varphi_{av}$ is the average difference of latitudes between the position of the above-mentioned ice boundaries on the meridians (in degrees); $\Delta\lambda$ is the longitude difference between meridians of the eastern and western sea boundaries (in degrees); φ_{av} is the meridian-averaged latitude of both ice boundaries; and 12,350 is the coefficient which transfers the latitude and longitude degrees into area dimensionality (km^2).

Figure 3.15 shows the average position of the indicated boundaries and their corresponding standard deviations (for 1954–1991). Based on these data, the value of ice exchange between the Laptev Sea and the Arctic basin, for the winter half-year, on average amounted to 328,000 km^2 (61% of sea area), which is only 10% higher than the value calculated from the atmospheric pressure difference. There is no reason to believe that the calculations of ice exchange areas for other seasons contain more significant errors.

The question of evaluating ice exchange between the 'supply' seas, to which the East Siberian and Chukchi Sea belong, with the Arctic basin appears to be much more difficult. In the winter season, onshore directed winds are often recorded here, which should result in the supply of a great amount of ice from the Arctic basin to these seas. The intensity of this process increases from west to east. However, the analysis of changes in the ice cover area in the eastern East Siberian and Chukchi Seas within the winter half-year shows that in the considered period ice export from these seas to the Arctic basin prevails (Gudkovich and Nikolayeva, 1963).

This conclusion is also confirmed by results of present-day studies: the design area of ice cover exported to the Arctic basin within the winter season is only 38,000 km^2, whereas its value determined from changes in position of ice boundaries, considered above, exceeds 440,000 km^2. Of them, about 100,000 km^2 are assigned to the area north of the New Siberian Islands and should mainly characterize additional ice export from the Laptev Sea. However, in this case, ice export from the East Siberian Sea (about 45% of its area) also represents a significant contribution to replenishing the Arctic basin with ice from the offshore seas. However, it should be remembered that dynamic divergence of ice cover near the northern boundary of the sea, often recorded under the prevailing cyclonic baric fields in the summer season, can have a significant effect on the position of the boundary of predominance of old ice at the end of winter and, hence, overestimate the value of ice export to the north determined by applying the described procedure.

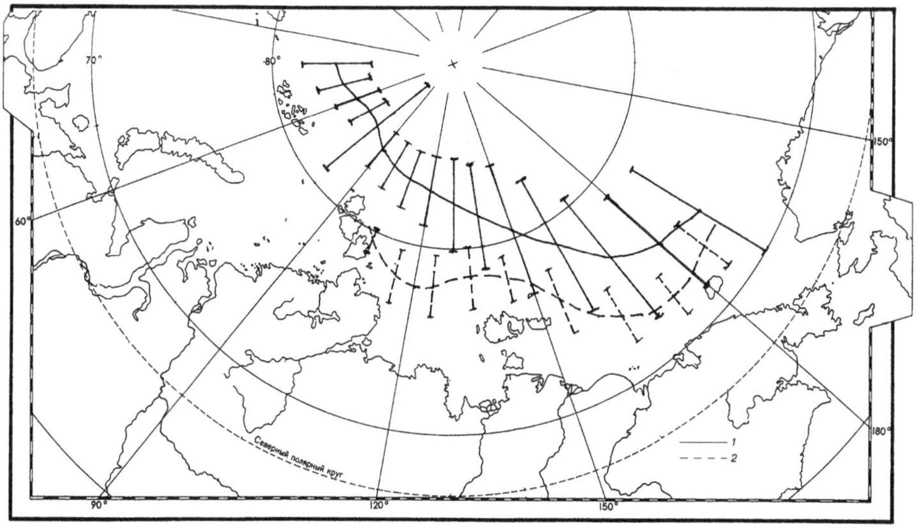

Figure 3.15. Average (1954–1991) location of the boundaries of old ice dominance in March (1) and of close residual ice in late September of the preceding year (2). Segments at meridians characterize corresponding standard deviations.

In the summer half-year, reiteration of exporting winds in the East Siberian Sea increases noticeably, which supports the design values of ice exchange between this sea and the Arctic basin (on the average, 103,000 km^2 under standard deviation of 122,000 km^2) as more reliable.

If it were not for the current bringing water through the Bering Strait to the north and north-west, as well as the influence of coasts restricting ice supply from the north, the Chukchi Sea at the end of winter, as a rule, would be blocked up with multiyear ice brought from the Arctic basin and Beaufort Sea. Average values for the areal extent of ice supplied to the Chukchi Sea within the winter half-year calculated from the Wrangel Island–Point Barrow pressure difference is 305,000 km^2; and within the summer half-year, 124,000 km^2. However, the calculations based on comparing the positions of the above ice cover boundaries at the end of summer and winter show that on average over 222,000 km^2 of ice (about 60% of sea area) are exported from this sea to the Arctic basin within the winter half-year. It is impossible to give a grounded estimate of ice exchange between this sea and the Arctic basin in the summer season due to the fact that under the strengthening Pacific Current in the Bering Strait at this time and heat advection in the atmosphere a significant part of the ice cover in this sea commonly thaws within the sea. In certain years, when a baric depression sets in for a long time period in the adjoining parts of the Arctic basin, 'ice collapse' can occur, due to which a large amount of ice cover from the north-west and north is supplied to the Chukchi Sea.

The above estimates demonstrate that only within the winter half-year over 1 million km^2 of ice cover are exported from the Siberian seas to the Arctic basin, which is commensurable with modern estimates of the area of ice annually exported from the Arctic basin to the Greenland Sea (Koesner, 1973; Mironov and Uralov, 1991; Vinje and Finnekasa, 1986; etc.). Taking into account characteristic ice thicknesses, an approximate estimate of ice volume exported to the Arctic basin within the winter half-year can be obtained (1,500–2,000 km^3). This value is much lower than the available estimates of ice export to the Greenland Sea in winter (Vinje and Finnekasa, 1986; Alekseyev *et al.*, 1997), which is quite reasonable taking into account the time of ice movement to the Fram Strait, ridging processes, etc.

Due to the fact that the thickness of ice cover involved in ice exchange is constantly changing, the most correct approach to determine estimates of the corresponding ice volumes should be based on dynamic–thermodynamic models of ice cover evolution. For resolving the above tasks, model calculations shall be performed over long time periods (years). The resultant drift velocity becomes commensurable with the value of systematic calculation error in this case (Gudkovich and Doronin, 2001) and the models used should be improved. Of particular importance is the possibly of a more precise registration of gradient currents and rheological properties of the ice cover.

If the ice cover of the Arctic was motionless, ice thickness distribution by the end of winter would be determined only by the conditions of ice accumulation (i.e., it would depend on air temperature, heat content of water, and heat exchange conditions between air and water). Ice motion significantly changes the distribution of ice cover thickness and its entire meso and macro-structure. As a result of ice exchange with the Arctic basin, a 'rejuvenation' of ice cover in the Siberian offshore seas occurs. Calculations show that, on average, at the expense of this process in the eastern Arctic seas, the volume of ice by the end of winter diminishes by 10–15%; and in the western seas (Laptev and Kara Seas), by 20–30%.

In the results calculated above, account was taken of the thickness and areas of non-deformed ice. During ridging, which is also associated with dynamic processes, the volume of ice per surface unit, which is sometimes called ice cover 'thickness' (Gudkovich and Romanov, 1970; Zubov, 1945), increases significantly. According to P. A. Gordienko, ice thickness due to ridge concentration in the majority of Arctic seas is, on average, almost 30%, and in the Chukchi Sea even 80% higher than ice thickness under quiet accumulation. According to the estimates of other authors, the thickness of first-year ice in the Arctic basin is, on average, 17% higher than that of even ice of the same age.

3.7 SEA ICE AND CLIMATE

Studies of long-term (climatic) changes of ice cover thickness and area (ice cover extent) in the Arctic seas are aimed at solving both economic and purely scientific tasks. Ice cover extent and ice thickness are associated with navigation conditions in the seas having ice cover. The presence of the latter also affects the construction of ports and other coastal structures as well as the operation of platforms producing hydrocarbon fuel and its transportation to the continent. On the other hand, without investigating long-term changes in ice cover thickness and areal extent it is impossible to understand the basic mechanisms of climatic changes on the Earth, and, thus, predict these changes.

3.7.1 Role of sea ice cover in climatic system

As is known, climate is a statistical regime of long-term fluctuations of atmosphere conditions (Khromov and Mamontova, 1974). Sea ice cover is comprised into this notion as one of the geographic conditions determining the occurrence of climate-forming processes on the Earth. The association of sea ice conditions (its thickness, area and other parameters) with climate (air temperature and other climate characteristics) has long been known (Budyko, 1969; Zakharov, 1996). However, the *climate-forming role* of sea ice is determined by a *feedback* (positive and negative) between ice cover conditions and processes in the atmosphere. The existence of such associations has been ascertained by both theoretical investigations and the analysis of observation data. The extent of their action is rather diverse: from local to global.

Dependence of the Earth's climate on sea ice cover is determined by some of its properties changing the space–time distribution of the intensity of thermodynamic and dynamic processes of interaction between the atmosphere and ocean. Among such properties are:

- a high reflectivity (albedo) of the snow–ice surface with respect to incoming solar radiation;
- an isolating effect, strongly restricting the flows of heat, moisture and gas exchange between the ocean and atmosphere; and
- a high specific latent heat of ice formation–melting, slowing down changes in the temperature of the atmosphere and the ocean under the impact of radiation and turbulent heat flows.

The above ice cover properties not only reduce air temperature in the Arctic by several dozens of degrees, but also have a major effect on temperate latitudes (Budyko, 1969). As a result, global gradients of air temperature between low and high latitudes increases, having a marked effect on atmospheric circulation and the prevailing trajectories of cyclones intensifying zonal air transfer at moderate latitudes and the hydrological cycle on the planet. It should be noted that the presence of ice cover results in self-oscillations in the ocean–ice cover–atmosphere system and corresponding climatic changes (Shuleykin, 1953; Alekseyev, 1976; Nikiforov and Shpaykher, 1980; Zakharov, 1996; Gudkovich and Kovalev, 2002a,b).

3.7.2 The stability of sea ice cover in polar regions

In Budyko (1969, pp. 24–25) the conclusion 'on the possible existence of two climatic regimes at high latitudes due to the presence or absence of polar ice' was reached. Both regimes are 'unstable, owing to which ice cover can appear and disappear under minor alterations of climate-forming factors and even in the absence of these changes as a result of self-oscillating processes in the atmosphere–ocean–polar ice system.'

However, Doronin (1968), and later Zakharov (1977, 1981, 1996) drew attention to the association of ice cover propagation with the impact of the underlying freshened surface water layer, below which the halocline layer occurs, in which water density quickly increases with depth. The presence of this layer, on the one hand, restricts the content of heat in the water, which accumulates in the summer season; and, on the other hand, strongly reduces the arrival to the surface (as a result of winter convection and vertical turbulent exchange) of heat supplied by sea currents from lower latitudes. All this promotes the growth of ice thickness and areal extent in winter, diminishing the possibility of thawing next summer. The authors of the cited papers believe that fluctuations in the area of freshened surface Arctic water results from a disturbance of the freshwater balance in the Arctic Ocean. The incoming part of this balance is made up of precipitation, continental runoff and inflow of low-salinity water through the Bering Strait; and the outgoing part, of freshwater evaporation and runoff to the Atlantic Ocean. Sea ice growth and melting exerts some influence on the freshwater balance of the surface layer also. In our opinion, a significant role in this process is also played by such factors as the volume and salinity of water supplied to the Arctic Ocean – mainly relatively saline Atlantic water (Appell and Gudkovich, 1984). Hence, in this matter, it would be insufficient to consider only the freshwater balance; account should also be taken of the salt balance in the upper layer of the ocean. It should be emphasized that providing grounds for the role of the halocline in sea ice cover formation in the Arctic Ocean rules out the presumption that a slight increase in the incoming part of the heat balance in the Arctic can result in a relatively quick disappearance of the Arctic ice.

3.7.3 Long-term ice cover thickness variations in the Arctic in the 20th century

Regular ice thickness measurements were conducted at polar stations. They started only in the late 1930s. The analysis of these observational data point to a close dependence of the thickness of this ice on the thermodynamic processes (mainly, on air temperature in the near-surface layer of atmosphere), which enables the consideration of this indicator as a reliable index of climatic change. However, a linear trend of ice thickness at polar stations located in the Siberian offshore seas, determined for the period from 1936–2000, averages at only +0.2 cm/year and statistically does not differ from zero. This is also proved by results of analysing long-term ice thickness variations in fast ice of the Arctic seas (Buzuev and Dubovtsev, 2002), which '... do not confirm climate warming in the Siberian offshore area within the 1960s–1990s ...'. However, the calculation of a linear trend of average thickness of fast ice within 30-year periods has demonstrated a positive tendency during 1940–1973 (+0.35 cm/year) and a negative one during 1973–2000 (−0.52 cm/year), which agrees with the earlier revealed (Karklin *et al.*, 2001) half-century cycle of ice cover extent fluctuations in the Arctic seas. Let us recall, that in addition to the half-century cycle in climatic fluctuations of ice cover extent in these seas, cycles with durations of about 20 years (Gudkovich and Kovalev, 2002b), about 10 years (the latter cycle is associated with the phenomenon of the Arctic fluctuation) and shorter cycles (Karklin and Teytelbaum, 1987) are also distinguished.

Noticeable climatic ice thickness fluctuations were recorded in the Arctic basin. Comparison of observational data of the expedition on *Fram* and *G. Sedov* ice-breaking vessels (Buynitsky, 1951) has shown that within the 44 years separating these observations the thickness of non-deformed ice in the near-Atlantic part of the Arctic basin decreased by 20%. Correspondingly, air temperature increased (by 6.5°C) as did the maximum temperature of the Atlantic water (by 0.55°C). As is known, in the late 1930s, the first warming in the Arctic in the 20th century reached its peak.

Much greater changes in ice thickness in the Arctic basin were recorded by some American scientists. From the analysis of sonar observations accomplished in the last third of the 20th century from American submarines the conclusion was drawn (Rothrock and Maykut, 1999) that within the period from the mid-1970s to the early 1990s ice thickness in that area decreased by 1.0–1.5 m (40% of its average value). Despite the fact that some other scientists (Shy and Walsh, 1996; McLaren *et al.*, 1994) did not confirm such changes, and still others (Wadhams, 1994, 1990) account for this phenomenon by ice ridging on the northern approaches to Greenland, the idea of polar ice thinning under the impact of anthropogenic greenhouse gas emissions and the corresponding climate warming was until recently actively discussed at different scientific meetings and international political conferences.

The research conducted in the last several years at the Antarctic and Arctic Research Institute (AARI) indicates that the cause of the revealed anomalies is not thermodynamic processes resulting from global anthropogenic climate warming, but mainly dynamic processes. The latter are associated with relatively short-term fluctuations of atmospheric circulation determining the processes of ice cover advection, ridging and divergence (Losev and Gorbunov, 1995; Porubaev, 2000; Makshtas *et al.*, 2001; etc.). The calculations accomplished by S. V. Klyachkin and given in the paper by Gudkovich and Kovalev (2002a) show that the results of advection depend not only on the anomalies of the ice drift velocity fields, but, very largely, on the initial ice cover distribution (position of the edge of residual ice, ice massif boundary, etc.). In compliance with the above calculations, under a reduced residual ice cover extent in seas east of Severnaya Zemlya, as a result of ice drift within 1–2 years to the sub-polar area, where 3–4 m and thicker multiyear ice commonly occurs, 1.5–2.5 m thick 1–2-year ice is supplied. Due to this, there is a marked ice thickness decrease, which within 2–3 years gives place to restoration of the ice cover conditions typical of this area. Alternation of anticyclonic and cyclonic regimes of atmospheric circulation is accompanied by changes in reiteration of convergence and divergence processes in the ice drift, due to which ice cover concentration and ridge concentration, and, hence, its average thickness change. The role of climatic changes in this case can only determine the probability of the onset of the corresponding conditions (initial ice distribution, cyclonicity degree of baric fields, etc).

Such a conclusion is also confirmed by investigative results (Rigor and Wallace, 2004), when significant changes in the sea ice distribution north of Alaska recorded at the close of the 20th century are associated with the phenomenon of the Artic oscillation (Proshutinsky and Johson, 1997; Polyakov *et al.*, 1999; Thompson and Wallace, 1998) and the corresponding changes in the area of multiyear ice.

Another possible reason for the changes in sea ice thickness in the Arctic basin could be a noticeable temperature increase of deep Atlantic water recorded during warming periods in the Arctic, which are accompanied by increasing surface layer salinity (Alekseyev *et al.*, 1998). These factors promote the growth of vertical heat flows in water, which can result in diminishing ice cover thickness. Such changes in the thermohaline structure of the water, coinciding in time with the maximum of the second half-century warming in the Arctic in the 20th century, exceeded in their intensity the above-mentioned changes in the first-period in the area of the *Fram* and *G. Sedov*'s drift. A possible cause of such an anomaly could be the large spatial specific features of the shows of these anomalies which were revealed: the largest deviations from the standard are recorded in the zone adjoining the archipelagos of Spitsbergen, Franz Josef Land and Severnaya Zemlya from the north. Away from this zone, toward the centre of the basin, the values of water temperature and salinity anomalies diminish noticeably. Thus, there are no conclusive proofs for assuming that two major warmings recorded in the 20th century have a fundamentally different nature, as Johannessen *et al.* (2004) believe.

3.7.4 The causes of ice cover change in the Arctic Ocean

The above facts indicate that the contribution of different processes to the changes of sea ice cover conditions, on scales from a decade to a century, can depend on the specific features of a certain region. Variations of ice cover areal extent were most adequately investigated in the marginal Arctic seas, where the available observation series show a polycyclic nature in these changes. However, the contribution of cycles of different duration to the variability of ice cover extent in different seas is not the same. In seas of the near-Atlantic region the role of long-period cycles (20, 50 years and more) is significant. In seas east of Severnaya Zemlya, short-period fluctuations are more important. The use of mathematical and physicostatistical models of ice cover evolution in marginal seas shows satisfactory results.

The question of ice cover variability in the central (deepwater) part of the Arctic basin is more complicated. Here, no long-term observations over changes in the ice cover area were conducted. The available irregular observations point to sporadically occurring ice thickness anomalies in certain water areas. There is no clarity as to the ratio of contributions to these anomalies, on the one hand, of the advection of ice of different age from the neighbouring areas; and on the other hand, of changes in the conditions of vertical convection bringing heat from depth, where relatively warm water of Atlantic origin occurs.

The causes of the above-mentioned cyclic fluctuations of different timescales are, as yet, a scientific hypothesis; therefore, further in-depth studies are required. A detailed review of the current notions on this matter would call for significant time and space. Obviously, the majority of cyclic changes recorded in the ice cover conditions are due to the processes in the atmosphere and ocean, which are affected by both external and internal factors.

Among the external factors are, as a rule, helio-geophysical effects (solar activity, tidal and nutational phenomena, Earth rotation speed, fluctuations of the solar constant under the influence of changes in the distance between the Earth and solar centre of mass, fluxes of energy and charged particles from space, etc). Internal factors are, on the one hand, natural processes of geological and biological nature as well as self-oscillatory phenomena related to the interaction in the lithosphere–ocean–atmosphere–sea ice cover–glaciers svstem. On the other hand, more and more often the attention of some scientists is attracted by an increasing anthropogenic impact on the climate of greenhouse gases (carbon dioxide, methane, etc.) accumulated in the atmosphere as a result of hydrocarbon fuel combustion and reductions in forested areas. The presence of a negative linear trend of ice cover extent (or a positive trend of air temperature) is regarded as a proof of anthropogenic warming. Meanwhile, it is known, that the assessment of a linear trend strongly depends on the time interval for which it is given. Under cyclic fluctuations, for assessing a trend, which is independent of them, not only the length of a series, but also the choice of origin relative to the phase of cyclic fluctuations is an important factor (Gudkovich and Kovalev, 2002a). It is quite inadmissible, as is done in the paper by Johannessen *et al.* (2004) and some other authors, to use for mathematical model calibration the observational data for time intervals between the neighbouring maxima and minima of cyclic fluctuations. In the above paper, the models from which changes in ice cover conditions in the 21st century were calculated and calibrated on the basis of observational data for the last 25 years, within which an extreme reduction of ice cover extent was recorded associated with a half-period of the last warming cycle in the 20th century.

Moreover, it is necessary to disclose a surprising fact, that it is the second-half of the 20th century, unlike the first-half, that was noted for the minimum linear trend of ice cover extent in the Arctic seas, which disproves the opinion on the decisive impact of greenhouse gas concentration on climate. As is known, this concentration in the indicated period increased exponentially (the notorious 'hockey stick' graph). Therefore, one cannot fail to agree with Academician Kondratyev (2004) that '... the observation data ... by no means contain a distinct confirmation of anthropogenically caused global warming ...' (p. 121), and '... the results of numerical simulation of climate providing grounds for the hypothesis of global greenhouse warming and, supposedly, correlating with the observation data represent only an adjustment to the observation data.'

A slight linear trend recorded in long-term variations of air temperature and ice cover extent in the Arctic seas can point to a certain anthropogenic impact, if the possibility of existence in climatic fluctuations of natural super-secular periods, which were repeatedly recorded in the Earth's history, is neglected. The above circumstances allow the statement that climatic changes occurring in the 20th century were mainly induced by natural causes, quite important among them being the factors regulating the supply of Atlantic water to the Arctic Ocean and the related processes of cyclogenesis in the Norwegian energy active zone of the ocean.

4

Oceanography

4.1 ORGANIZATION OF OCEANOGRAPHIC OBSERVATIONS AT THE 'NORTH POLE' DRIFTING STATIONS AND THE 'SEVER' AIRBORNE HIGH-LATITUDE EXPEDITIONS

Oceanological studies were the most important activities at the 'North Pole' (NP) drifting stations and during the 'Sever' airborne high-latitude expeditions (AHE). They were carried out at all the 'NP' and 'Sever' AHE stations without exception.

Observations at the 'NP' stations and during the 'Sever' expeditions are closely interrelated, primarily, scientifically. Long-term and, basically, year-round observations at a drifting station, in one quasiconstant point give a notion of the changes in oceanological characteristics with time; and the oceanographic surveying accomplished by a high-latitude expedition with aircraft landing on ice in many points demonstrates a quasisynchronous distribution of these characteristics in space, through the ocean water area. In certain years, oceanographic surveying practically covered the entire water area of the Arctic basin.

They are also related in respect of logistics – the 'NP' drifting station being the base for the 'Sever' AHE aircrew, where the members of the expedition can have a rest, hand in the samples taken for hydrochemical analysis, etc. In its turn, the second major task of the 'Sever' AHE is to provide the 'NP' drifting stations with everything required for long-term work on ice, as well as changeover of the station crew.

The above arguments provided grounds for forestalling the presentation of scientific results obtained on the basis of oceanographic observations at the 'NP' drifting stations and the 'Sever' expeditions by a description of the organization of oceanographic observations, the precision of determining the coordinates and immersion depth of devices, the precision of the devices, measurements of the depth in the area, the water temperature and salinity, the currents, water sampling, etc. Organization of each expedition, as a rule, comprises two stages: preparatory and field.

At the preparatory stage the scientific and organizational programme of each expedition was drawn up in compliance with the long-term program of investigations and approved by the Academic Council of the Antarctic and Arctic Research Institute (AARI). The task of the first 'Sever' expeditions and the 'NP' drifting stations was to get a rather complete notion of the Arctic basin floor relief, thermohaline structure of the water and the currents and distribution of hydrochemical characteristics. In the 1970s, the emphasis shifted toward studies of large-scale features of oceanographic and hydrochemical fields and the ocean and atmosphere interaction processes. During these years, in the winter–spring seasons (March–May), the 'Sever' expeditions accomplished large-scale oceanographic surveying of the Arctic basin. In the 1980s, scientific research of the 'Sever' expedition concentrated on the Siberian offshore seas.

Field stages of the 'Sever' expeditions lasted from March until June; however, in many years the expeditions were also organized in late autumn. Organization of the 'NP' drifting stations proceeded in the spring season by means of aircraft and in the summer season using ice-breakers. The duration of the 'NP' stations drift varied from several months to several years. In the last cases, the crews of polar explorers at the station changed after one year of work.

Oceanological investigations at the 'NP' stations and the 'Sever' AHE were carried out both on the basis of the standard observation programmes and special research programmes.

Within the standard programmes the following work and observations were accomplished: determination of geographic coordinates of the area, measurement of ocean depth, setting up deep-sea hydrological stations with measurements of water temperature and salinity, taking water samples for hydrochemical analysis and samples of bottom soil (sediments) and the hydrobiological sampling of benthos, zooplankton and phytoplankton.

Studies under special programmes were conducted to cover a broad range of oceanological problems – with each particular case having its own special programme.

Oceanological observations were conducted following the generally accepted procedure described in the manuals on hydrological (*Manual on the Hydrological Studies in Oceans and Seas*, 1977), hydrochemical (*Manual on the Procedures of Chemical Analysis of Sea Water*, 1977) and hydrobiological (*Manual on the Procedures of Biological Analysis of Sea Water and Bottom Sediments*, 1980) studies approved by the Hydrometeorological Service of the USSR. These manuals list the make-up of observations, observation period, procedures for preparation and undertaking of observations and other requirements necessary for ensuring a high level of the field information.

The main devices and equipment included: AARI bathometers (a modification of Nansen's bathometer with less capacity); VMM-type mechanical sea current meter (until the 1950s); self-recording mechanical current meters of Yu. K. Alekseyev's system (BPV-2, ESCM); Yu. K. Alekseyev's light-weight corer and piston corer of Kudinov's system (PGTP-54 corer); stand-alone digital meters of current, temperature, electroconductivity, and immersion depth (ATsITT) of R. A. Balakin's design; stand-alone digital current meter 'Vector-2' (in 2003–2004); and hydrological winch.

Transition from the technology of oceanographic measurements using bottle series to the CTD casting of water occurred in the 1970s during expeditions of the western states and in the 1980s in the Soviet sea expeditions. During the 'Sever' AHE, observations were mainly conducted using bottle series.

Throughout the entire period of operations at the 'NP' drifting stations (1937–1991), 1,760 deep-sea hydrological posts/stations were set up. During the 'Sever' expeditions from 1948–1993 – 5,150 were set up.

While processing the data of hydrological observations carried out by the Soviet and Russian observers at the 'NP' drifting stations and during the 'Sever' expeditions, sometimes, the question arose as to whether these data were comparable with the results of observations accomplished by foreign researchers in the Arctic basin. For checking the consistency of these observations, material from 1941–1993 was used for the deepest areas of the Canadian basin and the Nansen and Amundsen basins. For the Canadian basin, 238 Russian stations were used in 1949–1991 and 108 American stations in 1950–1985; for the Amundsen basin – 222 in 1948–1988 and 14 in 1991; for the Nansen basin, 86 in 1955–1982 and 14 stations in 1980–1991 in of the Russian and western expeditions, correspondingly (*Joint US–Russian Atlas of the Arctic Ocean*, 1997). The Russian observations resulted in a much larger data set and covered a much greater time span. Particularly large differences were recorded in the Amundsen basin – a fact that should be taken into account when analysing the results.

From the Tables 4.1–4.3 it follows that water in the selected areas differs essentially in its temperature and salinity. These differences are most pro-

Table 4.1. Mean values and mean-square deviations of water temperature (°C) and salinity (psu) based on the western and Russian data in the Canadian basin.
Joint US–Russian Atlas of the Arctic Ocean (1997).

	Temperature				Salinity			
	Average		Mean-square deviation		Average		Mean-square deviation	
Depth (m)	Russian	Western	Russian	Western	Russian	Western	Russian	Western
5	−1.67	−1.54	0.13	0.35	30.54	30.23	0.78	0.90
100	−1.40	−1.38	0.11	0.09	32.44	32.44	0.21	0.18
300	−0.02	−0.01	0.15	0.25	34.63	34.65	0.12	0.14
500	0.48	0.44	0.06	0.03	34.85	34.86	0.07	0.02
1000	−0.06	−0.10	0.05	0.06	34.90	34.91	0.06	0.02
2000	−0.41	−0.41	0.03	0.02	34.94	34.95	0.05	0.02
3000	−0.34	−0.35	0.03	0.02	34.96	34.96	0.04	0.02
4000	−0.29	–	0.05	–	34.96	–	0.05	–

Table 4.2. Mean values and mean-square deviations of water temperature (°C) and salinity (psu) based on the western and Russian data in the Amundsen basin.
Joint US–Russian Atlas of the Arctic Ocean (1997).

	Temperature				Salinity			
	Average		Mean-square deviation		Average		Mean-square deviation	
Depth (m)	Russian	Western	Russian	Western	Russian	Western	Russian	Western
5	−1.64	–	0.34	–	30.34	–	5.06	–
100	−1.60	−1.40	0.08	0.22	33.55	34.16	0.41	0.14
300	0.75	0.93	0.14	0.14	34.84	34.86	0.05	0.02
500	0.56	0.54	0.09	0.09	34.89	34.89	0.05	0.01
1000	−0.30	−0.29	0.06	0.07	34.90	34.90	0.04	0.01
2000	−0.74	−0.72	0.05	0.04	34.93	34.92	0.04	0.00
3000	−0.79	−0.77	0.04	0.01	34.94	34.94	0.04	0.01
4000	−0.69	−0.68	0.03	0.01	34.95	34.94	0.04	0.00

Table 4.3. Mean values and mean-square deviations of water temperature (°C) and salinity (psu) based on western and Russian data in the Nansen basin.
Joint US–Russian Atlas of the Arctic Ocean (1997).

	Temperature				Salinity			
	Average		Mean-square deviation		Average		Mean-square deviation	
Depth (m)	Russian	Western	Russian	Western	Russian	Western	Russian	Western
5	−1.85	–	0.06	–	33.76	–	0.62	–
100	−1.13	−1.29	0.91	0.52	34.35	34.36	0.19	0.11
300	1.72	1.62	0.39	0.13	34.91	34.90	0.06	0.03
500	1.27	1.04	0.28	0.13	34.94	34.92	0.06	0.01
1000	−0.20	−0.24	0.10	0.12	34.93	34.91	0.05	0.00
2000	−0.80	−0.80	0.03	0.04	34.93	34.92	0.06	0.00
3000	−0.79	−0.78	0.04	0.01	34.94	34.94	0.06	0.01
4000	−0.70	–	0.04	–	34.96	–	0.04	–

nounced in the Atlantic water layer – between horizons of 200 m and 800 m. For example, at the 300-m horizon, temperature in the Nansen basin is about 1.8°C higher than that in the Canadian basin; the salinity is, correspondingly, 0.25 psu higher.

A comparative analysis shows that the average annual values of both temperature and salinity, based on western and Russian data, show a good agreement. The mean values calculated from the western and Russian data best agree for the

Canadian basin. This can be easily accounted for by the fact that for this region the number of observations and observation period of the western and Russian data are most coordinated. For all three regions, for the least variable layer (1,000 m–floor), the average annual values show good agreement. Standard deviation values also match best for the Canadian basin. For other regions, particularly in the salinity field, this agreement is not so good. This can, primarily, be accounted for by the lack of the western data for this region both in respect to the observation numbers and periods.

The results of the comparison enable us to state that there is a similar level of quality for the Russian and western oceanographic measurements – generating the possibility of combining all the available western and Russian data into a single database and using the combined database for large-scale and climatic investigations.

4.2 DEVELOPING THE UNDERSTANDING OF SEABED RELIEF AND THERMOHALINE WATER STRUCTURE AND CIRCULATION: THE DISCOVERIES OF THE 1950s

The results of scientific studies on board the drifting ships *Fram* (1893–1896) and *G. Sedov* (1937–1940), at the 'NP-1' drifting station (1937–1938) and of the expedition on board the airplane 'SSSR N-169' (1941) allowed us to gain a first understanding of the natural conditions and life forms in the most inaccessible part of the Arctic – the Arctic basin. F. Nansen and his followers supposed that the Arctic basin bed presents a deepwater cup. In accordance with this, the water mass structure of the Arctic basin was considered in general the same as in the already investigated western part.

Another misleading conclusion was an understanding of the water and ice circulation in the Arctic basin. By 1947, few scientists doubted the existence of a strong current from east to west in the Arctic basin, which entrained surface water and ice from the Chukchi Sea to the strait between Spitsbergen and Greenland.

However, during flights to the North Pole in 1937 (to the 'Pole of Relative Inaccessibility' (Canadian hollow) in 1941), ice floes were encountered and aged at several decades (i.e., older than the duration of their movement from east to west from the Canadian sector of the Arctic). One could suggest that in addition to the Transarctic transport of surface water and ice from Alaska and Canada to Greenland, a different system of currents also existed in the Arctic basin.

In 1939, N. N. Zubov suggested a hypothesis for the possible system of an anticyclonic ice gyre in the Canadian–American sector of the Arctic. In 1940, Zubov and Somov, after specifying the famous 'isobaric ice drift rule' and using quite detailed pressure charts of the Arctic basin available at that time, suggested that the ice drift and the drift currents in the Amerasian part of the basin had a prevailing anticyclonic character (Zubov and Somov, 1940). But there was no direct evidence for supporting this hypothesis at that time. Organization of purposeful studies was needed.

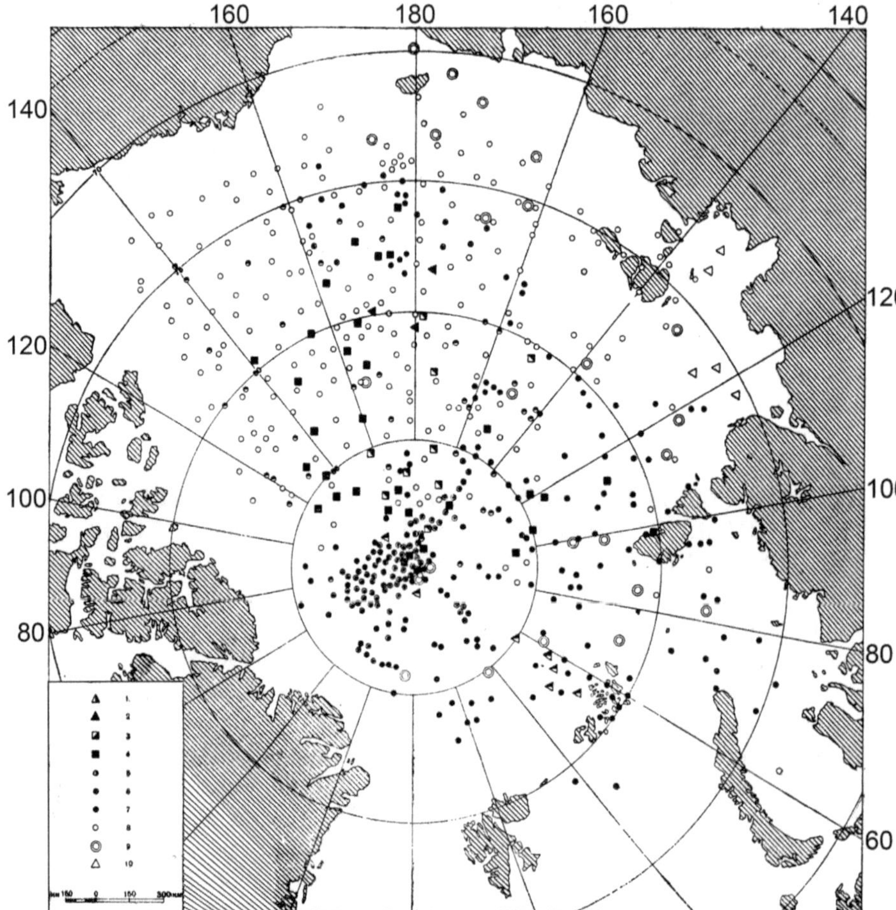

Figure 4.1. Airplane landing sites in the Arctic basin during 1937–1957.
From Laktionov and Shamontiev (1957).

Only in 1948, when scientific problems became a part of USSR northern territories development and were included in the state program of long-term Arctic investigations, did the first efforts to resolve this problem begin. Planned studies of the central Arctic began using the high-latitudinal airborne 'Sever' expeditions, and from 1950 the 'NP' drifting stations, the AARI being entrusted with their organization and implementation (Romanov *et al.*, 1997; Konstantinov and Grachev, 2000; Ugryumov and Korovin, 2004).

The first stage of these studies was full of discoveries. Observations made in the central Arctic Ocean from 1948–1956 literally turned the previous understanding of the ocean on its head. Figure 4.1 presents the areas of aircraft landings during the 'Sever' expedition operations and at the 'NP' stations, giving an understanding of the spatial coverage of observations in the Arctic basin. In general, the establishment in the 1950s of a modern knowledge of the seabed relief, structure of water masses of the Arctic basin, nature and causes of the Transarctic Current and the East (Canadian) anticyclonic water and ice gyre can be considered as the last great oceanographic discoveries on our planet.

4.2.1 Seabed relief

The first depth measurements in the near-Pole area by the 'Sever-2' expedition gave unexpected results. On 27 April, 1948, the oceanographic studies at a point 86°26′N and 154°53′E revealed a depth of 1,290 m – the minimum of all previously measured depths. This discovered seabed rise did not correspond to the existing opinion at that time of a deepwater hollow occupying the central Arctic basin. Upon their return from the expedition to Leningrad Timofeev and Gakkel revealed, for the first time, the underwater mountain system. Based on the data obtained, Gakkel prepared a bathymetry chart of the central Arctic basin with the projection of comparatively small depths on the previous charts to the north of the New Siberian Islands being extended to 86°26′N and 154°53′E. Timofeev, having analysed the additional hydrological observations of the 'Sever-2' expedition (1948), and of the previous expeditions, showed the detected seabed rise to be a component of the extensive sill, which crosses the entire Arctic basin separating into two parts – the sub-Pacific adjoining the Asian and the American continents and the sub-Atlantic adjacent to Europe and Greenland (Timofeev, 1960).

It should be noted that there was a similar line of argument from the US scientist Worthington. After analysing the data of the US oceanographic measurements of expeditions in 1951–1952 and data from Nansen, Worthington came to the conclusion that there should be an underwater ridge extending approximately from the Ellesmere Land to the New Siberian Islands, which separates the deep water of the Beaufort Sea from the rest of the basin – its depth around 2,300 m (Worthington, 1953). By that time the US scientists did not know about the measurements by the Soviet scientists.

In 1949, the 'Sever-4' expedition measured depths in the regions where, as suggested by scientists, the discovered seabed rise was suspected to continue. In fact these depths were even less. These measurements initiated the discovery and investigation of a vast underwater ridge extending from the New Siberian Islands across the North Pole and further to Grant Land. This ridge was called the Lomonosov ridge.

The operations of the 'Sever' expedition and depth measurements at the 'NP' drifting stations were commemorated with large geographical discoveries that, not only introduced significant changes to the geographical charts of the Arctic Ocean, but almost eliminated the 'white spots' (areas not covered by any types of observations) – total area of which comprised by 1939 about 1,800 thousand km^2. Legends of different mythical lands in the Arctic: the lands of Sannikov, Andreyev, Makarov, Peterman, etc., were dispelled. The understanding of the morphometry of the Arctic basin was radically changed. Instead of one deepwater oceanic hollow with a level seabed relief, as was considered before 1948, investigators faced an abyssal area in the basin with a complicated structure, the whole mountainous country dissected by underwater ridges and seabed rises into numerous depressions.

The discoveries were made so rapidly that information about them was not quickly disseminated to other countries and scientists. As a result, some underwater ridges and hollows received the non-coincident Russian and American names, which naturally was rather inconvenient. The US scientists proposed to the Russian scientists to standardize the names of different seabed parts of the Arctic Ocean (Treshnikov *et al.*, 1967). Such unification was made and one common seabed relief nomenclature of the Arctic Ocean was developed (*Regulation of* ..., 1974).

4.2.2 Discovery of the eastern anticyclonic water and ice gyre

The second major result of oceanographic studies made on the basis of full-scale information obtained at the 'North Pole' drifting stations and by the high-latitudinal 'Sever' expeditions was that the earlier hypothesis on the existence of the anticyclonic water and an ice gyre in the eastern Arctic basin was confirmed on the basis of full-scale data as the second dominant method of transports for the surface water layer of the basin.

The oceanographic observations carried out during the 'Sever' expeditions in 1948 and 1949 provided an experimental proof of the anticyclonic gyre's existence. On the dynamic charts of currents constructed by Treshnikov and Timofeyev, the anticyclonic circulation in the Canadian sector of the basin is quite pronounced (Timofeev, 1960). A total scheme of the general water and ice circulation of the Arctic basin appeared to be formed by two general dominants – first by the western Transarctic water and ice transport from east to west in the Eurasian part of the basin and second – the anticyclonic water and ice gyre – in the Amerasian part.

Worthington (1953), based on the analysis of oceanographic observational data during the 1951–1952 'Ski-jump' expedition, suggested that to the north of Alaska the entire water column from the surface to a depth of 600 m is entrained to the anticyclonic gyre. Thus, in the opinion of Worthington not only the surface water in the Canadian basin circulates clockwise, but also Atlantic water contrary to the earlier suggestion by Nansen of the cyclonic nature of Atlantic water spreading in the Arctic basin.

By 1960, the interannual variability of water and ice circulation in the Arctic basin and in the Siberian shelf seas was determined. The dynamic topography charts prepared by Gudkovich (1961b) from observations over 1949–1960 relative to a 200-dB surface showed that two circulation types designated as type A and type B can be identified.

At type A the circulation is distinguished by a strong development of the eastern anticyclonic water and ice gyre and, correspondingly by the displacement of the Transarctic Current core westward to 140–150°E, to the northern boundaries of the Siberian shelf seas. Export of water and ice from the seas to the basin predominates. At type B, the circulation of the eastern gyre is weakly developed while the

Transarctic Current core is displaced eastward and is remote from the seas. The cyclonic water gyres form in the seas contributing to the water inflow to the western areas of the seas.

The statement on the dominance and periodical replacement of two circulation types was then developed in the work of Proshutinsky and Johnson (1997). Based on numerical modelling, they confirmed that the cyclonic and anticyclonic regimes replace each other every 5–7 years and hence the complete period of oscillations comprises 10–15 years.

4.2.3 Surface layer

The formation of an understanding of the surface layer structure was most complicated, which by the definition of that time occupied the space from the surface to the upper boundary of Atlantic water (i.e., to the isotherm of 0°C).

According to observations from 'NP-1' drifting station and the ships *Fram* and *G. Sedov* in the sub-Atlantic part of the Arctic basin, Shirshov identified two water masses in the surface layer: surface Arctic water (0–50 m) and intermediate water (50–200 m). He suggested that intermediate water was formed genetically by mixing of Arctic water with Atlantic water (Shirshov, 1944).

New data of deepwater hydrological stations that covered now not only the sub-Atlantic part of the Arctic basin, but also the sub-Pacific region (see Figure 4.1) allowed depiction of the new features of temperature and salinity distribution in different horizons of the upper layer of the basin.

A full analysis of the observations made in surface waters was performed by Treshnikov (1959). Based on observational data obtained by the first 'Sever' expeditions and the 'NP' drifting stations, compilation of a 3-D picture of the thermohaline structure of the upper layer of the Arctic basin could be performed. In particular, the upper layer thickness changes from 50–100 m in the Fram Strait to 330 m at the centre of the anticyclonic gyre – the Beaufort gyre. Treshnikov explained the phenomenon of salinity decrease of surface water from the northern boundaries of the Siberian shelf seas to the deepwater regions of the Arctic basin.

Annual variations of mean water temperature and salinity for the 0–25-m layer in the observational data on board the ice-breaking ship *G. Sedov* and the 'NP-2' station showed the seasonal changes of water temperature to be insignificant, comprising tenths of a degree, while the salinity changes significantly increasing from summer to late winter by 2 psu. The freshening in the summertime is mainly determined by snow melting at the ice surface and by meltwater drainage under the ice, while salination is determined predominantly by ice growth.

Treshnikov constructed a chart of mean salinity for the 0–25-m layer for the winter period from which it followed that the salinity increased from west to east. The highest salinity zone is delineated in the form of a band from the Greenland Sea along the continental slope to the northern Laptev Sea. In this band due to the vertical winter circulation, a layer of uniform low temperature and higher salinity, compared with water located to the north of the indicated band, is created. This more saline and hence denser water gradually spreading northward sinks to a depth of 100 m.

Coachman and Barnes (1962), having supported Treshnikov's viewpoint and the suggestion of Nansen about the advective mechanism of occurrence and maintainance of the subsurface halocline layer with a water temperature minimum, considered the northern margins of the Barents, Kara and Laptev Seas as the primary sources of subsurface water. However, the genesis and spreading of the halocline with minimum temperatures, the so-called cold halocline, is still a present-day problem of polar oceanography (Steel and Boyd, 1998).

An important event in oceanographic studies in the 1950s was the discovery in the Arctic basin of water of Pacific origin beneath the upper mixed layer. Gudkovich (1955b) and Treshnikov (1959) were the first to point to the existence in the halocline layer of the water interlayer with increased temperature. In addition, Brodskiy (1956) during processing of plankton collections from the 'Sever' expeditions at the stations of the sub-Pacific region of the central Arctic, detected a large number of Pacific Ocean forms of plankton organisms in the 50–100-m layer. In 1961, Coachman and Barnes, analysing the Bering Sea water also made a conclusion about a wide spreading of Pacific origin water in the Arctic basin (Coachman and Barnes, 1961).

Gushchenkov (1964) also noted a water temperature decrease to −1.4 to −1.6 degrees at a depth of 140–170 m and connected this with the genesis of Pacific water during the winter period. Assuming the isotherms of −1.4° to be the lower and upper boundaries of Pacific water, he constructed the charts of Pacific water spreading (Gushchenkov, 1964).

4.2.4 Thermohaline structure of the Arctic Basin

Polemicizing with Nansen (1902) and Shirshov (1944), Timofeyev (1960) showed that three water masses and two intermediate layers could be identified in the Arctic basin. From water temperature and salinity, biological indicators and oxygen content, the following main and intermediate water masses were identified in the sub-Atlantic region of the central Arctic basin with the following characteristics (Figure 4.2).

Arctic surface water is characterized by a rather low temperature (−1.75 to −1.65°C) and low salinity (<32.00 psu). The water layer thickness in the sub-Atlantic region of the basin comprises 0–30 m and in the sub-Pacific region 0–100 m.

By origin, this water presents a mixture of river and Pacific Ocean water with seawater of the marginal Arctic seas and waters forming from snow and ice melting in the Arctic basin.

Atlantic water flows to the Arctic basin from the Greenland Sea. It has a positive temperature and a salinity higher than 34.60 psu. It is distinguished by a high temperature comprising 2–3°C in the sub-Atlantic region of the basin and 0.5–0.6°C in the sub-Pacific region and a relatively high (34.70–34.90 psu) salinity. The upper boundary of the Atlantic water layer (isotherm of 0°) in the sub-Atlantic region of the basin passes in depths of 50–200 m and in the sub-Pacific region in

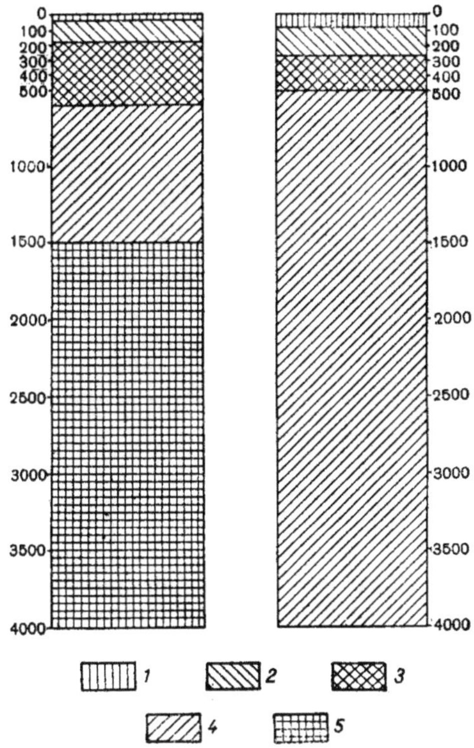

Figure 4.2. Vertical distribution of waters in different parts of the centre of the Arctic basin: (*left*) – the Eurasian sub-basin; (*right*) – the Amerasian sub-basin. 1 – the Arctic surface water; 2 – the intermittent upper water; 3 – the Atlantic water; 4 – the intermittent lower water; 5 – the bottom water.
From Timofeev (1960).

depths of 250–300 m with the lower boundary of this layer located at depths of 800–900 m.

Bottom water is characterized by a negative temperature comprising −0.80 to −0.85°C, and an almost uniform (34.90–34.96 psu) salinity. This water, filling the sub-Atlantic deepwater hollow of the Arctic basin is water of the north-eastern Greenland Sea. Bottom water can partly form during the ice formation period in the shallows of some Arctic seas, from where it sinks along the continental slope to the deepwater part of the Arctic basin.

Intermediate upper water is located between the Arctic surface and the Atlantic water masses, being characterized by mean transient temperature and salinity values of the aforementioned water masses.

Intermediate lower water is located between the Atlantic and bottom water masses and has transient temperature and salinity values of the aforementioned two types of the water masses. The upper boundary of this layer in the sub-Atlantic region of the basin is located at a depth of 800–900 m and the lower at 1,500–2,000 m.

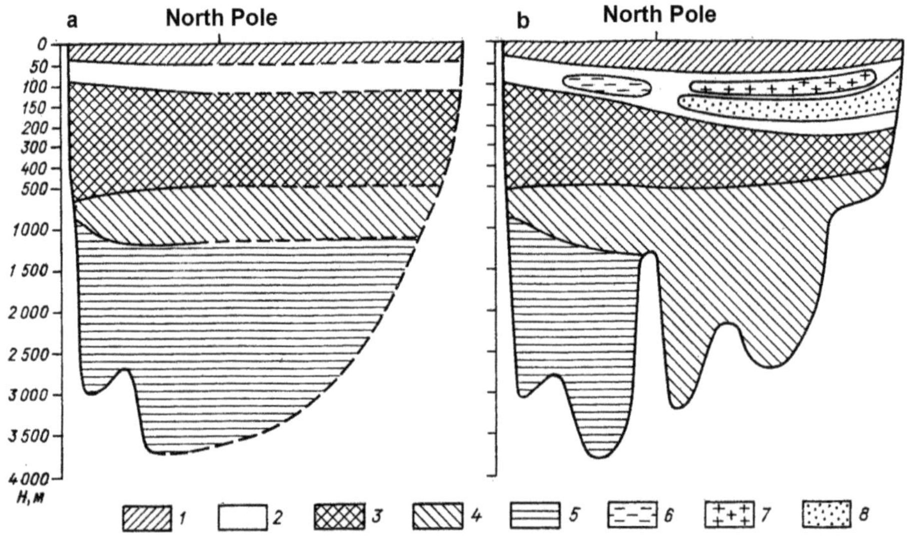

Figure 4.3. Evolution of our notion of the water structure in the Arctic basin before 1940 (a) and after the beginning of the 1940s to the end of the 1950s (b): 1 – the surface water of the Arctic basin; 2 – the intermittent upper water; 3 – the Atlantic water; 4 – the intermittent lower water; 5 – the bottom water; 6 – the intermittent water of winter formation; 7 – the summer Pacific water; 8 – the winter Pacific water.
From Nikiforov (1995).

In the sub-Pacific hollows of the basin, this water fills its entire deepwater area presenting bottom water here (an independent water mass type).

Evolution of the views of the structure of water masses is illustrated in Figure 4.3, which presents a transect from the Fram Strait to the Bering Strait. The understanding of bottom and intermediate waters of the basin has changed most of all. As can be seen from Figure 4.3, the Lomonosov ridge 'cuts' the coldest bottom water of the western area.

As to Atlantic water, the prevision of Shirshov, who suggested that Atlantic water flows not only eastward but also widely spreads in the central Arctic basin and, probably, fills it entirely, was fully justified (Shirshov, 1938). Timofeyev made maps of the Atlantic water characteristics and constructed a chronological scheme of deep Atlantic water spreading. Figure 4.4(a) presents a scheme of motion of Atlantic water in the central Arctic basin. In accordance with his estimate, warm water of Atlantic origin passes from the strait between Spitsbergen and Greenland to the Beaufort Sea over a period of 6 years.

Coachman and Barnes in 1963 published a chart of the Atlantic water circulation in the Arctic basin, which slightly differs from the chart of Timofeyev in the Beaufort Sea and above the Chukchi rise (Coachman and Barnes, 1963).

Nikitin (1969), having processed the data of deepwater hydrological observations for the period 1948–1956, constructed charts of geostrophic currents for different layers of the ocean. Figure 4.4(b) presents a scheme of currents in the

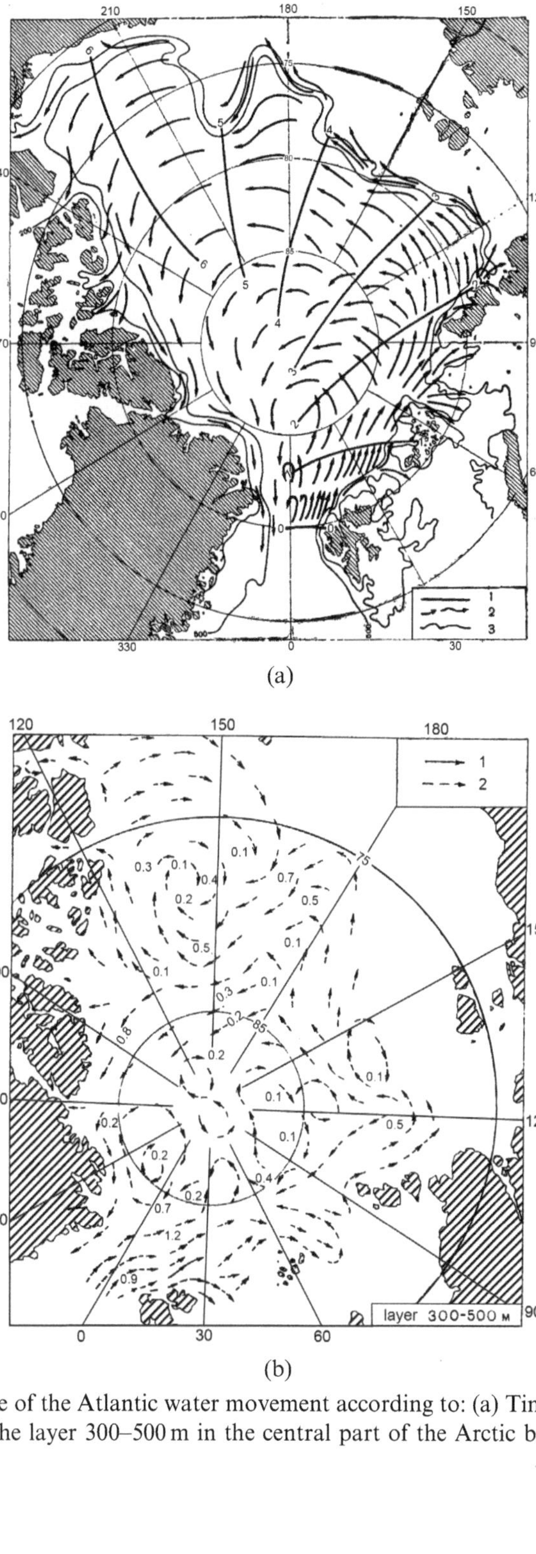

Figure 4.4. Scheme of the Atlantic water movement according to: (a) Timofeev (1960) and (b) Nikitin (1969) in the layer 300–500 m in the central part of the Arctic basin.

300–500-m layer, which reflects the Atlantic water circulation pattern. Comparing the schemes of Timofeyev and Nikitin, one can see that they coincide in general for the sub-Atlantic region of the Arctic basin and differ significantly for the sub-Pacific region. It is noted that the circulation pattern of Atlantic water of Nikitin is quite close to the circulation pattern, presented in the US *Oceanographic Atlas of Polar Seas* (1958).

4.2.5 Currents in the Arctic basin

A generalized analysis of observations of currents was performed by Nikitin (1969), who used 4,883 daily cycles of current measurements for the period 1949–1958. Nikitin was the first to give the most complete characterization of the regime of non-periodic currents of the Arctic basin – total (observed), wind, constant and residual.

The speeds of total currents measured in the Arctic basin vary over a wide range of values from close to zero to 86 cm/s. The speeds from zero to 6–8 cm/s have the largest frequency of occurrence (20% or more).

The total currents with the largest speeds from 45–86 cm/s were observed in regions located to the north-east of the New Siberian Islands, and at the approaches to the strait between Spitsbergen and Greenland.

Small current speeds are typical of the area of the anticyclonic water gyre and the near-pole areas where the largest measured speeds at the stations comprised 17 to 34 cm/s.

A scheme of the constant currents of the Arctic basin in the 3–10-m layer is presented in Figure 4.5 based on the data of instrumental measurements of currents and the results of current calculations using a 'dynamic method'. The scheme clearly presents two dominating water flows for the surface layers of the basin – the Transarctic Current crossing the basin in the general direction from east to west from the region to the north of the Chukchi Sea to the passage between Spitsbergen and Greenland and the anticyclonic water (and ice) gyre in the Amerasian part of the basin.

The water circulation shown in the scheme preserved its main features to a depth of 50–100 m. It is also stable in its direction of flows interannually. However, the speeds of currents are subjected to interannual fluctuations reflecting a corresponding variability of all processes in the atmosphere and the sea.

According to Nikitin's (1959) estimations the constant currents in the surface layers of the Arctic basin have prevailing speeds of 0.3–4.0 cm/s, in the strait between Spitsbergen and Greenland – 3.5–12 cm/s and in the area to the north-east the New Siberian Islands – 3.0–10.0 cm/s. In the 200–300-m layer, speeds of 0.3–1.5 cm/s predominate and in the strait between Spitsbergen and Greenland and near the New Siberian Islands – from 1.0–2.1 cm/s. In the 300–500-m layer where the speeds were obtained from the dynamic calculations, their values range from 0.1–0.3 cm/s.

The observations showed that strong currents episodically occur in the subsurface layer (50–300 m) of the Arctic basin. The subsurface currents were first observed

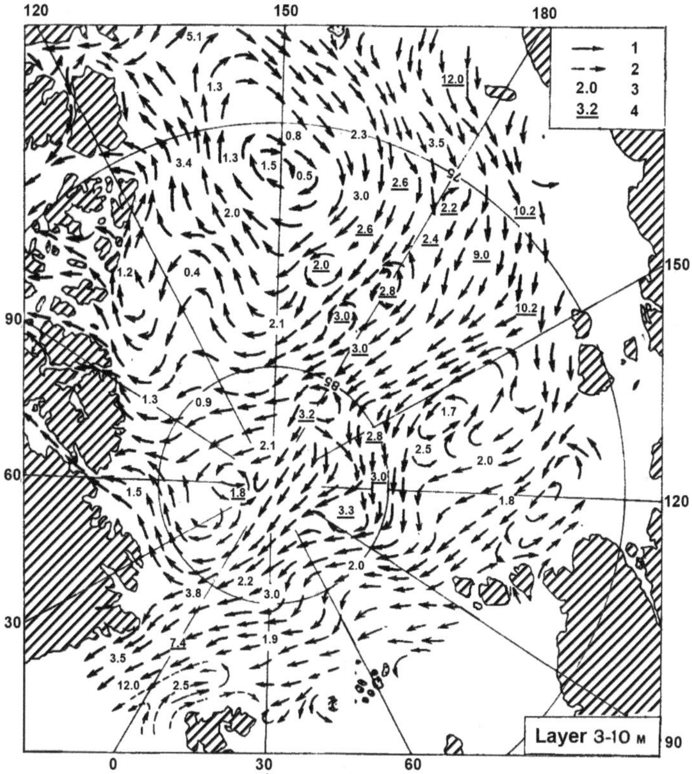

Figure 4.5. Scheme of the constant currents in the Arctic basin in the 3–10-m layer; 1 – the cold current, 2 – the warm current, 3 – velocity of the steady-state gradient current (cm/sec); 4 – instrumentally measured current velocity (cm/sec).
From Nikitin (1969).

by Shirshov in 1937 during observations at the 'NP-1' drifting station (Shirshov, 1944). Then the subsurface currents were twice recorded at the 'NP-2' drifting station, where they achieved speeds of 60 cm/s during 2 days and 21 cm/s during 6 days.

Brown and Grary investigated the subsurface currents in detail during the operation of the US 'T-3' drifting station in 1952–1955 (Brown and Grary, 1958). An analysis showed the currents to have a predominant direction from the high atmospheric pressure area to the low atmospheric pressure area. The authors considered the dynamic changes of sea level due to divergence and convergence of the Ekman flow as the main cause for occurrence of these currents.

In the summer of 1965, Galt observed (at the 'T-3' drifting station) subsurface currents whose speeds comprised 58 cm/s (Galt, 1967). At the time of observations, the drift speed was not greater than 4 cm/s. Belyakov superimposed his observation data and data of Galt in Figure 4.6 (Belyakov, 1974; Belyakov and Grishenko, 1980). As can be seen from this figure, the current speed core may not coincide with the depth of the largest seawater density gradient.

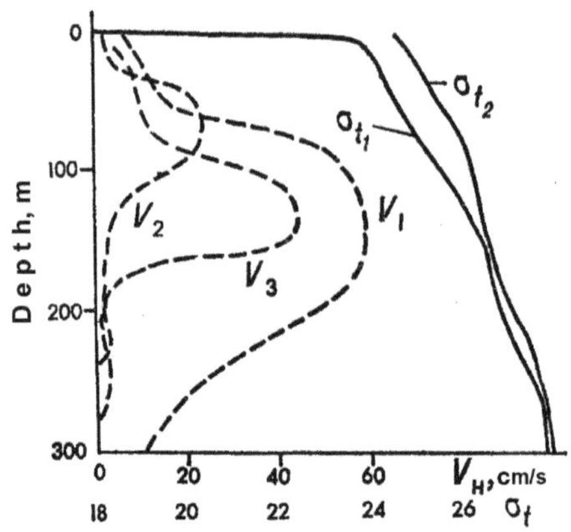

Figure 4.6. Vertical structure of maximum velocities of the subsurface currents (V_S) and apparent density (σ_t) at the moments of current measurements: V_1 and σ_{t_1} according to the data of Galt (1967); V_2, V_3 and σ_{t_2}, according to the data of Beliakov (1974).

The present-day importance of investigating the mechanism of occurrence of subsurface currents, which are the strongest of all types of currents observed beneath the ice cover of the Arctic basin, is obvious.

4.3 STUDY OF LARGE-SCALE OCEANOGRAPHIC PROCESSES

4.3.1 Motivations of the new direction of expedition studies of the Arctic Ocean

From 1954, the research activity at the 'NP' drifting stations significantly increased. Two to four stations drifted every year with a drift duration for each station comprising one to four years. However, there was some specific decrease in oceanographic observations performed by the 'Sever' expeditions from the late 1950s to the early 1970s. One of the main objectives of the 'Sever' expedition was the organization and support of the 'NP' stations.

However, by the early 1970s it became clear that studies of complex typical features regulating the dynamic and thermal phenomena in the ocean, and its interaction with the atmosphere, require observations over large areas and for a long times.

On the initiative of A. F. Treshnikov, the development of a program began that was called POLEX – the 'Polar Experiment' (Treshnikov, 1970). The POLEX region covered the area of the Arctic basin, the Arctic seas, the northern areas of the Atlantic and the Pacific Oceans and the extensive northern territories of Eurasia and the north. This program initiated a new direction for investigations of the Arctic Ocean. It is important to stress that a great deal of attention in this program was devoted to the observations to be carried out at the 'NP' stations

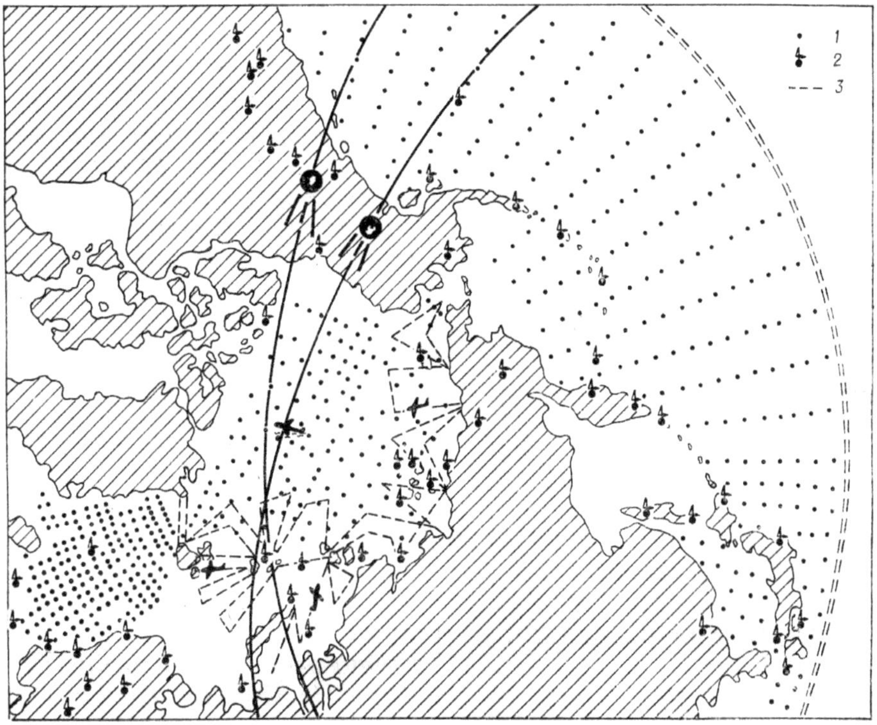

Figure 4.7. Survey of the POLEX–'Sever'. 1 – hydrological stations; 2 – air–meteorological stations; 3 – routes of flying observatories.
From Trioshnikov (1979).

and to the use of 'NP' stations as bases for arranging the work over the enormous high-latitudinal space using the 'Sever' expeditions (see Figure 4.7).

The implementation of POLEX began in the spring of 1973. Using the 'NP-21' station as the initial base, the 'Sever-25' expedition carried out deepwater oceanographic and meteorological observations at 182 ocean points. Execution of large-scale oceanographic surveys of the Arctic basin continued until 1979. Simultaneously, the 'NP' stations continued standard year-round observations and the hydrophysical studies under the special programs. This stage of the expedition studies can be, by right, called a new direction of research for the Arctic basin oceanography.

From 1981–1993, the oceanographic studies of the 'Sever' expeditions focused on the Siberian shelf seas, where teams performing the oceanographic surveys from drifting ice operated. The 'Sever' expeditions still performed work for supporting the 'NP' drifting stations and also carried out studies under special programs (Konstantinov and Grachev, 2000).

In 1991, the studies at the 'NP' drifting stations were stopped being resumed only in 2003. In 1993 the 'Sever' expeditions was terminated.

4.3.2 Surface Arctic water

Obtained data allow us to take a new look at the genesis and geography of the surface waters. Surface Arctic water is located from the bottom ice boundary or directly from the ice-free ocean surface to the pycnocline depth, which coincides with the thickness of the mixing layer. The layer thickness in summer comprises about 10 m in the northern areas of the Arctic seas and about 20–40 m in the central Arctic basin. In winter, the thickness of the layer at the northern boundaries of the Siberian shelf seas is 20–25 m increasing to 70–100 m (in some cases to 150 m) in the Eurasian sub-basin.

The regional peculiarities in the distribution of hydrological characteristics of surface Arctic water are clearly reflected on the charts of water temperature and salinity distribution at a depth of 5 m. Figure 4.8 (see colour section) presents the climatic charts of the average water temperature and salinity distribution for the period 1950–1989 in the 5-m horizon for winter and summer. We shall stress again those thermohaline structure peculiarities of the surface Arctic water layer that were earlier noted in Treshnikov (1959), Coachman and Barnes (1962), Nikiforov and Shpaikher (1980), etc.

The location of a zero isotherm at a depth of 5 m in the Arctic seas in summer is mainly determined by the ice cover edge and the ice concentration value. In the ice-free areas, the water temperature in summer is above zero with temperature increasing southward. In the central Arctic basin, the water temperature of the surface Arctic layer is close to the ice melting temperature.

In the wintertime, the water temperature of the surface Arctic layer is close to freezing temperature at a given water salinity value. Therefore, the spatial distribution of the water temperature correlates well with the salinity distribution. Let us pay attention to the zone of the lowest water temperatures of -1.8°C and below in the northern areas of the Barents and Kara Seas, western Laptev Sea and the adjoining area of the Nansen basin. Here the highest salinity values of the surface Arctic water mass are also observed.

The Arctic seas are the main source of brackish water for the Arctic basin, and on the salinity distribution charts both in winter and in summer the Kara, Laptev, East Siberian and Beaufort Seas are delineated as enormous marginal areas of freshened water. In summer, a large amount of freshwater continental runoff, primarily the river runoff, is added here to melt ice water (the contribution of atmospheric precipitation to the freshwater balance of the Arctic seas and the Arctic basin is insignificant comprising about 5–10% of the change in freshwater content due to the continental runoff). They supply a large-scale zone of low salinity in the anticyclonic gyre area of the Canadian hollow and a negative salinity anomaly in the Transarctic Current area along the Lomonosov ridge from the western East Siberian Sea and the eastern Laptev Sea towards the North Pole and further to the Fram Strait. These two zones of fresher surface Arctic water are delineated in the Arctic basin both in the winter and summer seasons on the climatic charts (Figure 4.8, see colour section).

The second peculiarity of the spatial water salinity distribution is the existence of two zones with large salinity values. The first zone is located in the Nansen basin adjoining the continental slope of Spitsbergen, Franz Josef Land and Severnaya Zemlya. The source of more saline water, which maintains a constant higher salinity of surface Arctic water in this region, is water of Atlantic origin. Atlantic water flows to the Arctic basin through the Fram Strait, directly underlying surface Arctic water and supplying it with salts from below. The second flow of Atlantic water, as can be seen from the bends of salinity and temperature isolines in Figure 4.8 (see colour section), intrudes to the Barents Sea and partly to the Kara Sea where Atlantic water undergoes significant transformation. First of all, a large heat loss by Atlantic water occurs. Its temperature at the northern boundaries of the Barents and Kara Seas becomes negative in the summertime, and in the wintertime, a zero isotherm at the surface is located much further to the south. Water of Atlantic origin in the Barents and Kara Seas in summer interacts directly with a zone of higher salinity of in the Nansen basin. Water of Atlantic origin in winter, at intense cooling and additional salination due to the salt release during the ice formation period, can submerge under surface Arctic water. Spreading further in the Arctic basin in the sub-surface layers, it serves as a source of salts and heat for the surface Arctic water.

The second zone with a relatively high water salinity is located in the Chukchi Sea, being connected with the flow of Pacific Ocean water to the sea through the Bering Strait, as can be seen from the bends of salinity and temperature isolines in Figure 4.9. Pacific Ocean water submerges under surface Arctic water and spreads in the Arctic basin as the interlayer of the so-called Pacific Ocean water of summer origin formation. Further cooling and salination of Pacific water in winter, in the southern Chukchi Sea, results in its sinking deeper than the interlayer of Pacific water of summer origin formation and spreading in the Arctic basin in the form of the so-called Pacific Ocean water of winter origin formation. The sub-surface summer and winter Pacific water is a source of salts for surface Arctic water.

In the summer, the frontal zone of the Transarctic Current is fuzzy, however, approaching the Fram Strait the contrasts of water salinity and temperature become more pronounced. In winter, the frontal zone of the Transarctic Current is clearer as can be seen in Figure 4.8 (see colour section).

An analysis of the distribution of oceanographic characteristics in the Arctic seas and the Arctic basin for individual years showed two main hydrological regimes that are identified by the formation conditions (Nikiforov and Shpaikher, 1980). One of them forms at the stable dominating influence of the Icelandic Low of the atmospheric pressure and the other – at a prevailing development of the Arctic High.

In the years dominated by the influence of the Icelandic Low, transport of water masses from west to east predominates in the Arctic seas resulting in the increased inflow of Barents Sea water to the western Kara Sea, outflow of surface Arctic water from the Kara and Laptev Seas to the East Siberian Sea and the attenuated inflow of the Pacific Ocean water to the Chukchi Sea. Correspondingly, the water outflow is formed in the Kara and Laptev Seas, and the inflow of surface water in the East Siberian and Chukchi Seas. The thickness of the surface water layer decreases in the Kara and Laptev Seas and becomes higher in the East Siberian and the Chukchi

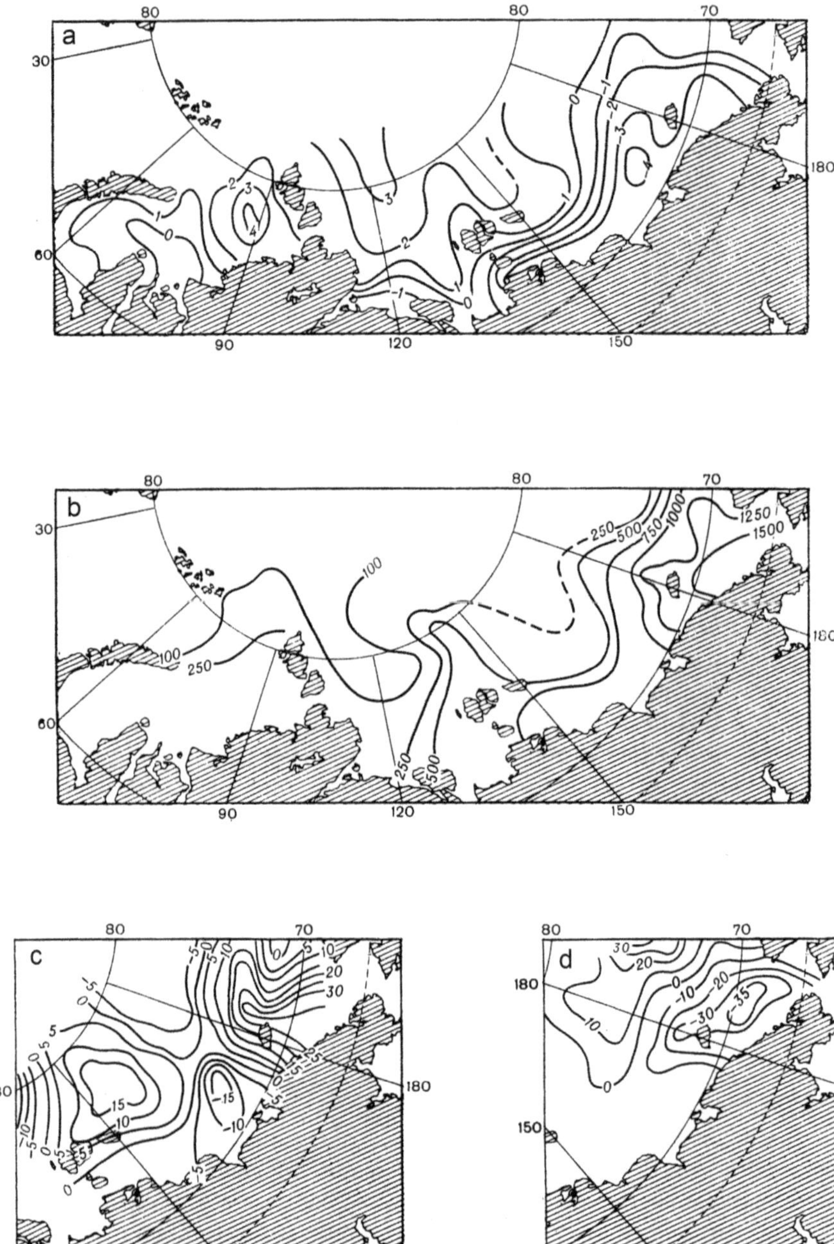

Figure 4.9. Distribution of hydrological characteristics by predominance of the Icelandic minimum: (a) anomalies of the water salinity (psu) in the surface layer (0–5 m); (b) silicon distribution (μcg/l); (c, d) anomalies of the layer thickness of surface Arctic and Pacific waters (m), respectively.

From Nikiforov and Shpaikher (1980).

Seas. The surface water salinity in the Kara and Laptev Seas increases and in the Chukchi and the East Siberian Seas, water freshens (Figure 4.9). The upper boundary of deep Atlantic water is located deeper than usual.

In the years of the predominant influence of the Arctic High, the transport of water masses from south to north is developed in the Arctic Seas, which determines the outflow of surface Arctic water from the East Siberian and the Chukchi Seas and the inflow of the Pacific Ocean water to them. The outflow of river water to the Kara and Laptev Seas significantly increases and the inflow of Barents Sea water to the Kara Sea decreases. The surface water salinity in the Kara and Laptev Seas is less, and in the East Siberian and the Chukchi Seas, it is higher than a multiyear average (Figure 4.10). In the eastern Laptev Sea, the East Siberian and the Chukchi Seas one observes water outflow and in the western Laptev Sea and in the Kara Sea water inflow. The thickness of the surface water layer in the Kara and Laptev Seas is greater and in the East Siberian and the Chukchi Seas is smaller than a multiyear average. Deep Atlantic water is located closer to the ocean surface than during cyclonic types of development (Nikiforov and Shpaikher, 1980).

4.3.3 Variability of intermediate water of the Arctic Basin

The oceanographic surveys of 1955 and 1956 supplemented with oceanographic observations at the drifting stations allowed, for the first time, the large-scale peculiarities of intermediate water spreading in the Arctic basin and the interannual variability of its distribution to be seen. An analysis of the vertical water temperature and salinity measured in different regions of the Arctic basin allowed identification of a number of new regularities of the thermohaline structure of intermediate water. It turned out that the intermediate water masses of the Eurasian and American sub-basins differ significantly by origin (Nikiforov *et al.*, 1977). This can be clearly seen from a temperature transect (Figure 4.11(a)), where cold water to the north of the Beaufort Sea between the 50 and 100 m depths is cut by a warmer water interlayer with a maximum in the 75-m horizon.

Below the layer of the summer Pacific Ocean water in the Amerasian sub-basin, water with a minimum temperature of less than $-1.4^{\circ}C$ is also observed almost everywhere. The analysis of the dissolved silicon concentration in this water allows the conclusion that this is also Pacific water, but of winter origin.

However, the traces of winter and summer Pacific water disappear at the approaches to the Lomonosov ridge. New data on the silicon concentration allows us to follow its concentration in the intermediate water of the Amerasian and the Eurasian sub-basins. It turns out that water enriched with silicon with concentrations up to 1,000–1,500 μg/l reaches not only the North Pole, but also the shores of north Greenland and the Canadian Arctic archipelago – then probably in the transformed form it is exported from the Arctic basin to the North Atlantic. That is, intermediate water of the Eurasian sub-basin consists, to a great extent, of strongly transformed water of Pacific origin. On the other hand, as shown by the temperature transect (Figure 4.11(b)), cold and saline water forming to the north of the Kara and Laptev Seas, sinks to depths and flows beneath the freshened surface water of the

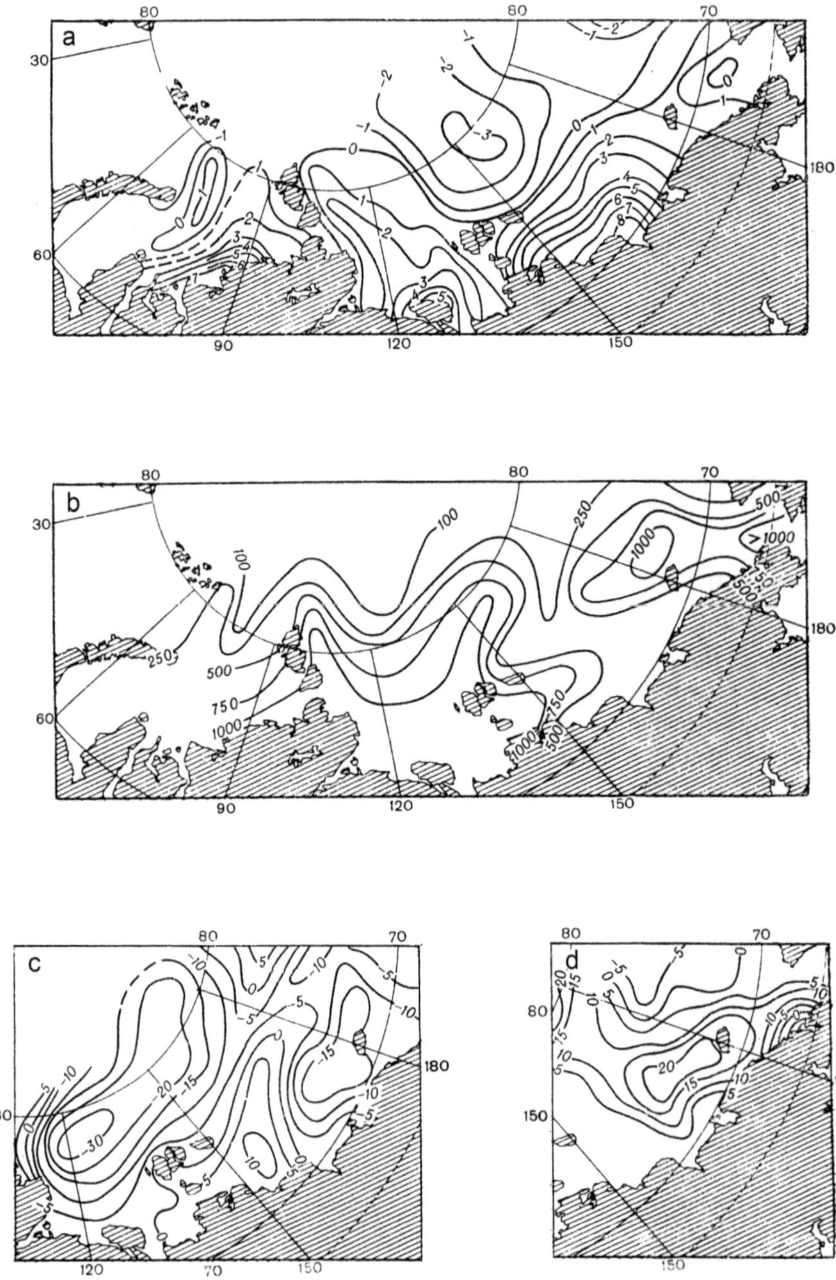

Figure 4.10. Distribution of hydrologic characteristics by predominace of the Arctic maximum: (a) anomalies of the water salinity (psu) in the surface layer (0–5 m); (b) silicon distribution (μg/l); (c, d) anomalies of the layer thickness of the surface Arctic and Pacific waters (m), respectively.
From Nikiforov and Spaikher (1980).

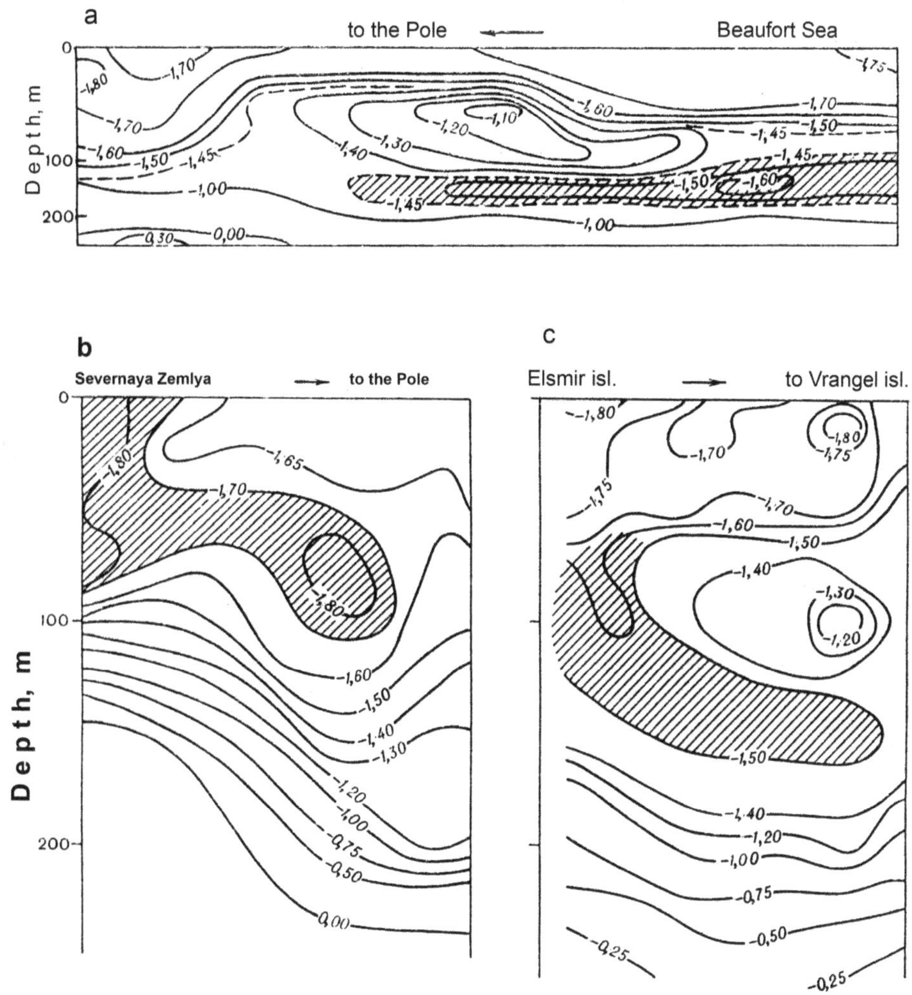

Figure 4.11. Vertical distribution of water temperature (°C) in the Arctic basin at sections from the Beaufort Sea to the North Pole (a); from the Laptev Sea to the North Pole (b); and from Ellesmire Island to Wrangel Island (c), in April 1973. Zones of the intermittent waters of the winter formation are marked by hatching.
From Nikiforov *et al.* (1977).

Eurasian sub-basin. This water spreads further in the form of winter intermediate water approximately in the same horizons as winter Pacific water differing from it only by its decreased silicon concentration. Thus, intermediate water of the Eurasian sub-basin consists of strongly transformed Pacific Ocean water and of winter intermediate water of shelf origin (Nikiforov *et al.*, 1977).

The observation materials of the 'NP-4' (1954/1955), 'NP-8' (1959/1960), 'NP-12' (1963/1964) and 'NP-20' (1970–1972) stations were used by Blinov and Vorobiev

Table 4.4. Mean monthly temperatures (°C) for a year in the core of summer and winter Pacific Ocean water, based on the observations at the 'NP-4, 8, 12 and 20' stations.
From Blinov and Vorobiev (1976).

Months	Summer water	Winter water
January	−1.23	−1.53
February	−1.30	−1.55
March	−1.36	−1.60
April	−1.31	−1.58
May	−1.19	−1.54
June	−1.18	−1.50
July	−1.19	−1.53
August	−1.29	−1.54
September	−1.30	−1.54
December	−1.21	−1.59

for investigating the variability of characteristics of Pacific origin water, since these 'NP' stations drifted to the north of the Chukchi Sea in the area of spreading of Pacific origin water (Blinov and Vorobiev, 1976).

For revealing the seasonal variability, all temperature observations, in the core of summer and winter Pacific Ocean water, available in the chosen regions were averaged by months. The results of averaging are presented in Table 4.4.

As follows from this table, the temperature of summer and winter Pacific Ocean water during the year changes over a significant range. In seasonal temperature variations of summer Pacific Ocean water one can easily detect two pronounced maxima (in June and December) and minima (in March and October). The amplitude of temperature variations of summer Pacific Ocean water within a year comprises about 0.20°C.

The changes in temperature of the winter Pacific Ocean water also indicates the presence of two maximums (in winter and summer) and two minimums (in spring and autumn). However, the seasonal temperature fluctuations of winter Pacific water are slightly less compared with summer: their amplitude is approximately 0.10°C. Thus, the intra-annual changes of temperature characteristics of summer and winter Pacific water manifest the same half-year periodicity with its extreme values for both water types being observed at the same time (Blinov and Vorobiev, 1976).

Blinov and Vorobiev, assuming as boundaries of summer and winter Pacific Ocean water the depth of the isotherm of −1.40°C, calculated the thicknesses of the layers and the depths of the upper and lower boundaries of summer and winter Pacific Ocean water. The thickness of the layer of summer Pacific Ocean water in the given region varies from 35–55 m. The thickest layer is observed from May–July, and the thinnest in November–December. The thickness of the layers of winter Pacific Ocean water of all the months of the year is slightly higher than in summer ranging between 55–75 m. The maximum thickness of the winter water layer is achieved in

August–September and the minimum – in March. Thus, there are pronounced annual variations in the change of thickness of both types of Pacific water with slightly displaced maximums and minimums relative to each other.

Whereas the thickness of the layers of summer and winter Pacific water during the year changes up to 20 m, their upper and lower boundaries experience significant fluctuations by depth comprising 40 m in some months. The upper boundary of summer Pacific water is the deepest in autumn and winter. The maximum changes by depth of its location are also observed at this time. This water is most close to the surface during the spring–summer period, which is also characterized by the greatest stabilization of the upper boundary: its observed motion in these months is not greater than 10 m. Hence, the annual variations of the upper boundary of summer water are in good agreement with the thickness change of its layer.

The location of the lower boundary of summer Pacific Ocean water, which is simultaneously the upper boundary of winter water, does not remain constant throughout the year, either. The annual variations of depth by location of this boundary are characterized by the presence of two maximums and two minimums. The maximums are observed in April and December while the minimums fall on the beginning of the year and August. The migration of the boundary of summer and winter Pacific Ocean water by depth during the year is not greater, on average, than 20 m. The exception are the winter months when the fluctuations of the boundaries can comprise more than 40 m.

The lower boundary of winter Pacific water barely experiences any fluctuations, although some deviations are possible in the wintertime (December–February). In December, it sinks as a rule and in January–February, it rises closer to the surface.

Study of seasonal changes of the Pacific Ocean water discharge through the Bering Strait (Fedorova and Yankina, 1963) showed that the inflow of Pacific water to the Chukchi Sea in summer is almost three times as large as in winter. This should primarily influence the thickness of the Pacific Ocean water layer in the Arctic basin and result in the rise of its upper boundary. In winter, a decrease of the Pacific Ocean water layer in the study region can result not only from the decreased water inflow through Bering Strait, but also from the increased advection of surface water from the Beaufort Sea, related to the wind regime peculiarities (Treshnikov, 1959). Thus, the annual variation in the considered characteristics of Pacific water is mainly connected with the inflow of this water through the Bering Strait.

Data on the interannual variability of thermal and dynamic characteristics of Pacific water are presented in Table 4.5.

Based on the data presented above, it can be said that the temperature and thickness of the summer and winter Pacific Ocean water experience significant interannual fluctuations.

The following dynamic factors contribute to the formation of cold intermediate water of winter origin in the Eurasian sub-basin (called 'winter intermediate water' above for short). First, the area of this water formation is located at the periphery of the Arctic High, which like a giant pump collects the lightest surface water at the centre. As a result, at the southern boundaries of the Arctic basin (especially at the outer margin of the Siberian shelf), the layer of surface Arctic water is either thin or

Table 4.5. Mean annual temperature and thickness values of the Pacific water layer in the study region.
From Blinov and Vorobiev (1976).

	Summer water		Winter water	
Year	Temperature in the core (°C)	layer thickness	Temperature in the core (°C)	layer thickness
1954–1955 (NP-4)	−1.33	40	−1.52	40
1959–1960 (NP-8)	−1.28	60	−1.48	55
1963–1964 (NP-12)	−1.12	50	−1.55	80
1970–1972 (NP-20)	−1.30	45	−1.60	55

absent in the wintertime. In all cases this water is strongly salinized as a result of an intense exchange with underlying water.

Second, the indicated region is situated at the left periphery of the Transarctic Current, therefore, due to a cross circulation of this current, less dense surface water is driven to the central regions of the Arctic basin, and more saline water from the lower layers replaces it.

As a result, in winter during the development of intense convective mixing caused by water cooling and salination at ice formation, water forms of increased salinity and decreased temperature equal to the freezing temperature at a given salinity.

The process of this water formation makes the stable geostrophic balance of the layers weaker or maybe destroys it completely, which results in water moving from these regions although along the corresponding isopycnic surfaces to the central regions of the basin. Thus, a layer of winter intermediate water forms underlying the surface water layer, which is characterized by the presence of the core with temperatures lower than of water located at the top and the bottom.

A similar situation is also observed in the other peripheral area of the Arctic basin, in particular to the north of Greenland, the Canadian Arctic archipelago and Alaska. It means that in winter one can expect the formation of cold and comparatively saline water in these regions that due to the aforementioned reasons can penetrate deep into the Arctic basin (Nikiforov *et al.*, 1979). There was an earlier suggestion that the formation of winter intermediate water can also be expected in addition to the European shelf slope of the Amundsen and Nansen depressions to the north of Greenland, the Canadian Arctic archipelago and Alaska. The oceanographic surveys of the 'Sever' expedition made under the 'POLEX-North' program and the oceanographic observations at the 'NP' drifting stations confirmed this suggestion and allowed us to specify the boundaries of formation and spreading of winter intermediate water. Figure 4.12 presents median values of the zones of formation and spreading of winter intermediate water of the Arctic basin, and Table 4.6 presents the characteristics of its spreading in the sub-basins.

A jet of cold intermediate water of winter formation can be clearly seen at the

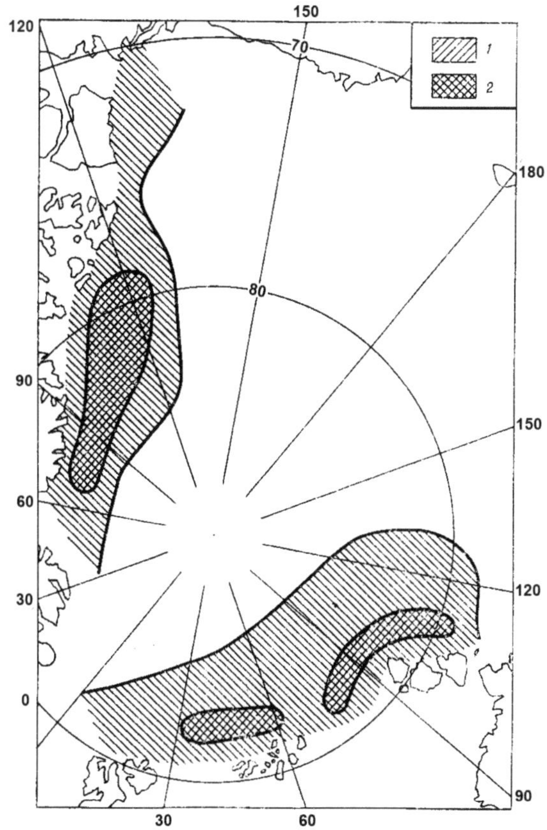

Figure 4.12. Median boundaries of the zones of formation and propagation of the winter intermittent waters of the Arctic basin (averaged for the winter period). 1 – areal extent of water; 2 – zones of water formation.
From Nikiforov *et al* (1979).

temperature transect (Figure 4.11(c)) oriented along the axis of the main flow of this water originating to the north of the Ellesmere Land.

We shall note one important peculiarity of water under consideration in the American sector of the basin: similar to the Pacific Ocean water, it is distinguished by the increased silicon concentration although its maximum level is approximately 1.5 times less than in Pacific water.

The interlayer of summer Pacific water in 1973 was especially pronounced only in the near-Pole area, while the water itself formed a strong jet with a core of water of increased temperature (up to -1.1°C) (Figure 4.11(a)).

Thus, the new data obtained show that the intermediate water mass of the Arctic basin presents water of Pacific origin, which comprises the main portion of intermediate water in the Amerasian sub-basin and winter intermediate water comprising the largest portion of intermediate water in the Eurasian sub-basin. Winter

Table 4.6. Characteristics of spreading of the winter intermediate water of the Arctic basin by years and sub-basins: Eurasian (Esb; km^2), Amerasian (Asb; km^2) and for the basin in general (Ab; in thousands km^2) on average over winter.
From Nikiforov *et al.* (1979).

	Regions			
Year	Esb	Asb	Ab	%
1955–1956	950	–	–	–
1972	380	–	–	–
1973	490	735	1,225	24
1974	1,100	490	1,590	32
1975	1,200	735	1,935	39
1976	1,470	610	2,080	42
1977	1,600	1,125	2,725	54
1978	1,100	610	1,710	34
Average	1,040	720	1,760	35
Amplitudes	1,220	635	1,855	37

intermediate water can play an important role in the formation of intermediate water masses in both sub-basins.

Data of the survey in April 1973 allow us to present one more important argument. The core of water of winter origin in both sub-basins is formed by waters that differ little in their salinity, so that in higher and lower lying water, the vertical salinity gradients are much higher compared with winter intermediate waters. Therefore, from the vertical salinity distribution, it is possible to identify them as a separate sublayer, which contains all the features typical of the fields of other elements.

The presence of the water interlayer, with a temperature less than in the above and lower lying layers, creates such inversion conditions under which the heat fluxes are directed to the layer with the lowest temperature both from beneath Atlantic water and the top of surface Arctic water. So the cold intermediate layer is a kind of locking layer (i.e., a layer through which the heat transfer from the depths of Atlantic water to the ice cover is difficult). This locking layer is comprised both of cold water of shelf origin and of winter Pacific Ocean water.

As can be seen from the Table 4.6, this water occupies about 35% on average and in some years more than half of the Arctic basin area. The interannual variability of the area of spreading of winter intermediate water is comparable with an average area for eight years (Nikiforov *et al.*, 1979).

4.3.4 Atlantic and bottom water

The hydrological surveys of 1955–1956 and data of observations at the drifting stations were used for the analysis of the maximum temperatures of Atlantic water. As can be seen in Figure 4.13, 'spots' of anomalously warm water are

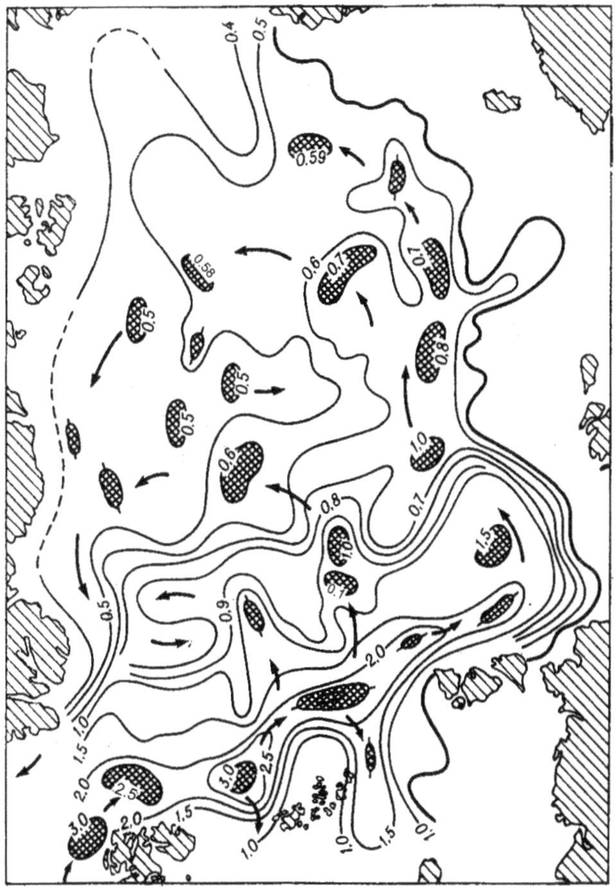

Figure 4.13. Schemes of circulation of the deep Atlantic waters according to the data on the spatial water temperature distribution (°C).
From Nikiforov and Spaikher (1980).

identified in the field of temperatures. These spots probably result on the one hand from the interannual fluctuations of the heat content of Atlantic water at the entry to the Arctic basin. On the other hand, it is clear that they are located along the water motion trajectories. Hence, this fact can be taken into account for the construction of water circulation in the Atlantic layer.

Nikiforov, Blinov and Lukin assessed the variability of the heat content of Atlantic water and its maximum temperature in the Arctic basin and its regions (Nikiforov *et al.*, 1979). It follows from Table 4.7 that the heat content variability (both interannual and multiyear) of Atlantic water is very large. The multiyear variability of the heat content of the deepwater layer in the sub-Atlantic part of the Arctic basin comprises 26 kcal/cm^2 a year and in some regions of the basin it is much greater – up to 40 kcal/cm^2 and more. The interannual variability of the heat

Table 4.7. Anomalies of the heat content (q, kcal/cm^2) of Atlantic water and its maximum temperature (T_M, °C) in the Arctic basin and its regions.
From Nikiforov *et al.* (1979).

	q				T_M			
Year	A	I	II	III	A	I	II	III
1950	5.2	6.9	2.3	−4.2	0.01	0.00	0.03	0.02
1955–1956	5.4	15.4	1.0	4.8	0.05	0.13	0.04	0.01
1973	0.6	−0.5	2.6	5.4	0.02	−0.06	0.06	0.04
1974	2.2	0.6	7.9	2.2	0.01	0.03	0.01	0.02
1975	2.5	2.7	0.4	0.4	−0.02	−0.05	−0.02	0.00
1976	−4.8	−7.5	−3.0	−2.9	−0.03	−0.04	−0.04	−0.02
1977	−7.1	−11.2	−7.9	−3.6	−0.02	0.02	−0.06	−0.04
1978	−3.9	−6.2	−3.5	−2.1	−0.02	−0.03	−0.02	−0.03
Maximum interannual changes	7.4	10.2	7.5	3.3	0.03	0.09	0.05	0.02
Maximum multiyear changes	12.6	26.6	15.8	9.6	0.08	0.19	0.12	0.08

Note: A – entire basin; I – Eurasian sub-basin region adjoining Eurasia; II – Eurasian sub-basin region adjoining Greenland; III – Amerasian sub-basin. The boundary between regions I and II passes approximately along the line connecting the middle of the Fram Strait and the De-Long Islands.

content of the layer occurs predominantly (not by less than 80%) due to the change of the layer thickness rather than due to the change of temperature of the water comprising it. The multiyear variability is on the contrary determined by water temperature changes by more than 40%, and in the sub-Atlantic part of the basin by more than 60%.

Both in the entire basin and in its separate regions, the water heat content and the maximum temperature decrease from the 1950s to the 1970s. The amplitude of multiyear variations of both characteristics naturally decreases with the mean heat content value of the layer – maximum in region I and minimum in region III.

It is important to note that from the 1950s to the 1970s, the Arctic High area increased in parallel so that the water and ice circulation in the basin became 'more closed' with the attenuated water exchange between the basins of the Arctic Ocean, and between the area of the anticyclonic water and ice gyre in the Canadian-American sector of the Arctic basin and ambient water. Since the area of the Arctic High began to increase only three years after the beginning of decreased Atlantic water inflow through the Fram Strait, it is doubtless that the increased size of the anticyclone of such a climatic character is caused by the decreased heat flux with the currents in the Arctic basin.

On the other hand, the more closed is the character of the circulation in surface and deep layers of the Arctic basin, the better conditions are created for its exchange with the north European bottom water. This is confirmed by data collected by the airborne expeditions in the first region and by the research vessels in the second region of the Arctic Ocean.

It is known that as early as 1898, a famous polar explorer F. Nansen detected that more than 80% of the volume of the Norwegian Sea is occupied by cold bottom water with a temperature up to −1.3°C. Now we know that this is the coldest and densest bottom water in the entire World ocean. The mechanism of its formation is still under discussion, however, it is clear that it is mainly formed by cooling of warm (probably of return Atlantic) water. A large amount of heat is released at this point to the atmosphere.

Later observations (e.g., in the 1950s) detected water with a temperature not less than −1.1°C, so that even doubts appeared about the reliability of the data of F. Nansen, more so as there was a firm belief in World oceanography about a large inertia of deep and near-bottom water, in particular with respect to heat. It was even supposed that temperature fluctuations of this water were not beyond the observation accuracy (±0.02°C). However, it was revealed in the 1970s that the temperature of bottom water began to rapidly decrease, dropping by 1977 to −1.4°C. The heat content of bottom water with a layer thickness of more than 2 km decreased by more than 45 kcal/cm^2.

Figure 4.14 presents both curves of the temporal temperature variations – the maximum temperature of Atlantic water in the Arctic basin and the minimum temperature of bottom water in the Norwegian Sea. The mean temperature of bottom water in the Amerasian sub-basin also decreased simultaneously. It is noted that for approximately 20 years the temperature of bottom water in the Norwegian Sea decreased by 0.3°C, the maximum temperature of deep Atlantic water – by 0.1°C and the average temperature of bottom water in the Amerasian sub-basin – by 0.05°C.

As can be seen, the cooling in the Arctic developing from the 1940s covered not only the atmosphere, but also the deep layers of the Arctic Ocean, 2,000 m or more from the surface. These are the first data in the World literature indicating that climatic changes entrain such large oceanic depths and that hence the entire ocean strata from the surface to the bottom at least in the zones of water formation participates in the fluctuations of climatic character and has such significant scales (Nikiforov *et al.*, 1979).

4.3.5 Value of oceanographic observations at the 'NP' drifting stations and in the 'Sever' expeditions

On the basis of oceanographic observations at the 'NP' drifting stations and in the 'Sever' AHE, extensive studies were carried out, numerous articles written and different reference, methodological and other manuals prepared. However, such observations were probably most fully and efficiently used in preparation of different atlases illustrating natural phenomena and processes occurring in the Arctic Ocean. Moreover, it would be impossible to make such atlases at all without the observations of the 'NP' stations and the 'Sever' expeditions. The most large and interesting national atlases prepared with a wide use of data of observations at the 'NP' stations and in the 'Sever' expeditions over the period up

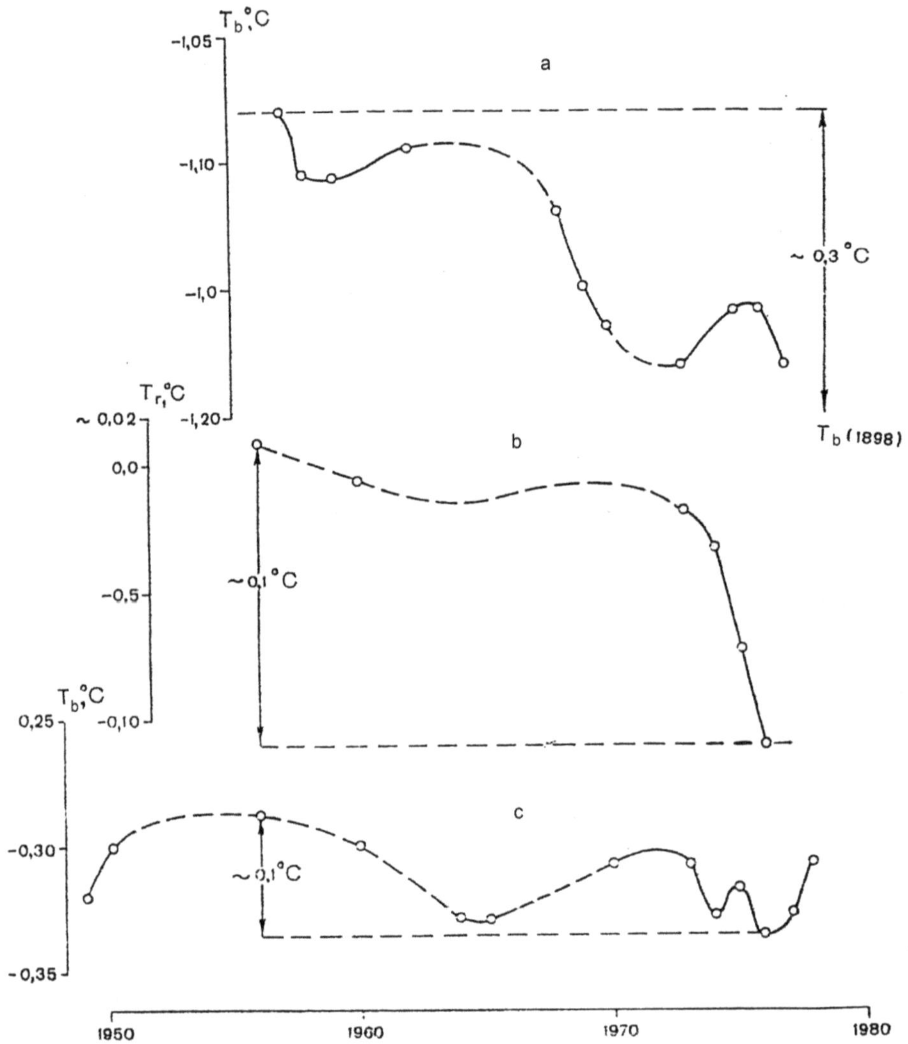

Figure 4.14. Multi-year course of minimum temperatures of the bottom waters in the Nordic seas (a), maximum temperature of the Atlantic waters in the Arctic basin (b) and mean temperature of the bottom waters in the Amerasian sub-basin of the Arctic basin (c). The temperature minimum of the bottom waters of the Norwegian Sea registered in the beginning of the 20th century (according to F. Nansen) is shown in Figure 4.14(a) at the right near the curve.
From Nikiforov *et al.* (1979).

to 1975 are the *Atlas of the Oceans: The Arctic Ocean* (1980) and the *Atlas of the Arctic* (1985).

The significance of the published atlases is enormous. The encyclopedic character of these publications is still valuable to-date. However, the development of science and practice set new challenges that require new efforts and approaches in the field of generalization of our knowledge of the northern polar area. The earlier published atlases contain predominantly information on the characteristics of media averaged over the entire observation period. But annual monitoring constantly provides information about the diversity of the environmental states and a large variability of meteorological, ice, oceanographic and hydrological processes in the individual key regions of the Arctic. The development of the climate theory and models of climatic processes require new climatic information in different presentation formats and use of this information.

The next step in the assimilation of data of oceanographic observations was the Russian–US electronic 'Oceanographic Atlas of the Arctic Ocean' (*Joint US–Russian Atlas of the Arctic Ocean*, 1997, 1998). During preparation of this atlas, the most complete oceanographic database of the Arctic Ocean from 1948–1993 was generated at the AARI on the basis of all available data of Russian and foreign oceanographic measurements. It should be noted that 80% of deepwater oceanographic stations for the Arctic basin are comprised of the observations in the 'Sever' expeditions and at the 'NP' drifting stations. Figure 4.15 (see colour section) shows the layout of deepwater hydrological stations made at the 'NP' drifting stations for the period 1937–1991 and by the 'Sever' expeditions from 1941–1993.

Using the objective analysis methods, the climatic fields of water temperature, salinity and density at regular grid points in standard horizons were obtained for some decades and in general over the entire period. In 1997, a CD-ROM of the atlas for the winter period was issued and in 1998 for the summer period (*Joint US–Russian Atlas of the Arctic Ocean*, 1997, 1998).

Oceanographic data are a fundamental base of information on the state and variability of the complicated natural system that is the Arctic Ocean. Observations record the instantaneous, or averaged for a specific time interval, values of oceanographic characteristics at discreet spacial points. The observations by means of probe zonds demonstrate a continuous variability of the ocean parameters at the specific spatial ranges. The entire set of data gives the evolvement of oceanographic phenomena and processes of different spatial and temporal scales. This non-uniform scale and the multi-level character of the results of field observations make handling primary data rather difficult. Therefore, for the success of the analysis of empirical material for determining natural regularities of different levels, the use of primary observation data for different types of calculations and obtaining an end product depend to a great extent on the system of database arrangement and processing techniques of primary observation data.

The oceanographic database prepared for generation of the Russian-US electronic 'Oceanographic Atlas of the Arctic Ocean', was used for analysing long-term changes of water of the Arctic basin (Swift *et al.*, 2004).

Pokrovsky and Timokhov (2002), using a method of spectral expansions made a reconstruction of the temperature and salinity fields for March–May. As a result, a continuous series of characteristics at grid points with a step by space of 50 km in standard horizons from 1948–1993 was obtained.

The oceanographic data of the 'Sever' expeditions and the 'NP' drifting stations were supplemented with observational data from other expeditions and used for the analysis of climatic changes of Atlantic water in the Arctic basin. For 100 years, two clear periods of Atlantic water warming were established – from the late 1920s to the 1950s and in the late 1980s–1990s, and also two cold periods – from the start of observations to the 1920s and in the 1960s–1970s (Polyakov *et al.*, 2004).

The scientific oceanographic observations and studies carried out at the 'NP' drifting stations and in the 'Sever' AHE made an outstanding contribution to the establishment and development of polar oceanography. The oceanographic database of the Arctic basin generated for the period 1937–1993 has not become obsolete. Its value is still very large for the scientific community for two reasons. First, these observations recorded the state of the Arctic basin in the past that can help us to reconstruct the climatic changes of thermohaline structure. Second, these data, especially the large-scale oceanographic surveys of the Arctic basin in 1973–1979, are unique pieces of information allowing us to adjust, test and check both the ocean circulation models and the Arctic climate models.

4.4 HYDROPHYSICAL INVESTIGATIONS

At the end of 1950s active field research work continued at the 'NP' drifting stations, their annual number varied from two to four. Since 1956 hydrophysical and radiophysical investigations occupied a significant place in research.

As it has been already mentioned, observations according to special programs were conducted for a wide range of oceanographic problems. In each case these programs were formulated by specialists, whose scientific interest was connected with that particular research area. Measurements of temperature and salinity were carried out at many drifting stations and in 'Sever' expeditions in fractures and on polygons under the ice cover; observations were performed for melting of ice floes from above, below, laterally, etc., hydroacoustical and hydrooptical investigations were also conducted. These underwater observations allowed us to fix specific features of the salinity distribution to the periods of freezing and currents directly under the ice, etc.

4.4.1 Freshening of the under-ice layer of the water by ice melting

Freshening of the under-ice layer of the water develops from water volumes formed as the result of snow and ice melting on the upper surface of ice floes, ice melting from below and lateral ice melting. Input of atmospheric precipitation is not great, in summer it equals 5–10%.

The first data on ice cover melting was obtained by ice levelling and measuring of snow and ice thickness. Analysis of these observations allowed us to reveal specific features of the melting process and transformation of the ice surface relief. These are, first and foremost, spatial irregularity of the snow and ice melting that depends both on climatic conditions, as well as on ice floe relief.

According to data observation obtained at 'NP-13' (Nazintsev, 1971b), 'NP-22' (Kochetov, 1981) and 'NP-29' (Grishchenko, 1987) mean ice thickness decrease in the hummocks for the summer period is on average 30% higher than for level parcels of ice. Ice melting in puddles is even more intensive. The total melting here for the summer period is twice as high as that of the level parcels of ice. As the puddles occupy 20–50% of the ice cover surface, taking them into account is obligatory. If we sum all these measured values, we notice that a layer of snow and ice of 40–50 cm melts from the upper surface of the ice cover in the central part of the Arctic basin and up to 1 m at its periphery (Golovin *et al.*, 1995).

The ice melting significantly varies from month to month and from year to year. This is true both for duration of melting period and for intensity.

Ice melts from its bottom surface also rather actively. According to measurement data at the stations 'NP-13', 'NP-16', 'NP-22', 'NP-29', the 10–50-cm layer melts from below during the summer period. This significant variability is determined by differences in thickness of the ice and snow, presence of puddles, oscillations of the solar radiation penetrating through the ice, outflow of freshwater through cracks and thaw holes and its freezing, and income of warmer water from fractures.

The penetrating solar radiation can be considered as one of the important factors determining ice melting from below. Special experiments performed at the station 'NP-24' demonstrated that variation of ice melting from below in first-year ice follows variations of radiative balance with a lag of 4.5 hours – the melting rate is 1.0–1.5 mm/day. The radiation weakly penetrates through thick ice with the melting rate decreasing to 0.6–0.7 mm/day.

According to observations at the drifting stations and on fast ice, the ratio of the ice melting from below to melting from above for the whole melting period was 0.47 at station 'NP-13', 0.54 – at 'NP-22' and 0.54 – on fast ice in Yenisei Bay. This ratio is typical mainly for first-year ice with a thickness of 1–2 m. Multiyear ice with a thickness exceeding 2 m melts from below less intensively, and thin ice – significantly rapidly.

The lateral melting for the summer period is 0.8–1.0 m. The process of lateral melting develops irregularly. In particular, at station 'NP-22' this melting significantly exceeded its typical value.

If we want to estimate its share in the total ice melting, it is necessary to take into account the ratio of the amounts of ice melted from above and below to the lateral ice melt.

For an ice floe with a size of $1{,}200 \times 800$ m the ratio of ice volumes melted laterally and from above is 1 : 36, for an ice floe with the size 27×27 m – 1 : 1, for an ice floe with the size 10×10 m – 3 : 1.

4.4.2 The thermohaline structure of water in fractures

The first measurements of vertical temperature and salinity profiles in a fracture were carried out at drifting station 'NP-2'. These field research works were continued at drifting stations 'NP-13' (Nazintsev, 1971b), 'NP-14' (Shpaikher *et al.*, 1968),

'NP-16' (Alekseev and Buzuev, 1973), 'NP-22', 'NP-29' and 'NP-31' (Golovin *et al.*, 1993).

Multidisciplinary and continuous observations at 'NP-31', as well as generalization of the results of the previous experimental investigations, allows us to understand the evolution of the spatial thermohaline structure of a fracture more accurately (Golovin *et al.*, 1993).

The period of intensive snow melting continues from 4 to 10 days. Water warming and avalanche-like freshwater flow from the surface of ice floes cause a rapid temperature increase of the surface layer of the fracture from −1.4 to −1.7°C to 2–5°C and salinity decrease from 30–33 psu to 4–19 psu (Figure 4.16). The fracture begins to fill up, therefore, maximum freshening and water temperature increase are observed near the fracture edge bank.

The period of intensive ice melting within the limits of 25–30 days is characterized by maximum development of the freshened layer in the fracture.

A maximum radiative warming-up is observed in this period. The water temperature in the fracture usually increases by 3–5°C, and in the case of 'NP-31' by up to 6–70°C. The maximum water freshening is also observed at this time. The upper layer salinity can decrease to 0.2–0.5 psu.

The thickness h_{max} of the freshened layer in this period reaches its maximum. The lower boundary reaches on average a depth of 100–150 cm at latitude 83–85°N ('NP-13, 16') and 180–280 cm at latitude 72–75°N ('NP-14, 31'). Longitudinal difference of the maximum thickness of the freshened layer (h_{max}) is also observed at this time. The value of h_{max} on 'NP-2' did not exceed 100 cm to the east off the meridian 180° at latitude 78.5°N (Nazintsev, 1971b), however, the h_{max} value of 180 cm was registered at 'NP-14' to the west of meridian 180° at latitude 76°N (Alekseev and Buzuev, 1973).

The *Brunt-Väisälä* frequency in the pycnocline reaches a value of 0.125 sec^{-1}. This value is 5–10 times higher than typical values of the *Brunt-Väisälä* frequency calculated for the seasonal pycnocline in the Arctic basin.

The ice on the fracture surface reaches 20–40 cm at the time of thinning of the freshened layer. It should be mentioned that the ice grows not only at the expense of phase change of seawater directly underlying the ice bottom surface. The ice thickness increases also at the expense of frazil ice that forms in the region of the pycnocline and then floats to the surface. Slices of the frazil ice with widths up to 10–15 cm were observed at 'NP-31' as 'corals' at ledges of the ice floe at the pycnocline depth during the whole summer period.

The process of freshened layer thinning ends after a phase of regelation of the surface ice and crusts of frazil ice. After that the under-ice layer in the fracture becomes completely vertically uniform – typical for winter conditions.

Summing up, the freshened layer thickness in a fracture in summer is about 1–2 m, and its lower boundary in the Arctic basin is, as a rule, not deeper than the draft of the ice floes.

Turbulent exchange practically does not exist between the freshened layer and underlying active layer of the ocean in the summer period. Great stability is observed at the interface, the *Brunt-Väisälä* frequency is as high as 0.12–0.18 s^{-1}.

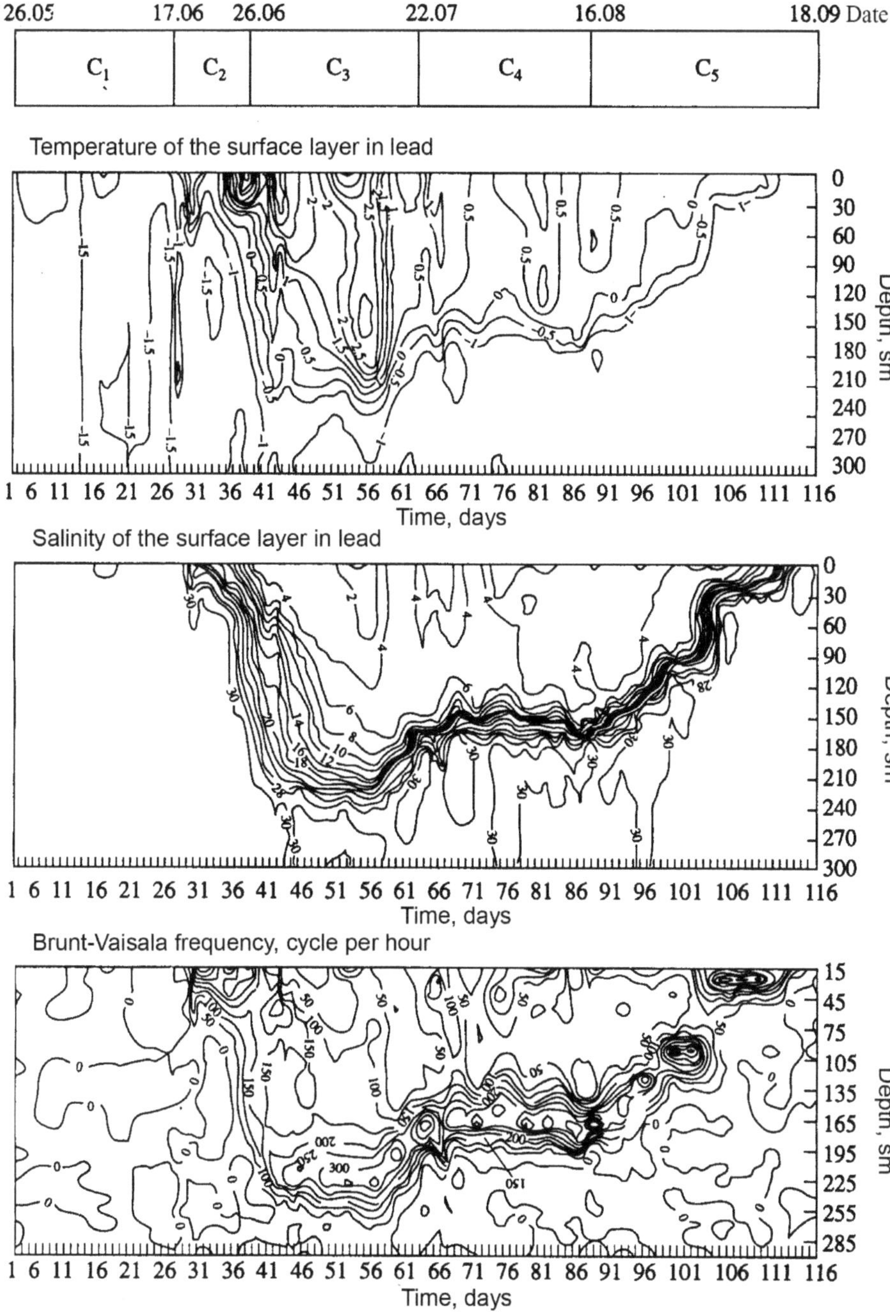

Figure 4.16. Spatial–temporal variability of the temperature, salinity and *Brunt-Väisälä* frequency in the surface layer of a fracture in summer for five stages C1, C2, C3, C4, C5. From Golovin *et al.* (1993).

Crystals of frazil ice are formed at the boundary zone between the fresh, warm waters of the freshened layer and the salty, cold surface Arctic waters. Optical effects of a blink are observed here.

The spatial thermohaline structure and thickness of the freshened layer in a fracture depend on the dynamics of large-scale ice concentration.

4.4.3 Formation of water salinity and density inversions under the ice in the winter period

One of the reasons of mineralization intensification of the under-ice layer in the Arctic basin in the cold period of the year is brine rejection out of growing ice. Using the data of Tsurikov (1976) on the dependency of the sea ice its their growth rate, it is possible to estimate approximately the amount of incoming salt in the under-ice layer from different intensities of ice formation. The maximum salinity gradients in the under-ice layer should be observed in the places with the maximum rate of the ice formation. In the Arctic basin such centres of seawater salinization are in the regions of young and thin ice. Observational results under water approve this proposition. A typical view of salinity vertical distribution under the bottom surface of undistorted young ice with a growth rate of 1.0–1.5 mm/hour is shown in Figure 4.17.

However, according to underwater observations the governing factor determining the presence of the halocline under the bottom surface of the growing ice is not only intensity of ice formation, but to a greater extent the dynamic state of the under-ice layer (Grishenko and Kochetov, 1989). The salinity gradient disappears everywhere due to the velocities of the relative under-ice current exceeding 3–4 cm/sec resulting in turbulent mixing (Figure 4.17).

A sharp salinity increase of this layer in the cold period of the year is often observed in the Arctic basin, besides the abovementioned factor of salinization of the under-ice layer caused by the process of natural ice growth. The mechanism of this phenomenon is as follows. It is well known that at low temperatures the majority of salt in the sea ice is in crystalline state, and the liquid phase is dispersed in small closed cells. With sharp ice temperatures increasing the mass of the liquid phase increases many times – intensive gravitational flow of concentrated brine into the under-ice layer occurs. In the conditions of the Arctic Ocean this warming of separate ice segments at the winter period is possible by: ridging, when cold ice fragments sink into or submerge in the water with temperatures 1.8–1.9°C below zero; by formation of cracks and fractures if the warm (in respect to the ice) sea water warms the ice at the region of the edges of the new cracks; and by a sharp and significant increase in the air temperature by 15–20°C (e.g., by cyclone passage).

Build-up of ice in the form of hollows and thin walled tubes (named also stalactites) form around concentrated fluxes of the cold brine incoming to the seawater with temperatures close to freezing point. Underwater observations of these stalactites, partially visualizing the process of intensive salinization of the under-ice layer, allows us to determine the scales of this phenomenon and to reveal its main specific features (Figure 4.18). Salinization intensity of the under-

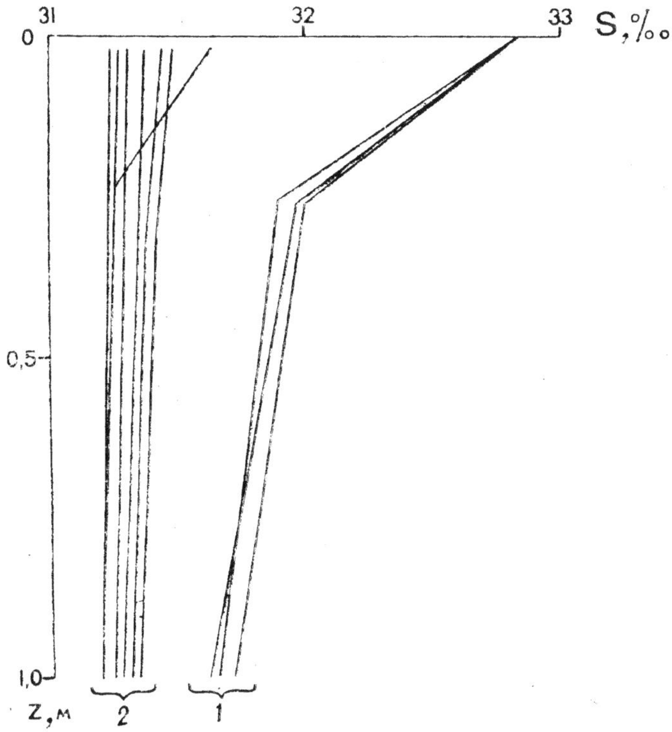

Figure 4.17. Salinity distribution in the under-ice layer of thin first-year ice in October (the Arctic basin): 1, 2 – velocity of the under-ice current up to 3–4 cm/s and above 5 cm/s, respectively.
From Grishchenko and Kochetov (1989).

ice layer increases by ridging by five times or more if compared with conditions of regular ice growth (Grishchenko and Kochetov, 1989).

As the area of the new ridging in the Arctic basin can reach 7–8% or more, and the young and more salty first-year ice is subjected to ridging; increases of ice temperature should be considered as one of the main factors of the salinization of the under-ice layer in the winter period.

By the formation of new cracks and leads their edges warm at the distance of 2–3 m off the ice thickness, and an intensive brine flow is observed in this ice strip into the under-ice layer that continues for several hours to several days depending on the ice thickness. It is natural that the intensity of this process decreases moving off the ice edge (Figure 4.18(b)). Taking into account the fact that the phenomena of ice cover disturbances in the Arctic Ocean in the winter period occur often, this described salinization mechanism of local segments of the under-ice layer should be considered as important.

A sharp air temperature increase results in the salts being warmed and rejected in ice comparable with the thickness of first-year ice. For example, the air temperature

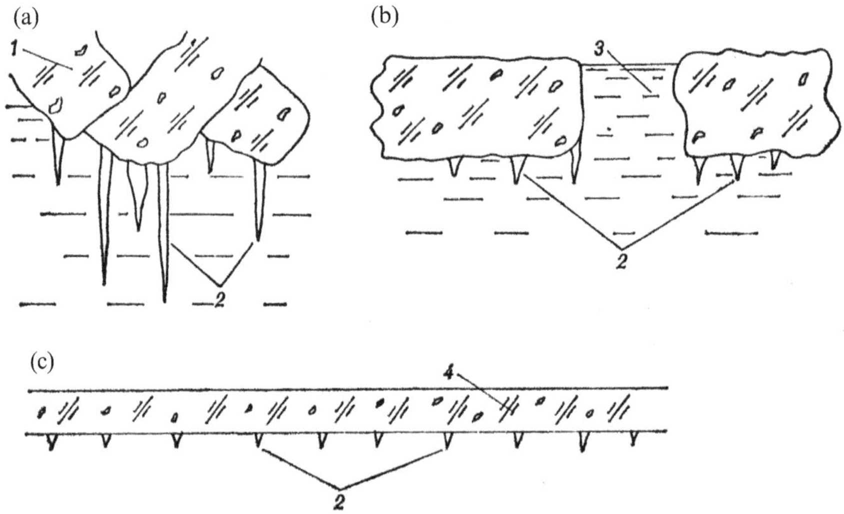

Figure 4.18. Scheme of ice tube location in places of brine rejection by warming of ridged ice (a), ice in a fracture (b), young ice (c). 1 – new ridge, 2 – ice tubes, 3 – new crack, 4 – undeformed ice.
From Grishchenko andKochetov (1989).

increased from −35°C to −17°C during several hours in April in the region of 'NP-22'. The amount of liquid phase in the young and first-year ice increased 5 times and was equal to about 20 g/kg having a salinity of about 200 psu. The result of it was the sharp salinization of the under-ice layer under ice of small and medium thickness. This process has been observed more than once in the Arctic basin and Arctic seas. Formation of small ice tubes demonstrating brine rejection was observed even on the bottom surface of nilas with thicknesses of 9–10 cm resulting from air temperature increases during 4–5-hour periods from −12°C to 0°C (September 1974, NP-22, the Arctic basin) (Figure 4.18(c)).

An inverse type of salinity distribution under the ice cover accompanied by water cooling can be a reason for an inverse type of density distribution under the ice in the winter period. Lukin and Timokhov have analysed data from nearly 1,000 stations of standard observations for temperature and salinity under the ice cover. Analysis results demonstrate that the negative gradient of water density under the ice cover in the winter period can often be observed. The layer thickness is comparable with the thickness of the boundary layer. Maximum values for the under-ice layer thickness within the density inversion are 50–100 m. Gradients of apparent density for the entire layer can reach several tenths of the apparent density, and this value exceeds the stochastic differences caused by observational errors (Lukin and Timokhov, 1989).

On the whole, a density inversion has been observed at each fifth station in the winter period for the entire array of hydrological stations. Gradients have exceeded 0.04 arbitrary units. Vertical profiles with the density inversion in the winter period are irregularly spatially distributed. According to the data of selected hydrological stations with station frequencies of 150–200 km, the inversion is alternated with the stable stratification in the under-ice layer. Sometimes these stations are grouped in regions with a positive or negative gradient under the ice cover. Sizes of these regions are about 300–600 km. There is no regularity between these stable and unstable stratifications, and we can suppose a spatially intermitting character of the inversion under the ice in winter. Variations of the inversion thickness layers between oceanographic surveys and observation results in the layer of more dense water demonstrate that this inversion layer episodically collapses.

Analysis of the vertical density distribution shows that, in some cases, the density inversions are observed not directly under the ice, but at 10–50 m. Such an anomalous distribution can be caused by the fact that the inversion layer separated for some reason from the ice cover and submerged at the moment of observations to a fixed depth.

A significantly smaller number of observations have been conducted for temporal development of stratification conditions. At one hydrological station the weakly unstable stratification was fixed for 40 hours, the stable stratification continued for about 200 hours and then the new inversion layer was formed. It testifies that the inversion appearance under the ice surface in the winter period is also intermittent in character.

Revealing the factors exerting influence on the inversion conditions has demonstrated the following. The closest correlation of the apparent density for the inversion layer has been registered with air temperature and ice thickness. Statistical analysis has revealed inverse negative dependency of the values of the apparent density on the ice thickness and air temperature over the ice cover.

Influence of drift velocity on the value of the gradients of apparent density has not succeeded unambiguously. The density inversions have been observed, as well as by ice drift absence, by significant drift velocities of the order of 30–50 cm/s. At the same time, an increased number of cases with negative density gradients is registered in dynamic regions. For example, ice circulation in the Arctic basin in 1975 was more intensive compared with 1976. As a result, the negative density water gradients under the ice were registered in 1975 10% more often than in 1976 (Lukin and Timokhov, 1989).

4.4.4 Investigations of the fine structure of water

A significant achievement of the up-to-date oceanography radically changing our notions of the ocean, was revealing of the general presence of a wide spectrum of internal waves, synoptic eddies and the thin structure of water. The Polar regions of the World Ocean are, in this sense, not excluded. Investigations of these phenomena in the Arctic basin, practically simultaneously with the American specialists, began at the end of the 1960s.

With the help of fast-response devices developed in the AARI together with the Morphyspribor Institute and Kazan Aviation Institute, such as thermistor chains, thermal sounding devices and cyclical measurement devices of sound speeds,

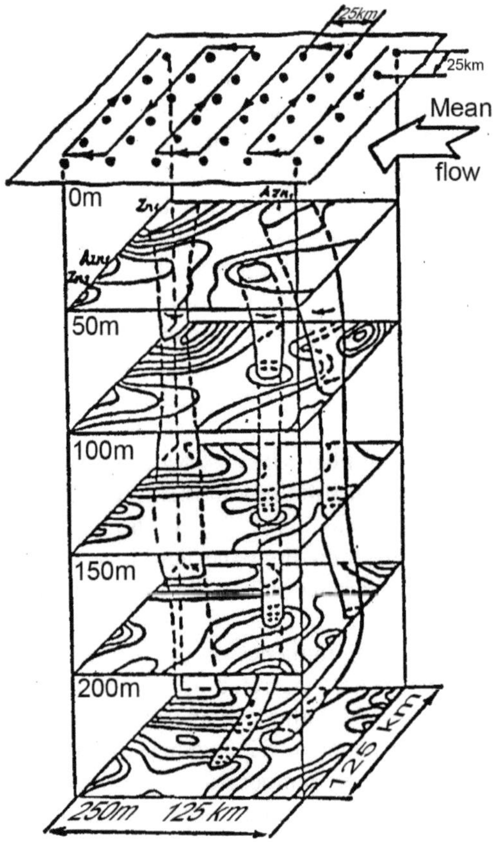

Figure 4.19. 3-D structure of temperature fields under ice.
From Lebedev (1999).

information on small and mesoscale spatial and temporal variabilities in the hydeo-physical fields of the upper layer of the Arctic basin have been identified (Bogorodskiy *et al.*, 1977; Bogorodskiy *et al.*, 1980; Bogorodskiy *et al.*, 1982b). Experimental investigations allowed us to reveal the presence of water layers up to 70 m with vertically constant temperatures divided by interlayers with thicknesses up to several metres and temperature gradients up to 0.2°C/m.

Investigations conducted in the sub-polar region on a polygon (125 × 125 km) allowed us to reveal the cellular structure of the temperature sound speed fields in the horizontal plane. Supposing that the character of these fields is caused by an eddy structure, it is possible to affirm that the horizontal sizes of these eddies are 10–40 km, and that the intervals between these eddies are 20–50 km (Bogorodskiy *et al.*, 1982b; Lebedev, 1999). An example of a 3-D temperature field obtained by this field research work is demonstrated in Figure 4.19.

Investigations of these phenomena allow us to reappraise both the pure oceanographic processes (genesis and transformation of waters and frontal interfaces, and heat exchange between separate layers of the ocean), as well as specific features of propagation of sound and optical waves.

Use of a system of thermal sensors spatially distributed from 10–450 m for long-term observations and collected information with a discreteness of 20 seconds allowed us to reveal short-period waves propagating in the ocean as wave trains. Their magnitudes are of the order of 1–1.5 m for wave periods of 20–60 minutes.

The internal gravitation waves generate in the ice cover so-called slow waves, and their investigation is possible using tiltmeters placed on the ice cover surface (Smirnov, 1972b; Bogorodskiy *et al.*, 1982a).

4.4.5 The under-ice drift currents

Detailed measurements of current velocity in the under-ice layer showed that the microstructure of these currents depends on the character of the bottom surface relief. Under ice with a flat bottom surface the current velocities depend on the velocity of the under-ice flow (Beliakov and Grishenko, 1980). Observations have been made by divers at selected points in the under-ice layer injecting a colorant of neutral buoyancy. The thickness of the under-ice layer for the entire range of observed velocities from 5–55 cm/s did not exceed 4–5 m, and with velocities of 5–10 cm/s thicknesses did not even exceed 1 m. Thickness of the water layer adhered to the ice oscillated from tenths of millimeters when velocities were 50–60 cm/s up to 1–2 cm with minimum velocities. This adhesion of the under-ice layer has been observed under the bottom surface of smooth, melting ice in the summer period. In the case of formation of growing crystals on the bottom ice surface in the winter period, the adhesion layer thickness somewhat increases, however, it does not exceed several centimeters even under velocities of 40–50 cm/s in the under-ice layer.

Turning of the current vector in the under-ice layer reached 30–40° clockwise for current velocities of 10–15 cm/s and then decreased with an increase of the ambient velocity. The most complicated structure of the under-ice layer dynamics is registered in the regions of ridge keels, by rafting and near edges of ice floes of great thickness, especially near flowing glacier bits. On the one hand, this macroheterogeneity of the ice bottom surface exerts an influence on the under-ice layer in the water column up to several tens of meters, on the other hand zones are observed under the ice cover of the so-called ‘dynamic shadow’. These zones under the ice are characterized by insignificant current velocities and even by complete immobility. The last circumstance sometimes exerts a decisive influence on the character of biological processes, distribution of hydrochemical and other water characteristics, rate of ice formation and destruction under the water.

The drift of under-ice currents have many common features with the sea wind-induced currents in the ice-free areas. Velocity of the under-ice current rapidly decreases with depth: at 5 m it is equal to 90–100% of the ice drift velocity and at 25 m – to 25–30% according to observations made at ‘NP-1’. The current vector at 25 m deviates to the right off the current vector by 25–30° (Shirshov, 1938).

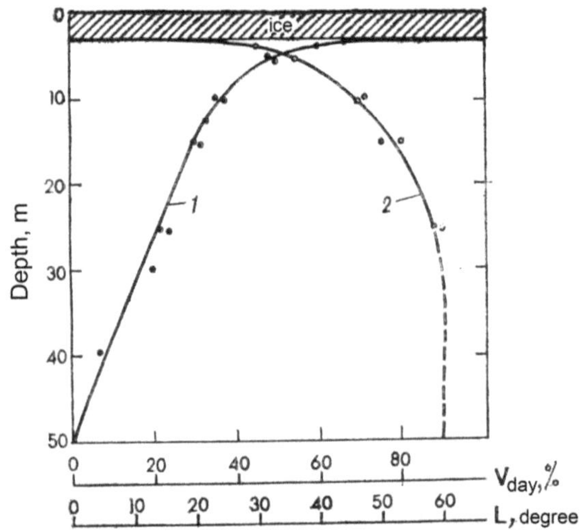

Figure 4.20. Distribution of velocity V (1) and direction L (2) of the drift under-ice current at different horizons in the Arctic basin.
From Belykov and Grishchenko (1980).

L. N. Beliakov generalized the results of his observations for the drift currents under the ice cover and measurements performed by other specialists. These averaged data allow us to assume that distribution of the current velocity and direction demonstrate a steady-state character of the ice drift and accompanying drift current (Figure 4.20).

The current velocity at 3.5 m is only 66% of the ice drift velocity. Repeated measurements of the current velocity in the vicinity of the ice bottom surface showed that the current velocities had been 65–70% of the ice drift velocity. Vertical distribution of the true drift velocity and direction is given in Figure 4.20.

The drift current rotates to the right with depth and generally corresponds in structure to the Ekman spiral. At 3.5 m the current turns to the right of the ice drift vector for 30°, maximum for 58°. With an increase in depth the turning angle varies less, and it demonstrates an inconstancy in the turbulent friction with distance variation from the ice bottom surface.

The obtained data showed that the drift ice in winter entrained for a 25% greater water mass than in summer, and the drift current velocity increased nearly twice in the period of a significant freshening of the under-ice layer if compared with the winter one (Belyakov, 1974).

4.4.6 The hydroacoustic investigations

These observations were made regularly in the Arctic since the organization of a hydreoacoustic laboratory in the AARI in 1956.

The first guidelines of these investigations was related to the division into districts of hydrological and acoustic measurements of the central part of the Arctic basin and Arctic seas, determining in a statistical aspect specific features of the under-ice distribution of the acoustic oscillations in different seasons of the year and creating a physical basis for their propagation by solution of practical problems.

The second important guideline of these investigations determined as a priority by AARI specialists in polar hydracoustics is related to the study of spatial–temporal and spectral–energetic characteristics of the under-ice noise field for different conditions of its formation and dynamics.

These investigations have formed the basis for the creation of special purpose databases, as well as for the solution of the inverse problems of hydroacoustics.

The level of these investigations for development of fundamental science and a solution of applied problems was rewarded the State Prize of the USSR for hydroacustical investigations in the Arctic in 1969. Among the prize winners were specialists in the department of ice and ocean physics at AARI (V. V. Bogorodskiy and A. V. Gusev).

This success was based on investigations of the physics of sound formation by the deformation and destruction of the ice cover (Bogorodskiy *et al.*, 1970; Gavrilo *et al.*, 1970) by intra-ice friction (Bogorodskiy *et al.*, 1974), interaction of snow storm fluxes (Lebedev and Popov, 1973; Lebedev and Popov 1974) and thermal fracturing of the ice cover surface (Bogorodskiy *et al.*, 1972). Acoustic signals radiated by cetacea were studied as a characteristic of the noise field of the Arctic seas (Bogorodskiy and Lebedev, 1978). Characteristics of these complicated signals and the means by which they were generated are, by all means, of great interest for the biology of the northern seas.

The study of statistical dependencies of sound level in the seawater and in the ice cover on wind speeds and ice drift, air temperature, deformation and divergence of the ice cover and their derivatives allowed us to obtain connections between them (Gavrilo and Gusev, 1975).

Using this information we can determine deformation or divergence of the drift velocity vector from measured values of the sound pressure, air temperature and wind speed, thus characterizing stress state of the ice cover over the areas determined by radiuses of spatial correlation of the processes under consideration.

The other practical solution related to investigations of the physics of sound formation associated with the processes of ice destruction at different stages of its state and with processes of its movement was the development of a control method of ice conditions at local river segments for forecasting jams by the analysis of the spectral–energetic characteristics formed by their (jams) noise field (Bogorodskiy and Gavrilo, 1976).

A further development of this 'passive' acoustic method of investigation of ice cover deformations was realized by the radio hydroacoustic method in the middle of the 1970s and experimentally tested in the Arctic basin. This method allowed us to estimate qualitatively the mesoscale (up to tens of kilometres) characteristics of compacting and diverging of the ice cover using the radiation of acoustic signals in a sound range (Bogorodskiy *et al.*,1975b). These investigations demonstrated that variations of differences between the specially marked ice floes, being at the distance of 30 km, were measured to an accuracy of 6 m.

Organization of such a system of acoustic control for the regions of construction and exploitation of offshore structures in the Arctic seas has now become especially real, as the collection of information on the dynamics of ice conditions in the vicinity of a structure allows us to diminish the risk of its exploitation and to take opportune measures to avoid dangerous situations.

It should be noted that the acoustic method of studying growth/melting rate of the ice bottom surface (Bogorodskiy and Gavrilo, 1980) allows us to obtain remote information over vast areas, using the principles of echolocation of the ice bottom surface for selected points, and transferring information using a radio channels. The detailed description of the above-mentioned acoustic methods of ice cover study for the solution of different hydrophysical problems and estimation of variability of the acoustic fields is presented in the monograph of Vagapov *et al.* (1993).

The third direction of hydroacoustic investigations in the Arctic basin were measurements of specific features of acoustic signal propagation (up to 1,000 km). These investigations conducted with the use of wideband noise and explosive signals allowed us to obtain information on magnitude fluctuation, phases and time of signal propagation and to reveal the main oceanographic parameters determining these processes.

Investigations carried out in the Arctic basin in 1984 at 'NP-25' and 'NP-26' at a distance of 880 km showed (Andreev *et al.*, 1995) that a strong intra-modal and inter-modal signal dispersion is observed in the underwater sound channel of the Arctic waveguide. The inter-modal dispersion is the reason for mode division, and it is encouraging for the solution of problems of acoustic monitoring.

Results of these investigations and the work of the following scientific collectives: Acoustic Institute of the Russian Academy of Science, Institute of General Physics of the Russian Academy of Science, Institute of Oceanography of the Russian Academy of Science, Institute of the Applied Physics of the Russian Academy of Science, allowed us to formulate the problem of practical realization of the acoustic monitoring of climatic variability of the Arctic Ocean proposed in the scientific project of Mikhalevskiy *et al.* (1991).

The fourth direction of study is the acoustic investigation of ice cover. One of the main characteristics of the sea ice cover from the point of view of analysis of the acoustic signal structure reflected from its bottom surface is ice relief.

Ice fathometers were, at first, used for relief study – mounted on submarines. The physical basis of their functioning is based on the dependencies of angular, frequency and energetic characteristics of acoustic backscattering (boundary reverberation) on the physical nature and geometric properties of illuminated areas.

Investigations carried out at 'NP-13' in 1965 for ice at different stages of development: nilas, first-year and multi-year (Bogorodskiy and Smirnov, 1975) demonstrated that the mean value of the coefficient of backscattering depends on ice thickness and increases wich its growth, irregularity of the backscattering correlates with the variability of the ice thickness.

Field measurements of the speeds of ultrasound wave propagation in the ice cover and in specimens of sea ice have been conducted on the drift ice in the Arctic basin since 1958 ('NP-4'). Annual cycles of these measurements were performed at the drifting stations 'NP-13', 'NP-20' and 'NP-26', and they described seasonal variations of the sound speed in the ice cover column (Bogorodskiy and Gusev, 1973; Gavrilo *et al.*, 1980).

An important place in these investigations was occupied by the analysis of energy weakening of the elastic waves in the ice cover related both with scattering on the crystal structure, as well as with the thermal adsorption caused by the viscosity of the brine and air entrapped between the ice crystals (Lebedev and Suhorukov, 2001). Physically justified relations between some mechanic and acoustical ice characteristics, manifesting themselves in the process of its deformation and delayed fracture, were determined in the results of these investigations (Gavrilo *et al.*, 1978). Parameters of acoustic signals can be an objective criteria of ice strength and degree of destruction at all stages of its deformation by constant compressive stresses exceeding the creep strength.

4.4.7 Hydro-optical investigations

One of the important guidelines for field research carried out at the drifting stations was the study of the optical characteristics of snow, ice and water exerting influence on transportation of radiation in the atmosphere-ice-ocean system and thus determining the characteristics of the under-ice optical layer and related processes of photosynthesis and luminescence, plankton migration and coloring of biological objects. Such optical characteristics as the indices of weakening, backscattering, adsorption and phase function of light diffusion are directly related with the distribution in the seawater of suspended and dissolved matter. Their distribution is determined by the density structure and dynamic processes. The field of refraction index is unambiguously determined by the density field. The luminous flux penetrating in the water column exerts an influence on its biological productivity, and is an important factor in the thermodynamic process. In this aspect, investigations of the optical characteristics of waters conducted in the Arctic basin guided by B. Ya. Gaitskhoki (Gaitskhoki, 1974; Gaitskhoki *et al.*, 1978; Gaitskhoki and Ivashnev, 1980) allow us to consider the optical methods as good instruments for studying the oceanographic characteristics of water in the presence of ice cover hindering the transfer of mineral particles from the air into the ocean.

Undoubtedly the significance of hydro-optics is in the solution of many problems: optimal designing and use of the underwater TV, photography and filming, optical location and communication, measurement methods of oceanographic characteristics using the effect of the combinational scattering, etc. Solution of these problems demanded a multidisciplinary approach to the study of natural illumination of the ocean surface in the Arctic, radiation transfer through the snow and ice cover, light field in the ocean covered by the ice and the optical characteristics of water.

An important place in these investigations has been occupied by the analysis of the regime of natural illumination in the Arctic determined by specific features of the physical–geographical peculiarities of the region. These investigations allowed us to obtain dependency of illumination values on the solar height for three typical gradations of cloudiness (Gaitskhoki, 1974).

Within the framework of these investigations a photometric model of the snow and ice cover was developed (Gaitskhoki, 1970) which allowed us to calculate the coefficients of diffusive transmission and reflection in the snow and ice cover with sufficient accuracy; specific features of variation of snow and ice optical properties according to their structure, content, spatial and temporal anisotropy of their properties giving a possibility to make approximate optical classification of the ice cover (Gaitskhoki, 1975; Bogorodskiy and Gaitskhoki, 1979; Bogorodskiy and Volodin, 1982).

4.5 HYDROCHEMICAL OBSERVATIONS AT THE 'NP' DRIFTING STATIONS AND DURING THE 'SEVER' AHE

4.5.1 Introduction

Life in the oceans and the seas, its diversity and well-being of the ecosystems depend on the hydrochemical conditions and especially on the fluxes of nutrients such as carbon, nitrogen, phosphorus, silicon, etc. The distribution and temporal variability of the concentrations of nutrients are closely connected with the life cycle of phytoplankton and the sea's dynamics. The hydrochemical measurements not only supplement well the hydrological observations, but also allow a detailed study of the water structure and specification of the place and time of formation of different water masses.

Systematic studies of the hydrochemical regime of the Arctic basin at the drifting stations and in the 'Sever' AHE began in 1948. Over half a century of Arctic basin studies at the drifting stations and in the high-latitudinal expeditions have been undertaken, more than 2,300 oceanographic stations were occupied during which the hydrochemical parameters were determined. This period can be divided into two stages by the number of publications generalizing the observational materials.

In the first stage from 1948–1956, the results of observations at the 'NP-2', 'NP-3', 'NP-4' and 'NP-5' drifting stations and from three high-latitudinal expeditions of 1948, 1949 and 1950 were generalized by Musina (1960).

The results of hydrochemical studies of the Arctic Ocean from 1960–1979 were generalized in the monograph by Rusanov *et al.* (1979). During this period, the main portion of data on the distribution of dissolved oxygen, silicon, phosphates, pH values and alkalinity in the Arctic basin was collected. Intense hydrochemical studies of the water column, ice, and the assessment of water, ice and snow contamination in the Arctic basin were continued in the 1980s. The accumulated information was systematized in the electronic (on CD) *Hydrochemical Atlas of the Arctic Ocean* (2002) and was used for preparation of yearbooks and reviews of chemical contamination of the marine environment and for implementation of planned scientific projects by AARI (Smagin, 1995).

4.5.2 Results of the first stage of hydrochemical studies in the Arctic basin (1948–1956)

Over the time of operation of the AHEs and drifting stations from 1948–1956, about 3,000 determinations of chlorinity of seawater, more than 1,500 determinations of oxygen, more than 2,000 determinations of pH values, about 1,000 determinations of alkalinity, more than by 1,000 determinations of phosphates and silicates and about 500 determinations of nitrites and nitrates were carried out. In addition, a full chemical analysis of the salt composition of 20 samples of seawater collected in different regions of the central Arctic basin was performed.

In those years, a great deal of attention was paid to investigating the carbonate system (pH value and alkalinity) in seawater. The distribution of specific alkalinity (alkalinity-to-salinity ratio) was widely used in oceanography as a criterion for identification of water masses of different origin. Based on the materials of the first high-latitudinal expeditions, charts of specific alkalinity distribution at the surface of the Arctic basin and in other standard horizons were constructed. First data showed the specific alkalinity in surface water to change over a wide range (700–750 conventional units), while its average value (722 conventional units) was much higher than the values typical of the surface water of the North Atlantic (670–680 conventional units). High values of specific alkalinity were explained by the influence of continental runoff. In addition, changes of specific alkalinity in the Atlantic water masses in passing to the Arctic basin and in the troughs of the Arctic Seas (Guryanova and Musina, 1960) were established.

The hydrochemical materials collected in the first expeditions allowed characterizing in general the hydrochemical conditions in the Arctic basin and establishing their close relation to the hydrochemical regime of the Arctic seas. The hydrochemical conditions were described in accordance with an understanding of the water masses of the Arctic Ocean formed at that time. Chemical characteristics of surface Arctic water, which is rich in oxygen and poor in nutrients, are given. The entire layer from the lower ice surface to a depth of 25 m and in some cases to 50 m was considered as a surface layer.

The layer from a depth of 50 m to the upper boundary of the Atlantic water masses (zero isotherm) was considered an intermediate layer. The largest difference in the distribution of oxygen, pH value and nutrients in the basins separated by the Lomonosov ridge were observed in this layer. In the sub-Atlantic region, the oxygen concentration and pH value smoothly decreased with increasing depth without forming the minimum while the concentrations of nutrients also smoothly increased. In intermediate waters of the central (near-Pole) and the sub-Pacific regions, the extremes of hydrochemical parameters in the vertical distribution were revealed.

The Atlantic water in the Arctic basin filling the intermediate depths (from 200 to 750 m in the sub-Atlantic region and 300 to 750 m in the sub-Pacific region) differed from higher and lower lying waters by a slightly increased oxygen concentration and decreased specific alkalinity. It was determined that Atlantic water moving from west to east gradually changed not only regarding temperature, but also its chemical properties – especially noticeable from the changed concentration of dissolved oxygen. The difference in the oxygen concentration in fresh Atlantic water and transformed water comprised 0.4 ml/l.

A typical feature of the deepwater of the Arctic basin was a high concentration of dissolved oxygen. It was believed that the presence of the Lomonosov ridge making the water exchange between the depressions of the Arctic basin has a significant influence on the hydrochemical regime of deepwater. First data showed however that the deepwater of the sub-Pacific region is fresher (81% of oxygen saturation) compared with deepwater of the sub-Atlantic region (77%).

4.5.3 Results of the second stage of hydrochemical studies in the Arctic basin (1960–1979)

The interannual variability of hydrochemical elements in the Arctic Ocean depends on the physical–chemical and biological processes and also on the hydrological regime, which are in turn under the influence of the climatic system of the Arctic Ocean. The distribution of chemical elements in surface waters is determined by the water exchange between the Arctic basin and the Arctic seas (Pivovarov, 2000), ice formation and melting and to a great extent by the inflow of Atlantic and Pacific waters.

The hydrochemical indicators began to be actively applied as indicators of the water masses, in particular, the use of silicon allowed scientists to trace the Pacific Ocean water spreading into the Arctic basin. In subsequent years, the hydrochemical observations significantly expanded. Their number and the list of ingredients increased and their quality was improved due to application of modern methodologies. The materials of these observations were analysed in Rusanov *et al.* (1979). This material was also used for plotting the charts of distribution of hydrochemical elements in the ocean in the 'hydrochemistry' section of the *Atlas of the Oceans* (1980) and the *Atlas of the Arctic* (1985).

The oxygen distribution in the surface layer of the Arctic basin depends not only on temperature and salinity, but also on the zonal redistribution due to surface currents (Pivovarov *et al.*, 1997). Water most rich in oxygen is observed in the areas of the anticyclonic gyre of the Canadian basin – more than 9 ml/l, whereas the lowest oxygen concentration is recorded in the regions bordering the Arctic seas – less than 8 ml/l. In the same region in the near-bottom horizons, an unusually low oxygen concentration is noted, as compared with the other regions, comprising less than 4 ml/l.

The silica distribution allows tracing the Pacific Ocean waters spreading over the Arctic Ocean. It should be noted that one can determine the area of this water spreading by the silicon concentration more accurately than by the hydrological indicators, which change much more rapidly than the hydrochemical characteristics. Thus, the Pacific Ocean interlayer is traced by the silicon concentration to the North Pole and even to Greenland, while by the hydrological indicator, this interlayer is identified only within the Canadian basin. Thus, the work showed the availability of using the hydrochemical indicators for revealing the structure, distribution and transformation of water masses in the Arctic Ocean.

4.5.4 Results of the third stage of hydrochemical studies in the Arctic basin (from 1980)

Further work based on a more complete database contained in the *Hydrochemical Atlas* with the use of the computer software 'Ocean Data View' (Schlitzer, 2003) allowed performing calculations and constructing charts of the distribution of hydrochemical elements. As a result of data analysis, the spatial distribution of hydrochemical elements was obtained for the period 1960–1989, corresponding to a low Arctic Oscillation Index and for the period 1990–2001 with a high Arctic Oscillation Index. The interannual variability of hydrochemical elements for the period 1973–1979 was also obtained (Smagin *et al.*, 2003).

In 1965–1989, the Pacific Ocean water was present in the Canadian basin practically up to the Lomonosov ridge (Figure 4.21, see colour section). In the area of its formation at the shelf slope of the East Siberian and the Chukchi Seas the thickness of the Pacific Ocean layer comprised 250 m (from 50–300 m). In the surface water of the Transpolar Current, the silicon concentration was more than 10 μmol/l, which distinguished this area from the other part of the Canadian basin, where the silicon concentration in surface water was less. One should note that the area restricted by the isopleth of 10 μmol/l includes the Siberian shelf. Therefore, surface water in addition to the Pacific Ocean water also included waters of the Siberian rivers. The concentration of dissolved silicic acid was higher than 50 μmol/l in the core of this water that had a lower temperature as compared with the temperature of higher and lower layers. The low layer presented warmer Atlantic water.

In the higher lying layer of the Canadian basin, the increase in temperature was noted in the 75-m horizon (Figure 4.22, see colour section). It is usually considered that this layer presents summer Pacific Ocean water that has a decreased silicon concentration and a higher temperature as compared with winter Pacific Ocean water. This layer exists due to a quasi-closed water system as a result of the anticyclonic gyre in this region where water can circulate for several years. With the Pacific Ocean water moving northward, its lower boundary rose to 150 m and higher due to the inflow of Atlantic water underlying the Pacific Ocean water (Figure 4.23, see colour section). As a result, the maximum concentration of dissolved silicon comprised 15 μmol/l in the 100 m horizon and more than 25 μmol/l in the 75-m horizon (Rusanov *et al.*, 1979).

In the 1990s when the Arctic Oscillation Index changed from a higher to a lower value, the concentration of dissolved silicon in the former layer of Pacific Ocean water decreased to a background value of less than 10 μmol/l. In reality, the Atlantic water replaced the Pacific Ocean water that was locked on the shelf slope of the East Siberian and the Chukchi Seas. Unfortunately, the expedition observations were too few during the new climatic period to construct good gridded fields similar to the previous years. However, the data obtained confirm a sharp change of the water masses behind the Lomonosov ridge in the Canadian basin, caused by a strong inflow of Atlantic water. Thus the interannual variability in the vertical distribution of temperature, salinity, oxygen, phosphates and silicon in 1973–1977, when the situation was quite stable, was insignificant in the North Pole area (Figure 4.24, see colour section).

The distribution of temperature, salinity, oxygen, phosphates and silicates significantly changed in the 1990s. For example, the silicon concentration in the Pole area comprised 40 μmol/l in the water layer up to 200 m in 1973–1977 while in 1998, the silicon concentration practically decreased to a background level.

Thus, investigation of the available data showed that the interannual variability should be considered in the framework of specific climatic periods. Thus, during the period with a low Arctic Oscillation Index, in particular from 1973–1979, the interannual changes of hydrochemical and hydrological indicators in the anticyclonic gyre were practically absent. In a more dynamic near-Pole area, these changes were also insignificant. A significant change in the spatial distribution of these indicators took place in the 1990s when the Arctic Oscillation Index changed from being low to high.

4.5.5 Ice chemistry

Sea ice as a multi-phase, heterogeneous and a multi-component system presents a complicated feature participating practically in all natural processes in polar areas. Therefore, a large number of studies are devoted to investigating the physical–mechanical studies of sea ice. The problems of ice chemistry are covered to a much lesser extent. At the same time many aspects of oceanography, ice research, marine chemistry and fishery cannot be resolved sufficiently fully without the knowledge of sea ice chemistry. In this respect, the work program of the drifting stations included observations of sea ice chemistry. It was established that the formation of sea ice composition is influenced by the following factors:

- temperature field distribution at the time of ice formation;
- temperature field variations in the ice strata;
- chemical composition of the underlying water;
- gravity brine drainage;
- selective removal of ions from the ice strata; and
- local conditions.

The full-scale data on the chemical composition of drifting ice of the Arctic Ocean were generalized and systematized for the first time by Gudoshnikov *et al.* (1987). Based on the use of cluster analysis of these data, evidence on the average chemical composition throughout the entire strata of sea ice of different ages was obtained according to the World Meteorological Organization Sea Ice Nomenclature (Table 4.8).

The values of the average chemical composition of sea ice of different ages allows us to take into account the quantity of salts and other substances included to the ice strata, in balanced calculations (Table 4.9). For example, it was obtained from the calculations that the total export of salts by ice of different ages through the

Table 4.8. Ion-chloride ratios for different ice types according to the WMO nomenclature.

		Ion-chloride ratios					
Ice type		Ca/Cl	Mg/Cl	Na/Cl	K/Cl	SO_4/Cl	HCO_3/Cl
Shuga	Avg.	0.022	0.066	0.550	0.021	0.140	0.007
	Std.	0.004	0.009	0.025	0.003	0.015	0.001
Dark nilas	Avg.	0.021	0.067	0.560	0.021	0.140	0.007
	Std.	0.003	0.010	0.021	0.003	0.014	0.001
Light nilas	Avg.	0.020	0.067	0.439	0.017	0.139	0.007
	Std.	0.003	0.008	0.019	0.002	0.016	0.001
Young grey, grey-white	Avg.	0.022	0.069	0.482	0.016	0.140	0.007
	Std.	0.001	0.003	0.016	0.003	0.008	0.001
Thin first-year	Avg.	0.021	0.067	0.554	0.018	0.148	0.011
	Std.	0.001	0.005	0.014	0.001	0.010	0.001
Medium first-year	Avg.	0.020	0.068	0.569	0.017	0.152	0.012
	Std.	0.003	0.007	0.017	0.002	0.011	0.001
Thick first-year	Avg.	0.021	0.067	0.570	0.019	0.154	0.012
	Std.	0.003	0.005	0.019	0.003	0.014	0.001
Second-year	Avg.	0.022	0.065	0.561	0.023	0.143	0.011
	Std.	0.004	0.009	0.023	0.002	0.013	0.002
Seawater	Avg.	0.022	0.066	0.550	0.021	0.140	
River ice (Ob′ Bay)	Avg.	0.097	0.268			0.315	
	Std.	0.008	0.032			0.018	

Note: Avg. – average value. Std. – standard deviation.

Table 4.9. Salinity and concentration of nutrients in Arctic sea ice based on data of the 'NP-22' and 'NP-23' drifting stations.

Ice age	Salinity (psu)	Phosphates (μg/l)	Silicon (μg/l)
First-year autumn	2.92	20	50
First-year winter	5.70	20	90
Second-year autumn	2.96	25	80
Second-year winter	5.80	29	60
Multiyear	2.44	24	50

Fram Strait from the Arctic basin comprises 6,120 million tons a year. The ingress of salts incorporated to the ice strata from the Arctic Ocean seas to the Arctic basin during a year is equal to 6,350 million tons. The total export of phosphates comprises 55,500 tons and the ingress – 37,500 tons. For silicon, these values are equal to 126,000 and 106,000 tons, respectively.

Based on the full-scale studies, the peculiarities of formation of the sea ice composition under the influence of local conditions – presence of ice ridges, ice formation in fractures and presence of contaminants were revealed. The most active influence on the formation of drifting ice chemistry is produced by oil, oil products and detergents belonging to a class of surface-active substances.

4.5.6 State of the chemical contamination of the Arctic Ocean

Due to a global spreading of contaminants, studies of the ocean environment contamination were carried out. The 'NP' drifting stations present a unique platform for the study of long-range transfer of contaminants by air and water masses and drifting ice, as they are located in difficult to access regions, remote from the main contamination sources. The observations included observations of such priority contaminants as oil hydrocarbons, heavy metals and chlorinated hydrocarbons (Smagin, 1995). Thus, at the time of the drift of the 'NP-22' station from the Arctic basin to the northern Beaufort Sea from 1977–1978, the levels and distribution of hydrocarbons in the samples of freshly fallen snow and ice were investigated (Dmitriyev and Pivovarov, 1981). The average concentrations of hydrocarbons in ice and snow comprised 0.038 mg/l and 0.017 mg/l, respectively. The fraction of biogenic hydrocarbons was within 1.3–25.0% in ice and 1.6–21.8% in snow.

In 1985, the levels of heavy metals, sulfates and organochlorine pesticides in snow were measured for the first time at 'NP-27' in the central Arctic basin (Smagin *et al.*, 1990). Given the specific synoptic situations it was established that the minimum levels of heavy metals in freshly fallen snow were observed in the case of the closed circulation of air masses in the station drift area and also at the transport of air masses from the North Atlantic regions. The increased concentrations of lead, manganese, nickel, copper and zinc were noted at the movement of air masses from the regions of Eurasia and South-East Asia. The baseline levels of sulfates in snow were observed during the period of closed circulation of air masses, at the transport of air masses from the North Atlantic regions and the northern Pacific Ocean. The increased level of sulfates in snow was noted at the movement of air masses from the regions of Eurasia. High levels of organochlorine pesticides in the freshly fallen snow were recorded at the movement of air masses from the regions of South-East Asia (Table 4.10).

4.6 BIOLOGICAL STUDIES AT THE 'NP' DRIFTING STATIONS

4.6.1 Introduction

The biological studies in the central Arctic Basin started in 1937, when four people from Papanin's crew landed on ice near the North Pole, among them, hydrobiologist P. P. Shirshov. During the drift of the 'NP-1' station from the North Pole into the Greenland Sea he conducted different observations of the composition and distribution of phytoplankton and zooplankton (Shirshov, 1938). After the end of the expedition, several ice samples were taken from Papanin's ice floe for studying of the species composition of microflora.

At approximately the same time, the drift of the *G. Sedov* ice-breaker vessel

Table 4.10. Concentrations of contaminants in the freshly fallen snow from 'NP-27'.

Sampling date	Heavy metals (μg/l)							Sulfates (μg/l)	Pesticides (ng/l)	
	Mn	Pb	Fe	Ni	Cd	Cu	Zn		α-HCH	γ-HCH
25 May, 1985	0.70	1.15	4.6	0.15	0.05	0.35	5.3	415	0.88	0.45
4 June, 1985	0.70	1.10	25.0	0.15	0.12	0.55	8.6	210	0.71	0.33
20 June, 1985	0.23	0.30	25.0	0.15	0.05	0.15	2.3	150	–	–
26 June, 1985	0.65	1.10	25.0	0.90	0.61	0.55	12.3	240	–	–
25 July, 1985	0.10	0.25	1.0	0.15	0.05	0.12	1.6	310	1.55	0.68
14 August, 1985	0.05	0.26	7.1	0.15	0.05	0.03	0.5	100	–	–
23 August, 1985	0.70	1.10	11.6	0.15	0.05	0.25	2.3	2780	–	–
3 September, 1985	0.05	0.05	4.0	0.15	0.05	0.09	2.5	140	1.68	0.57
13 September, 1985	0.20	0.45	2.8	0.15	0.05	0.58	1.5	410	1.35	0.87
3 October, 1985	0.15	0.22	2.0	0.15	0.05	0.12	0.9	220	3.23	2.43
13 October, 1985	0.20	0.60	4.1	1.0	0.05	1.00	5.7	2740	–	–
31 October, 1985	0.20	0.10	1.0	0.50	0.05	0.30	2.0	680	2.39	0.98
15 November, 1985	0.11	0.40	4.6	0.15	0.05	0.15	1.3	340	1.36	0.79
21 November, 1985	0.10	2.40	0.5	1.0	0.05	0.15	5.5	200	2.30	0.99
27 November, 1985	0.10	1.40	0.9	0.50	0.05	0.50	2.0	250	4.51	1.88

proceeded along the *Fram* drift route (October, 1937–January, 1940). In addition, the *Malygin* and *Sadko* ice-breaker vessels drifted with the *G. Sedov* until August 1938. On the *G. Sedov* complex oceanographic observations was accomplished, and hydrobiological samples were taken by V. Kh. Buynitskiy, whose tasks, in addition to this work, were also observations of birds and mammals (Buynitskiy, 1946).

In April, 1941, the first Soviet AHE was delivered to the central Arctic Basin by the 'USSR N-169' aircraft in the area of the Pole of Relative Inaccessibility, where three hydrological daily current observation stations were set up, and three layer-wise plankton catches were positioned at a depth of 2,000 m.

Processing of materials collected during the drifts of 'NP-1', *G. Sedov* and expeditions to the Pole of Relative Inaccessibility have shown that at high latitudes in the Arctic and in near-Pole areas, as well as in other parts of the World Ocean, zooplankton is dominated by copepods (Bogorov, 1946a,b). About 20 species were found dominated by *Calanus glacialis*, *C. hyperboreus* and *Metridia longa*. In addition to the copepods, plankton also contained radiolarians, siphonophores, ostracods, amphipods and appendicularians. P. I. Usachev, while processing the phytoplankton samples, revealed that the vegetation of algae in the central Arctic Ocean started in July, after the snow melted away, and the ice let through light in sufficient amounts for phytoplankton development. The duration of the vegetation is only about one month. In the near-Pole area, 73 algal forms were recorded (mainly diatoms) comprising both planktonic and benthic forms, the latter being predominant. P. I. Usachev assigned 24 benthic (littoral) forms of sea diatoms (Usachev, 1949) to typical cryophiles, which were constantly found in the ice mass.

G. P. Gorbunov and E. F. Guryanova, who processed the bottom fauna collected by the *Sadko* and *G. Sedov* expeditions, established the fact of disconnection between deep-sea benthos in the areas west and east of the New Siberian Islands and the surrounding shallow-water area (Gorbunov, 1946; Guryanova, 1957). This fact enabled them to presume, that there was a physical barrier preventing dispersion of deep-sea benthic organisms through the central areas of the Arctic Ocean. It is rather difficult to overestimate the importance of such a discovery for understanding the distribution of water masses in the Arctic basin and the distribution of its deep-sea bottom fauna and plankton. The most important aspects of this investigation are discussed in papers by Zenkevich (1966) and Guryanova (1970).

A review of hydrobiological and cryobiological studies conducted in the central Arctic basin in the middle of the last century shows that before starting systematic observations at the 'NP' drifting stations there was little evidence on the qualitative and quantitative composition and distribution of the main components of the deep-sea part of the Arctic Ocean, such as phytoplankton and zooplankton, benthos, ice biota and abyssal fishes. Thus, some general biological problems appeared not to be explored. This, primarily, concerns phytogeography and zoogeography, ecology and physiology of cryophile organisms, systematic studies of the basin pelagic zone, assessment of the productivity of its communities and other issues.

4.6.2 Phytoplankton

Collections of materials for the qualitative and quantitative analysis of phytoplankton were first carried out at the 'NP-1' drifting station (Shirshov, 1938). Pretreatment of the samples from the 'NP-1' station has shown that phytoplankton in the samples taken in July is poor both qualitatively and quantitatively; however, in August an abundant development of diatoms is recorded (Shirshov, 1938). Peridinians occur rarely. However, the species composition of phytoplankton is rather poor even in August. There are mass occurrences of *Chaetocerus socialis*. Usachev (1961), analysing the plankton flora based on 'NP-1' materials, showed that almost half of the species identified by him were typically planktonic, and the rest were mostly of benthic form. Phytoplankton from the 'NP-4' station also appeared to be poor (Virketis, 1957).

Phytoplankton samples from 'NP-22' were processed by Belyaeva (1980). It was shown that the flora in the study area was mainly composed of diatom algae. The role of dinoflagellates, silicoflagellates and blue–green algae is negligible. Overall, 52 diatom species were recorded, among which the number of pennate and centric algal species is approximately the same, the number of pennate diatoms almost not changing season-wise, whereas the number of centric algae varies extensively. Depth-wise distribution pattern of algae was revealed as well as the specific features of taxonomic composition and numbers of phytoplankton populations.

4.6.3 Zooplankton

In the 1950s–1970s, plankton was collected by specialists of the AARI on the assignment of the Zoological Institute of the USSR Academy of Sciences. During the period from 'NP-2' to 'NP-20', 1,638 zooplankton samples were taken at 257 stations (Pavshtiks, 1980). Catches were carried out in the Arctic and Atlantic bottom water, and intermediate and surface water masses in turn.

During the drift period of the 'NP-22' and 'NP-23' stations in 1975–1980, zooplankton was collected by specialists from the Oceanology Institute named after P. P. Shirshov (Melnikov, 1976a,b, 1978; Melnikov and Tsinovsky 1978). Catches were carried out layer-wise with and without locking of the net. Layers of catches are (in metres): 0–10, 10–25, 25–50, 50–100, 100–150, 150–200, 200–300, 300–500, 500–1,000 and 1,000–floor.

Seasonal variations of vertical and horizontal plankton distribution were studied as well as diurnal migrations, numbers and biomass of the mass species, age composition of populations, etc. The main results of these observations at the 'NP' stations were published in the following papers: Virketis (1959), Pavshtiks (1977, 1980), Heinrich *et al.* (1980), Kosobokova (1980), Marhaseva (1980), etc. The results of zooplankton studies in the central Arctic basin carried out at the 'NP' drifting stations were used to compile hydrobiological maps for the *Atlas of the Arctic* (Brodskiy and Pavshtiks, 1976). The materials collected at the 'NP' drifting stations also significantly enriched our knowledge on biodiversity of pelagic organisms in the Arctic basin.

4.6.4 Benthos

In section presents only the results of studying benthos at the *Sever* and 'NP' stations. Work was performed from 1950 to 1979 with several interruptions during this period.

At the first stage of investigations, in the 1950s, trawlings were carried out at 39 stations. The material was taken from a depth of 108–4,379 m; mainly from depths below 1,000 m.

General lists of bottom organisms caught at 'NP-2', 'NP-3' and 'NP-4' (1954–1955) were published by Koltun (1964).

During these catches, different representatives of bottom fauna of the Arctic basin were collected: foraminifers, sponges, coelenterates, nematodes and polychaetes, scaphopods, gastropods and bivalves, bryozoans, crustaceans (cirripedials, opossum shrimps, cumaceans, amphipods and isopods, shrimps) and echinoderms (mainly holoturians). For a long time, these materials served as the basis of our knowledge on the faunal composition and structure of bottom communities in the central Arctic and were used to prepare several scientific papers and monographs (Ushakov, 1982; Guryanova, 1957; Koltun, 1964). Based on the materials collected by the expeditions, a new shrimp species was identified – *Bythocaris curvirostris* (Kobyakova, 1957) and a mass holothurian species – *Elpidia heckeri* (Baranova, 1989). In many respects, due to the analysis of benthos collections taken from the drifting stations, our notions of zoogeography of the polar basin were defined more precisely (Guryanova, 1957, 1970). One specimen of a large sea pen collected at the 'NP' decorates the exhibition of the Zoological Museum in St. Petersburg.

From 1976–1979, specialists from the Oceanology Institute named after P. P. Shirshov collected benthos at 'NP-22' and 'NP-23' drifting stations (Afanasiev, 1978; Moskalev, 1980). As a result, 115 bottom fauna samples were taken: 53 trawl samples. In 1976–1977, benthos was sampled at a depth of 2,600–3,550 m in the eastern Canadian basin ('NP-22') and at a depth of 266–2,821 m ('NP-23'). Most of the collections contained foraminifers, sponges, bivalves, polychaetes and crustaceans. And in 1978–1979, the study area moved to the depth of 150–2,700 m north of the Beaufort and Chukchee Seas.

The principal forms of benthos were more abundant here: sponges, sea pens, actinias, alcyonarians, zoantharians, bryozoans, polychaetes, gastropods and bivalves, isopods, sea lilies, starfishes, sea urchins, basket stars and holoturians. On the basis of the data obtained in the areas investigated in 1978–1979, bottom communities and trophic zones were distinguished (Moskalev, 1980).

Overall, 10 biocoenoses were distinguished in the outer part of the Chukchee Sea offshore: *Umbellula + Heliometra + Bryozoa + Jasmineira + Gersemia*; *Gersemia*; *Trochostoma + Ctenodiscus*; Astarte; *Ctenodiscus + Pontaster*; *Kolga + Pourtalesia*; *Anthosactis + Bryozoa*; *Anthosactis + Astarte*; *Anthosactis + Ctenodiscus + Pontaster*; and *Ophioscolex*.

In subsequent years, organisms from the benthos samples taken at 'NP-22', were studied by specialists from the Oceanology Institute of the RAS, Zoological Institute of the RAS and the Institute of Marine Biology of the RAS. The first incomplete list of the species was published in a paper by Afanasiev and Filatove (1980). The materials on sea pens *Umbellula encrinus* and deep-sea holoturians *Kolga hyalina* collected at 'NP-22' were used in Sirenko *et al.* (2004).

Separate articles were published on some bottom fauna taxa: on bristle(-bearing) worms – polychaetes (Zhirkov, 1980: including the description of 4 new genera and 5 new species); on sea pens – pennatularians (Pasternak, 1980); on brachiopods (Zezina, 1980); on octopus (Kondakov *et al.*, 1980); on amphipods —freshwater shrimps (Kamenskaya, 1980: including the description of a new species), on opossum shrimps (Petryashev, 2004: including the description of one new species and a redescription of another species, earlier known only from one occurrence); on isopods (Kusakin, 1983; Malyutina and Kussakin, 1996: includes the description of 13 new species) and cumaceans (Vasilenko, 1988: including a description of a new species). Taxonomic determinations of benthic organisms collected at the 'NP' drifting stations served as the basis for drawing up a list of the species of free-living invertebrates of the Eurasian seas and the adjacent deep-sea parts of the Arctic edited by Sirenko (2001, 'Central Polar Basin' section).

4.6.5 Necton

On the assignment of the Zoological Institute of the USSR Academy of Science, ichthyological specimens were collected at seven 'NP' stations. Fish catches at different times were carried out from the surface to a depth of 25 m using different techniques. Based on the materials of 'NP-6', which drifted in the northern East Siberian Sea, the *Arctogadus glacialis* codfish species, new for the USSR fauna and very poorly studied, was identified. Overall, during the observation period from 1954–1977, over 10,000 fish specimens were caught. It was ascertained that in the near-Pole areas of the Arctic and under pack ice, two codfish species dwell (*Boreogadus saida* and *Arctogadus glacialis*), which at times form large shoals. For instance, at 'NP-16' in November 1968, within several days, several thousand fish were caught on hook in holes in the ice.

The first attempt of catches at depth was made at 'NP-23' in May–October, 1977. Catches of abyssal fish were carried out by bottom trawls and using a long line dropped through a 2×1 m hole in 5–6 m thick ice, undertaken using the 'NP' hydrological winch (designed at the AARI) with an electric drive. Overall, 25 trawlings and several dozens of catches with the long line were made at different depths (to 3,000 m). During trawlings, two specimens of *Paraliparis bathibius* Collett were caught and one specimen of *Licoides frigidus* Collett (Tsinovskiy, 1980). In the course of ice diving in the immediate proximity of the ice, above depths exceeding 2,000 m, two specimens of *Liparis koefoedi* Parr were caught (Tsinovskiy and Melnikov, 1980). The occurrence of this typical nectobenthos above great depths near the ice could be due to the fact that these fish were carried out by a strong drift current directed from the East Siberian Sea offshore to the central Arctic basin.

4.6.6 Microbiological investigations

In the areas of the 'NP-3', 'NP-4' and 'NP-5' stations drift, vertical distribution of saprophyte bacteria in water samples and soils was studied. Samples for studying the distribution of bacteria in water were taken from the standard horizons. Bacteria were discovered practically at all depths near the North Pole. Particularly numerous cells were recorded at the interface between the Atlantic and surface water. Noticeable seasonal variations were recorded in the numbers of saprophytic bacteria in the vertical and horizontal distributions in the central parts of the Arctic Ocean. Colonies of micro-organisms grown from water samples and silt comprised non-sporiferous and sporiferous bacilli, microbacteria and coccal forms. Yeast colonies were also distinguished (Criss, 1976).

4.6.7 Cryobiota

The first sea ice samples for the purpose of studying biological composition were taken at 'NP-1'. After the end of the drift (19 February, 1938), when Papanin's ice floe was carried out to the Greenland Sea, 10 samples were split off the ice ridge, each of them was divided into three parts: the upper, middle and lower layers. P. I. Usachev, who processed these samples (Usachev, 1949), revealed 73 forms, among them 64 diatoms, 3 peridinians, 1 silicoflagellate, 5 forms of infusorians, spores, etc. Among the diatoms were cells with well-preserved chromatophores, however, most of the cells were void. Among the diatoms, there were 25 plankton and 43 benthic forms, which corresponded to 37 and 63% of the total numbers of cells. Among the latter, P. I. Usachev recorded 24 forms of typical cryophiles, which accounted for about one-half of all benthic diatoms. These cryophiles were marked by P. I. Usachev as 'typical', since they occurred ubiquitously in collections from the central areas of the Arctic Ocean, and these particular forms were responsible for mass 'blooming' in the ice. V. I. Vernadsky, who commented on cryobiological investigations at 'NP-1' (Vernadsky, 1938), placed major emphasis on the biogeochemical role of ice diatoms in 'eating away' ice due to heat release during biochemical reactions in the course of their development. The subsequent observations of the composition and structure of biological communities in sea ice were only undertaken at the 'NP' drifting stations between 1975–1990.

In the course of three all-year-round observations at 'NP-22' (1975–1976), 'NP-23' (1977–1978), 'NP-22' and 'NP-24' (1979–1980), which drifted in the central areas of the Arctic Ocean, hundreds of samples of perennial and one-year ice were taken, field experiments were conducted with young growing ice, snow and meltwater samples were taken from puddles and seawater samples from the lower ice surface (Melnikov, 1976, 1978, 1989; Melnikov and Tsinovsky, 1978; Melnikov, 1981). Processing the samples showed that sea ice in the Arctic Ocean is a composite and time-stable ecological system. The presence of a mechanism regulating the average equilibrium thickness as well as a stable large-scale ice circulation, and a system supporting an equilibrium ice balance on the whole, ensures their stable existence within the Arctic Ocean. A simultaneous occurrence of freshwater green algae and sea diatoms was recorded within a 3–5-m ice thickness. Their distribution shows a clear vertical zonation, which is stably retained within the physiographic boundaries of the Arctic Ocean. Two functional stages are distinguished in the annual cycle of ice ecosystem development: (1) photoautotrophic, when the synthesis and accumulation of an organic substance proceeds, and (2) heterotrophic, when delivery of the organic matter to organisms wintering in the ice thickness and at the water–ice interface occurs. Ice flora exceeds 90% of the total product synthesized by all the algae (ice and planktonic). The materials received provided grounds for distinguishing two main flows of substances in the ecosystem: (1) supply of biogenic elements from water to the ice mass, where they are assimilated by microflora during photosynthesis (summer stage of evolution) and (2) a reverse one – from ice to water – movement of organic matter synthesized during summer (winter stage of evolution). It was shown convincingly, that the accumulation of organic matter in the ice during the summer development stage of plant communities was an important biogeochemical function of the ecosystem. Accumulation of organic matter during the photosynthesis of ice flora can be considered as an energy depot for the organisms within the entire trophic network of the Arctic sea ice ecosystem. Taking into account the specificity of sea ice substrate and of the organisms inhabiting it, the stability of links between them in the ecosystem, as well as with regard to the planetary extent of this natural phenomenon, it was proposed to distinguish the communities inhabiting the Arctic sea ice as an independent class of water communities and name it a *cryon* (i.e., existing in ice or associated with ice; and the space taken up by sea ice), and its population a *cryal* (Melnikov, 1989).

An important contribution to understanding the species composition of plant communities inhabiting the sea ice of the Arctic seas was made by Yu. B. Okolodkov, who took the opportunity of collecting materials during the AHEs of the AARI (AHE 39–42 in 1987–1990) (Okolodkov, 1989, 1992). For the first time, the appearance of freshwater and brackish water species in ice was recorded due to the influence of the river runoff on the sea ice population. In addition, for the first time, the predominance of peridinians over diatoms was recorded in certain geographic areas of the Arctic Ocean, the latter until recently being regarded as the dominant group in the plant community of the sea ice ecosystem.

4.6.8 Conclusion

The results obtained using the materials collected at North Pole drifting stations became significant. They were mainly received during the period, when no changes in the Arctic natural environment were recorded, and, thus, should be considered as the basis for comparison with present-day materials. As is known, currently, there is a marked reduction of ice cover thickness and area, surface water freshening and warming has occurred due to climatic changes in the Arctic Ocean. Comparison of data on the ice and plankton communities from the central areas of the Arctic Ocean, which were received in 1975–1980 in the Canadian sector, with similar data from the same geographic area obtained during the SHEBA expedition in 1997–1998 and 'Arctic-2000', has shown that in the last two decades there occurred major changes in the qualitative and quantitative composition of ice and planktonic communities (Melnikov *et al.*, 2000). Further observations at the 'NP' drifting stations will enable more precise determination and supplement our knowledge of the nature of the Arctic Ocean.

5

Geophysical observations

5.1 INTRODUCTION

Geophysical observations – a vital part of the explorations of the polar regions nature – began to be an important component in the scientific programs developed for the Russian ice drifting stations. The idea of using ice drifting floes as natural moving platforms from which extensive multidisciplinary scientific studies in the Arctic Ocean basin could be made was a very reasonable solution for Russia with its vast polar and sub-polar territory – occupying 10 time zones in longitude. During the drift of the 'North Pole-1' (1937–1938) future academician Eugene Fedorov performed an annual cycle of geomagnetic observations, which were a valuable contribution to more full understanding of the features of geomagnetic variations in the central Arctic. The scientific programs of the next ice drifting stations were expanded considerably by including other geophysical observations (mostly connected with studies of the polar ionosphere). Choice of the geophysical observations to be held at the ice drifting stations was determined by the necessity to enlarge our knowledge of the physical processes in geomagnetic fields, the Earth's ionosphere and propagation of radio waves through it, auroral events, cosmic rays as well as to solve some important pragmatic problems connected with the establishment of regular vessel navigation along the Great North Polar Sea Route. Several geophysical explorations made by Russian scientists were pioneering in their fields and had no analogies in other countries. In this connection such projects as the determination of spatial and temporal variations of the zone of anomalous radio wave absorption in the Arctic ionosphere (riometer observations), as well as detailed explorations of radio waves in the central Arctic (oblique sounding of ionosphere) should be mentioned.

In this chapter a general review of the geophysical observations made at the 'NP' drifting stations during the whole period of their operation is given. The main emphasis in this review has been placed on the physical essence of these observations and on the final results obtained from analysis of the measurements.

5.2 EXPLORATIONS OF GEOMAGNETIC FIELD

The study of the geomagnetic field is a traditional kind of geophysical research made in the Arctic and Antarctic Research Institute (AARI) over many decades. The irregular and intense character of geomagnetic variations in high latitudes attracted the attention of several generations of polar geophysicists.

Quite naturally, the first object of geomagnetic study in the Arctic was the exploration of distributions of the permanent geomagnetic field and its secular variations. Extensive reconnaissance of mineral resources in the Arctic and sub-Aarctic regions and the development of maritime and aviation navigation systems required the elaboration of detailed magnetic maps of the vast northern territories where magnetic observations have never been undertaken. Compilation of the geomagnetic data obtained by the 'NP-1' station crew together with similar data collected during previous arctic voyages of such vessels as the ice-breaker *Sedov* (1937–1940), *Mod* and *Fram* allowed us to obtain a rough estimation of the secular variations of the geomagnetic field in the Arctic Ocean basin. All these data, collected over different years, have let scientists of the AARI to attribute them to the only one year – 1940. As a result, the first magnetic maps of the Arctic Ocean basin were produced for the epoch of 1940. Evidently, accuracy of these maps was not very good. After World War II a series of airborne geomagnetic survey expeditions were performed in the central Arctic. The experts of the AARI, M. F. Ostrekin, P. K. Sen'ko and N. A. Milayev, were responsible for the scientific results of these expeditions. These data together with results of geomagnetic observations made at the ice drifting stations 'NP-2', 'NP-3' and 'NP-4' gave the possibility of producing detailed magnetic maps of the central Arctic for the epoch of 1955. Accuracy of these maps was high enough to satisfy all requirements of maritime and airborne navigation.

Analysis of the data of geomagnetic observations made first at the ice drifting stations showed that there were several local geomagnetic anomalies in the central Arctic. Their localization was made with definition to their sizes and intensities. Some speculations were proposed concerning the depths of the possible sources of these anomalies.

Rather intriguing results were obtained which showed a close connection between configuration of the Ocean bottom and character of the geomagnetic variations above the bottom elevations. For the first time reliable data about secular variations of geomagnetic fields in many areas the Arctic Ocean became available.

An important part of geomagnetic explorations is the study of an alternating component of the geomagnetic field, connected with the influence of external sources, preferably the Sun. Thorough analysis of the data of geomagnetic activity obtained at many high-latitude stations including 'NP' ice drifting stations made by one of the pioneers of Arctic geomagnetic observations – Dr. Alexey Nikolskiy

(1956) showed that the main peculiarity of the polar geomagnetic variations is a definite presence of three separate zones of disturbance: the morning ones, the daytime ones, and night-time ones. The relative magnitude of these disturbances depends strongly on the geographic position of the point of observation. Discovery of the second zone of geomagnetic disturbances located in the near-Polar region and observed in summer made by A. Nikolskiy was an outstanding scientific achievement – developed further by the next generations of polar geomagnetic scientists.

The most complete data on the spatial distribution of geomagnetic activity in high latitudes was obtained during observations in the International Geophysical Year (1956–1957) and the International Year of Quiet Sun (1964–1965). It was shown that the most intense geomagnetic and auroral disturbances were observed along branches of the auroral oval located around the geomagnetic pole in both hemispheres. Conception of the auroral oval as a form of global distribution of the high latitude geomagnetic disturbances was introduced into geophysical science by the Russian explorers O. Khorosheva and Ya. Feldstein (Troshichev, 1986) who analysed data of photometric auroral observations at high latitudes including data obtained at the 'NP' ice drifting stations. Configuration of the auroral oval is generally constant during both disturbed and quite periods. However, the oval is not of uniform formation, it has several centres of disturbances where the level of activity is higher than at nearby regions. These centres of activity are located in the daytime, morning and night-time hours and are concentrated at the following geomagnetic latitudes: 75°–78°, 69°–72°, 67°–69°. Conception of the auroral oval as a new coordinate system in polar geophysics was one of the main basic ideas, which determined the development of a new direction in science – solar terrestrial physics.

Analysis of the maps of the spatial–temporal variations of geomagnetic activity made in the AARI revealed their strong dependence on Greenwich (global UT) time. It was shown that geomagnetic activity increased significantly in the auroral oval at 16–18 UT. This increase of activity is especially evident during wintertime at that part of the oval where 16–18 UT coincided with local geomagnetic midnight. In the near-Pole region maximum geomagnetic activity was recorded during the summertime in periods when local geomagnetic noon coincided with the period 16–18 UT.

The very first geomagnetic observations in the Arctic showed the presence of the strong sinusoidal oscillations of the geomagnetic field with amplitudes of several hundreds of nanoTesla (nT) and periods of 100–500 seconds. These oscillations were called 'giant pulsations' and were classified as a special type of geomagnetic pulsation. It was found that the maximum activity of these pulsations depends on the geomagnetic latitude and on the solar activity level. Further studies showed that these pulsations are the result of the complex interaction between the magnetosphere and ionosphere at polar latitudes.

The beginning of permanent geomagnetic observations in Antarctica (International Geophysical Year in 1957–1958) gave start to systematic studies of geomagnetic disturbances in the geomagnetic conjugate points (i.e., in the stations located at the opposite ends of the same geomagnetic field line). Of special interest was the study of conjugate processes in the region of the open geomagnetic field lines the so-called polar cap. This region in the Arctic is located in the centre of the Arctic Ocean where drift trajectories of most of the 'NP' stations took place. The scientists of the AARI were the pioneers of such explorations. They used data of magnetic observations made at the ice drifting stations for these studies. Milyaev (1961) compared peculiarities of geomagnetic variations recorded at the ice drifting stations 'NP-6' and 'NP-7' operated in the northern polar cap with data of simultaneous geomagnetic observations made at the Antarctic near-Pole stations (Vostok and Mirny). It was shown that variations of geomagnetic fields in the opposite polar caps were almost similar. However, this similarity disappeared if the data of the polar cap stations were compared with data of lower latitude stations. It is worth noting that the full physical interpretation of the processes of conjugacy of geophysical events in the region of open geomagnetic field lines has not been made even today.

A significant contribution was made by the geomagnetic scientists of the AARI in the study of features of the Earth's electromagnetic field in the World Ocean. The data of the geomagnetic observations made at the 'NP' stations were experimental material for this study. Zhigalov (1960), analysing the character of geomagnetic variations in the central Arctic, discovered that the depth of the ocean beneath the geomagnetic station influenced strongly the variations of the vertical component of the geomagnetic field. Galkin (1970), using the ideas of Zhigalov showed that the data of geomagnetic observations at the 'NP' stations were connected directly with the relief of the ocean bottom.

Another discovery in geomagnetic studies made by the scientists of the AARI was the so-called 'Coast Effect' (Sen′ko, 1958). The essence of this effect are the specific variations of the vertical and horizontal components of the geomagnetic field, which are expressed in abnormally high amplitudes of variations of these components nearby a sea coast. Presumably this effect is due to induction currents in the sea and to a sharp gradient of electric conductivity in the boundary between the sea and coast.

The achievements of the Russian geomagnetic scientists have a pioneering character and these ideas put forward need to be developed on the basis of modern observational equipment. It is necessary to say that all the above-mentioned results were obtained using original geomagnetic devices designed and produced by the experts of the AARI.

5.3 EXPLORATION OF THE POLAR IONOSPHERE AND PROPAGATION OF RADIO WAVES

The epoch of the active advance of various branches of industry and mineral resource mining into the northern territories of Russia (including the Arctic) in the 1920s coincided with the beginning of extensive use of the high-frequency (HF) range of radiowaves for wireless communication on both regional and federal scale. Russian scientists and engineers were among the pioneers in using this new and progressive kind of radio communication. In conditions of disorder after the Civil War (1918–1922) the experts of Nizhniy Novgorod radio laboratory elaborated the original equipment for receiving and transmitting signals in HF range (wavelengths of 10–100 m). Radio waves of HF range propagate by sky ray by means of reflection from an ionized layer of the upper atmosphere (so called ionosphere) located at altitudes of 100–400 km. The theory of this radio wave propagation has not been elaborated in this period and the knowledge about it is based on accumulation of experimental data only. The study of propagation of HF waves in the Arctic had not been performed at that time, since the Arctic itself was an unexplored region of our planet. In previous years, radio communication was made at the middle radio waves (wavelengths of 300–500 m) by means of the spark radio transmitters for distances not more than 300–400 km. This equipment was technically unreliable and cumbersome, it consumed a lot of electrical power and its aerial fields occupied very big territories. Besides that, conditions of propagation of the middle radio waves depended crucially on a level of electric conductivity on the ground surface. They were very good for communication above open seawater but became worse for communication above the permafrost and soggy tundra. Quite naturally these peculiarities of the middle radio wave propagation seriously limited using this kind of communication for aviation, which was the only reliable kind of transportation for all the industrial and scientific activities in the Arctic where the distances between the few operating centres of activity were vast.

E. T. Krenkel organized the first HF radio unit in the Russian Arctic for permanent service at the polar station Matochkin Shar (Novaja Zemlja Archipelago) in 1927 (Krenkel, 1973). Several other similar stations were organized in the Arctic a little later. The HF radio operators, beside their professional obligations, established unprofessional connections worldwide with other colleagues. It was the beginning of very important scientific explorations into the study of the propagation of HF radio waves in the Arctic – undertaken by the enthusiastic volunteers. Analysis of these data accumulated by the HF operators during many years gave start to scientific explorations of conditions of HF radio wave propagation at high latitudes. It is worthy to note that E. T. Krenkel established the first radio communication between the Arctic and the Antarctic (Krenkel, 1973) – on 12 January, 1930 when Krenkel worked at the polar station Tikhaja Bay (Franz-Josef archipelago). A steady communication was established with the centre of the US Antarctic expedition (Little America station). If one takes into account the comparatively low level of development of radio equipment at this period this achievement could be considered as a significant event in the history of Arctic exploration. Among the other pioneers of studies of the HF radio wave propagation in the Arctic were the scientists of the AARI – B. F. Archangelskiy, F. Ya. Zaborshchikov, G. N. Egorov, V. M. Driatskiy, L. P. Kuperov, etc.

The crew of the first ice drifting station 'NP-1' transmitted all the information about their activities by means of HF radio communication – E. T. Krenkel was responsible for this (Krenkel, 1973). Besides that, regular transmitting of meteorological data from 'NP-1' made it possible to successfully accomplish the World's first direct transpolar aeroplane flights from Russia to North America by teams lead by

V. P. Chkalov and M. M. Gromov in 1938. Radio communication between 'NP-1' and the territory of Russia was made by using an intermediate retransmitting station located at Rudolf Island (Franz-Josef archipelago). This completely new method of organization of radio communication in the Arctic, by means of intermediate point, was widely used in following years.

5.3.1 Vertical sounding of the ionosphere

The first experimental observations of the HF radio wave propagation in the Arctic revealed significant irregular oscillations of the signal level at the receiver input produced by the corresponding variations in the polar ionosphere. In this connection a strong necessity to organize the regular ionospheric observations in the Arctic became evident. A specialized radar – ionospheric station of vertical sounding was elaborated, which made it possible to get information about ionospheric behaviour within a radius of 300 km from the point of observation. Sounding of the ionosphere is made in the frequency range of 1–20 MHz (wavelengths of 300–15 m). The receiver of the ionospheric station registers the signal reflected from the ionosphere. This process required an evaluation of the time (Δt) allowance for time spent travelling between the ground and the ionosphere. This time interval allows one to determine the height of the atmosphere from which a vertically radiated impulse is reflected (i.e., altitude of the ionospheric layer). It is expressed by the equation:

$$h_d = c.\Delta t/2 = 1.5 \times 10^8 \Delta t \text{ (metres)}$$

Where h_d is the so-called acting altitude of the ionosphere and $c = 310^8$ m/s – the velocity of light (velocity of propagation of a radio impulse).

In the real ionosphere a radio impulse propagates with a group velocity V_g, which is less than c. In this case the time of the radio impulse propagation is determined as:

$$\Delta t = 2h_o/c + \int dh/V_g$$

Where integration is made between the limits h_0 (altitude of the lower boundary of the ionosphere) and h_h (altitude of the radio impulse reflection).

Taking into account well-known expression (Rishbeth and Garriott, 1969):

$$V_g = ch_i = c(1 - 80.8N_e(h)/f^2)^{0.5}$$

where N_e is the density of the electrons in the ionosphere (cm^{-3}) and f is the frequency of the sounding (MHz) we get an expression for the value of the acting altitude of the ionosphere as a function of the frequency of sounding:

$$h_d(f) = h_o + \int dh/(1 - (80.8N_e(h)/f^2)^{0.5}$$

where integration is made again between the limits h_o and h_i.

The last expression determines altitude – frequency characteristics of ionosphere, so-called ionogram – the main product of the vertical sounding of the ionosphere.

Figure 5.1 shows an example of two ionograms of vertical soundings recorded at the drifting stations 'NP-8' on 21 June, 1961 and 20 September, 1961, taken at local midnight. The station was located in the near-Pole part of the eastern Arctic ($\varphi = 83°$N, $\lambda = 142°$E). The sticks at the horizontal axis of Figure 5.1 mark the frequency of sounding in MHz, and at the vertical axis – values of acting altitude of reflection (i.e., height distribution of the ionospheric layers). The bottom layer of the ionosphere (region E) is located at altitudes of about 100 km and besides the regular *E* layer the so-called sporadic layer *Es* can be seen frequently. The upper part of the ionosphere is called the *F* region. During summer this part of the ionosphere is divided into two layers – *F*1 layer located at height 180–220 km and the main layer *F*2 – used for distant communication. The electron density of the ionosphere required for reflection of vertically incident radio impulses at the frequency f is equal to:

$$N_e = 80.8/f^2 \text{ (cm}^{-3}\text{)}$$

All radio waves with frequencies of sounding exceeding the so-called critical frequency of the *F*2 layer equal to:

$$f_0F2 = (80.8N_e)^{0.5}$$

pass through the ionosphere without reflection and this range of frequencies is used for communication with Earth's satellites. The frequencies less than f_0F2 are used for communication with ground-based objects (by means of oblique incidence sounding). So, it is evident that the ionosphere's condition plays a crucial role in the reliability of all kinds of radio communication.

Systematic explorations of the ionosphere made mostly at middle latitudes allowed us to reveal the main peculiarities of the ionospheric variations in dependence on the level of ultra violet solar radiation – the main source of formation of the ionosphere. In the most general form these regularities are shown in Figure 5.2 (Johnson, 1965). There are two types of significant variations of electron density in the ionospheric layers: 'day–night' and 'maximum–minimum solar activity'. Such types of behaviour of the middle latitude ionosphere allows one to elaborate more or less reliable methods of prediction of ionospheric condition based on the solar activity indices. As it was indicated previously, the first experiments on the HF radio wave propagation in the Arctic revealed significant and irregular variations in the high-latitude ionosphere. Hypotheses about the crucial contribution of solar radiation in the formation of the ionosphere allow one to suggest that during the polar night, which is present for several months when the level of the solar radiation significantly decreases, the ionosphere in the near-Polar region could be absent completely.

The World's first experimental verification of this hypothesis was made during the operation of the ice drifting station 'NP-3' (April, 1954–April, 1955). The scientists and engineers of the AARI performed annual cycles of vertical sounding of the near-Pole Arctic ionosphere at this station. The station operated for 370 days, the

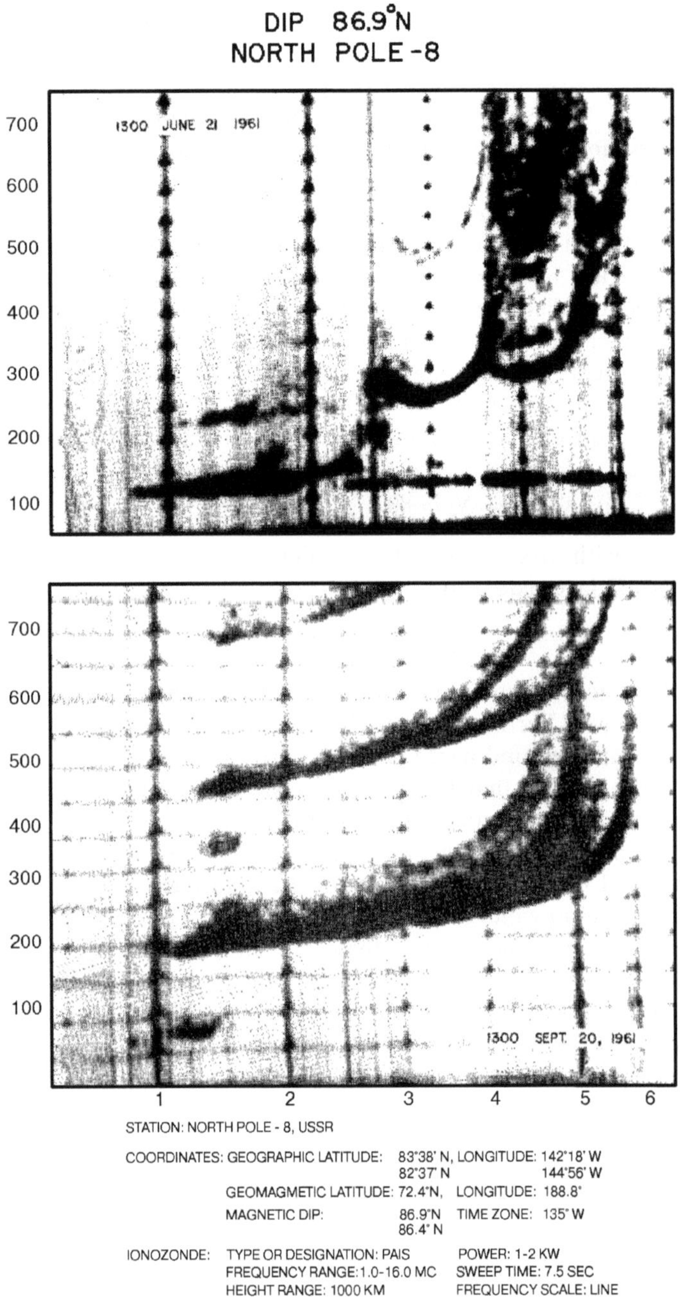

Figure 5.1. Examples of the daytime (*upper panel*) and night-time (*lower panel*) of the ionograms of vertical ionospheric sounding obtained at the 'NP-8' drifting station at 13:00 local time on June 21, 1961 and at 13:00 local time on 20 September, 1961 correspondingly. The station was located (during these time periods) at 83.6°N–82.6°N; 142.3°E–144.9°E.

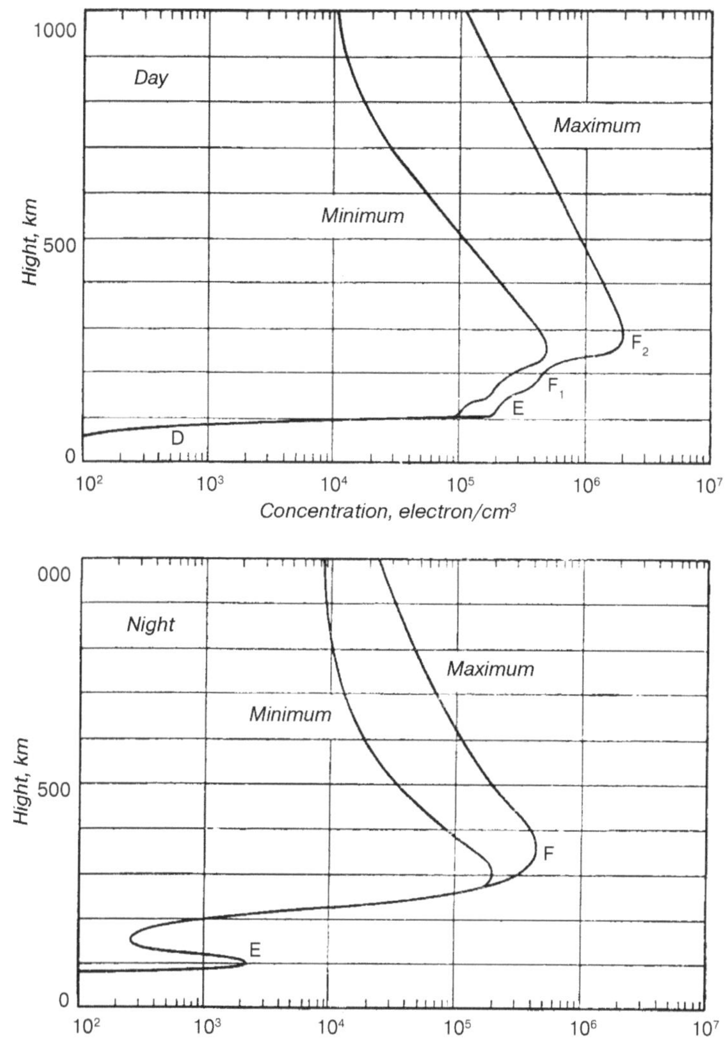

Figure 5.2. Vertical profiles of electron density in the ionosphere for daytime (*upper panel*) and night-time (*lower panel*) hours in the epochs of maximum and minimum solar activity.

length of its drift was equal to 1,860 km. The trajectory of the station drift could be presented as an arc on the great circle passing exactly through the North Geographic Pole with starting point coordinates of 87°08′N; 180°54′E and end point coordinates 87°00′N; 24°00′W. On 25 August, 1954 the station 'NP-8' approached the point of the North Geographic Pole at a distance of 30 km. So in accordance with the contemporary knowledge the whole drift of the 'NP-3' station laid at latitudes above 80°N (i.e., in the region of the open geomagnetic field lines) in the so called polar cap.

The scientists and engineers from the AARI already possessed some experience in performing ionospheric observations in the Arctic. The very first Russian vertical ionospheric station was put into annual operation at Tikhaja Bay Station (Franz-Josef archipelago; 80°30′N; 52°80′E) – the most high-latitudinal geophysical observatory at that time. Similar observations were started in Tiksi bay in mouth of the Lena River in 1944. However, the near-Pole part of the Arctic remained to be an absolutely unexplored region in the study of the ionosphere. It is worthy to note that results of ionospheric observations at the 'NP-3' station were unique having no analogous measurements elsewhere in the World. These data were obtained a long time before the International Geophysical Year (1957–1958) when our knowledge about the configuration of the near-Earth space were changed radically together with understanding of the physical processes controlling the condition of it. Launch of the World's first artificial satellite in October 1957, from the USSR, represented the start of the new era in space exploration – including studies of the near-Earth surroundings (magnetosphere and ionosphere). In this connection, achievements of the scientists of the AARI, in the studies of the polar ionosphere, obtained by analyses of the results of ionospheric observations at the ice drifting stations especially at 'NP-3', continue to be scientifically significant today.

The main conclusions derived from analyses of the ionospheric observations at the 'NP-3' station are the following (Driatskiy and Besprozvannaja, 1958). It was show that ionospheric layer $F2$ is permanently present in the polar ionosphere during polar night and its electron density is very high ($f_0F2 = 5$–6 MHz). It was unusually high for epochs of minimum solar activity corresponding to 1954. Since the level of the solar ultraviolet radiation at ionospheric altitudes decrease strongly during polar night periods, a hypotheses was introduced that a source of ionization for the polar ionosphere was corpuscular radiation of the Sun – so-called solar wind. Another important feature of the night-time $F2$ layer of the polar ionosphere was the formation of the so-called sporadic (irregular) ionization at heights of this layer. Specific peculiarities of these events were analysed including their dependence on geomagnetic activity levels.

This effect later attracted the attention of scientists from many countries – after about 30 years of the 'NP-3' drift. These ionospheric forms were called 'patches of ionization', which are moved by magnetospheric convection across a polar cap from the middle latitude ionosphere illuminated by the Sun. Ionospheric plasma inside the patches is strongly disturbed and therefore radio communication with satellites made through these formations became very complicated. Since the area of the patches of ionization is comparatively big (about 1,000 km) this problem turned out to be a serious one for control of satellite operation (Crowley, 1996). Character and peculiarities of the patches of ionization distribution in the polar ionosphere discovered later turned out to be very similar to the features obtained by analyses of ionospheric observations made at the 'NP-3' station.

During wintertime the ionograms of vertical soundings of the near-Pole ionosphere demonstrated the presence of many diffusive reflections when the correct interpretation of the critical frequencies of ionospheric layers was complicated and sometimes impossible. It is clear that organization of reliable radio communication in high latitudes requires knowledge of regularities of the distribution of these diffusive structures (*F*-spread events) in the Arctic ionosphere. The data of many years of ionospheric observations made by vertical and oblique incidence methods were used by the experts of the AARI to develop maps of the temporal/spatial distribution of the *F*-spread events in the Arctic ionosphere (Nekrasov *et al.*, 1982). The data of ionospheric observations made at the ice drifting stations contributed to this study significantly. The leading ionospheric scientists in the World evaluated this study positively. Figure 5.3 shows the maps of the *F*-spread distribution in coordinates 'geomagnetic latitudes – local geomagnetic time' for the different seasons of 1980 (maximum of solar activity). One can see from Figure 5.3 that diffuse reflections in the polar ionosphere are maximal in winter during night-time. Low-latitude belts of the *F*-spread events during wintertime coincide with a region of the main ionospheric trough – the region of very low electron density, which is a serious obstacle for radio communication at high latitudes. During other seasons, distribution of the diffusive reflections becomes more complicated with strong variability in their intensity.

Important contributions to the studies of the peculiarities of the polar ionosphere were made by observations of the lower part of the ionosphere (*E* layer) performed at the 'NP-3' station. The ionospheric station at 'NP-3' recorded very high activity in the appearance of various types of sporadic (irregular) ionization in the *E* region (*Es* layer). The flat types of *Es* layers with low electron density and different height positions were dominant in the polar ionosphere. Their dependence on the geomagnetic activity level was opposite to the *Es* layers typical to the middle ionosphere: probability of appearance of the polar types of *Es* layers decreases with increase in geomagnetic activity. These studies were very important for understanding of the physical processes in the polar ionosphere and they had no analogies in international studies of that time. Unfortunately all these achievements of the scientists of the AARI had very few reflections in international scientific publications. One can mention the sole paper devoted to this subject – Driatskiy and Besprozvannaja (1958). The further studies based on the analysis of ionospheric observations, made at the ice drifting stations, expanded our knowledge about various complex physical processes in the Arctic ionosphere not only in the near-Pole region but also in the zone of the maximum auroral activity (auroral oval area) located at lower latitudes.

Information accumulated in various ionospheric observations in the Arctic, especially at the ice drifting stations, was used by the scientists of the AARI for developing special methodical documents for the practical organization of radio communication in the Arctic as well as for the prediction of conditions favourable for performing it (Gorbushina *et al.*, 1969). These findings and methods allowed us to organize a special system of radio navigation in the Arctic, which was widely used by aeroplanes and vessels. This system was also used for tracking the positions of some ice drifting stations. For example, during operation of the 'NP-2' station (August, 1950–March, 1951) the radio transmitter at the station was permanently recorded by two remote bearing-finding stations of 'DAN-2' type. Determination of the station coordinates by this radio method was permanently made while the

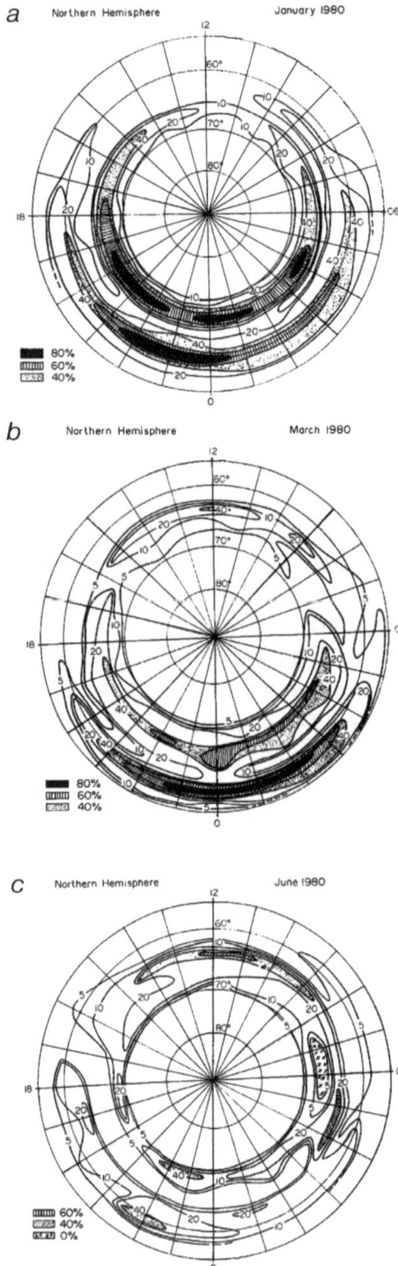

Figure 5.3. Spatial–temporal distribution of the *F*-spread events of the high-latitude ionosphere in the Arctic in winter (*upper panel*), summer (*lower panel*) and equinox (*middle panel*) for the epoch of minimum solar activity. The coordinates are geomagnetic latitude (concentric circles) – local geomagnetic time (external numbers at the axes). Diffusion intensity is shown by different shading.

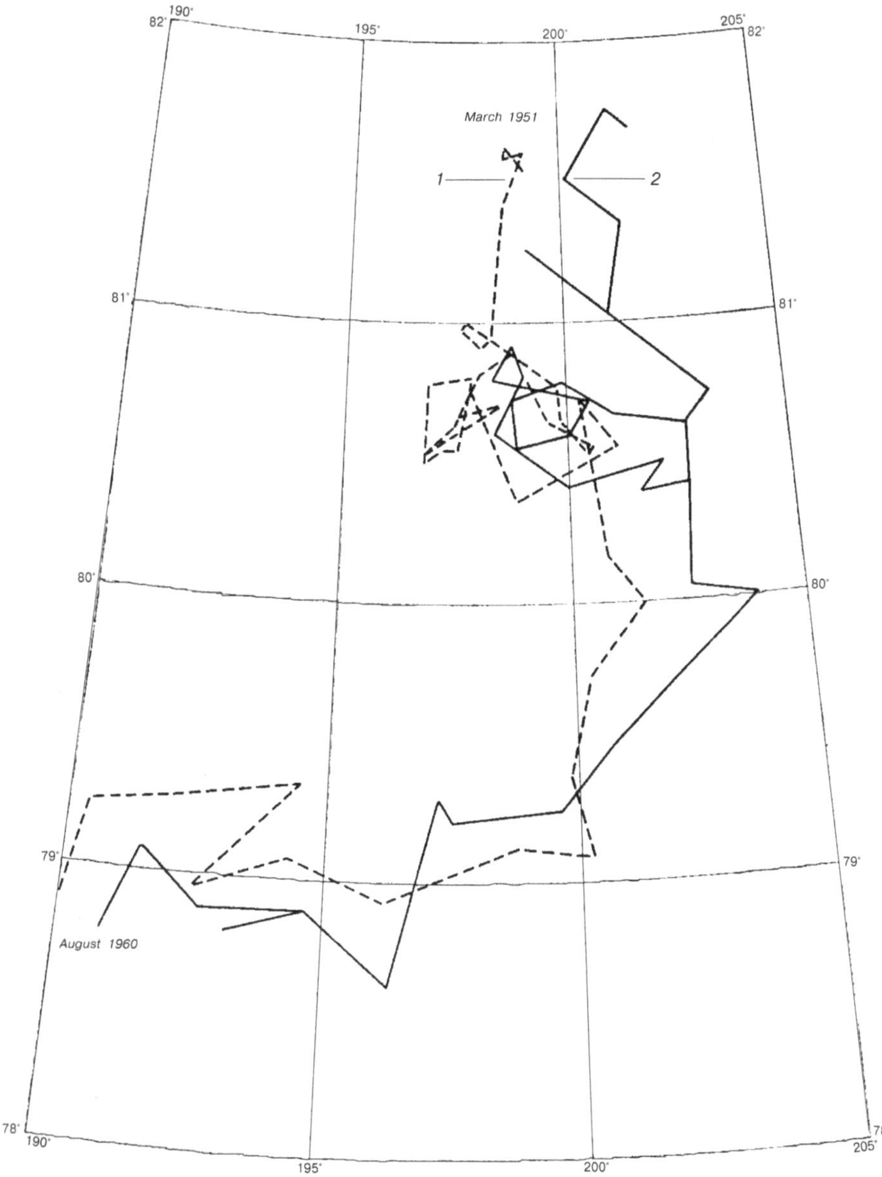

Figure 5.4. Trajectory of drift of 'NP-2' in August, 1950–March, 1951 obtained from astronomical observations (curve 1) and from radio navigation measurements (curve 2). Geographic coordinates are shown.

astronomical method (by the stellar observation) was possible only in cloudless periods. Comparison of these two methods showed that the accuracy of the radio method was of 1–1.5° (Driatskiy, 1956). Figure 5.4 shows the trajectory of the drift of the 'NP-2' station as determined by astronomical method (curve 1) and by radio method (curve 2). One can see that the both curves coincide pretty well even in their finer details.

5.3.2 Riometer observations

The data of ionospheric vertical soundings describe ionospheric conditions at altitudes above 100 km. It means that information about the lower layer of ionosphere – the *D* region (altitudes 50–100 km) must be provided by other technical methods. The lower ionosphere at high latitudes is under almost permanent impact by precipitating fluxes of charged particles (protons and electrons) of solar and magnetospheric origin. These fluxes produce, in the lower ionosphere, geomagnetic disturbances and auroral forms of various intensity, as well as abnormal ionization in the *D* region, exceeding the level of regular ionization at these altitudes by several orders of magnitude. A layer of abnormal ionization in the lower ionosphere effectively absorbs the energy of radio waves passing through that layer and the conditions of radio communication are significantly deteriorated due to this phenomenon. In some cases radio communication breaks down completely (effect of so called blackout). Vertical ionospheric stations do not register any reflections from the ionosphere in such cases. Information about ionospheric conditions during blackouts can be obtained by means of a riometer – an indicator of the opacity of the ionosphere. This technically simple, but very effective device, is actually a receiver recording the level of galactic cosmic rays – radiation at frequencies of 20–50 MHz which penetrate the Earth's ionosphere. If a layer of abnormal ionization exists in the lower ionosphere then the level of receiving signal decreases considerably. The degree of this decrease is a measure of the absorption of radio waves, which is proportional to the value of electron density in the layer of abnormal ionization.

Absorption magnitude indicated as A can be calculated by the following expression:

$$A = -10\lg I/I_0$$

where I_0 is the level of the received signal at the receiver input under a quite period and I is the level of received signal during a moment of measurement.

Riometer absorption can be expressed through a special equation, which includes an electron density value N_e and a frequency of electron collisions of ν:

$$A = 0.46\int N_e\nu\, dh/[\nu^2 + (\omega \pm \omega_h)^2]^{0.5}$$

where h is the length of radio wave in the layer of absorption; ω is the angular frequency of the operating radio wave of the riometer; and ω_h is the electron gyrofrequency.

So, it is possible to transform a magnitude of riometer absorption into a value of electron density in the layer of abnormal ionization in the lower ionosphere. This value of electron density obtained from riometer measurements can be compared with calculated values of electron density derived from a condition of electrodynamic equilibrium:

$$N_e = (q/\alpha_{\text{eff}})^{0.5}$$

where q is the rate of ionization from a special source (solar UV radiation, corpuscular radiation, etc.) calculated by various methods and α_{eff} is the coefficient of effective recombination in the lower ionosphere.

The experts of the AARI developed the original type of riometer, which was the first unit of its kind in the USSR and was used it for ionospheric measurements at high latitudes (Driatskiy, 1974). The same experts developed the original methods of reduction and analysis of the riometer data for both pragmatic and scientific purposes. The comparatively simple operation required for riometer measurements and its very low consumption of electrical power allowed one to use a riometer widely in ionospheric studies at the 'NP' ice drifting stations. The very first regular riometer observations were performed at the 'NP-10' station in 1963–1964. The main types of abnormal absorption of radio waves at high latitudes are the so-called auroral absorption produced by the fluxes of magnetospheric electrons with energy $E_e > 30$–40 keV and the so called absorption in the polar cap (PCA) produced by the solar cosmic rays (mainly protons with energy $E_p > 5$–50 MeV).

The balance of ionization in the D region of the polar ionosphere changes greatly during the appearance of both types of abnormal ionization. During quiet (undisturbed) conditions, balance of ionization of the polar D region (shown in Table 5.1) is similar to the corresponding conditions of the lower ionosphere in middle latitudes due to the same sources of ionization: solar UV radiation in the Lyman-α line ($\lambda = 121.6$ nanometers) and galactic cosmic rays. Joint impact of these two sources determines significant changes in the ionization balance during a day (ionization is greater during daytime than during night-time) and not so great in the cycle of the solar activity (ionization is greater during an epoch of high solar activity).

The most frequent type of abnormal absorption of radio waves in the polar ionosphere is the auroral absorption (AA). The main source of ionization in this

Table 5.1. Balance of ionization in the undisturbed ionospheric D region during a mimimum of the solar activity at a geomagnetic latitude of 60°N.

H (km)	Rate of ionization (el/cm^{3}s): Solar radiation in the Lyman-α line	Rate of ionization (el/cm^{3}s): Galactic comic rays	Total rate of ionization (el/cm^{3}s)	Electron density (el/cm^3): Daytime conditions	Electron density (el/cm^3): Night-time conditions
40	–	2.0	2.0	–	–
50	–	5.4×10^{-1}	5.4×10^{-1}	–	–
60	–	1.8×10^{-1}	1.8×10^{-1}	20	–
70	6.0×10^{-2}	5.0×10^{-2}	1.1×10^{-1}	80	–
80	2.0×10^{-1}	1.1×10^{-2}	2.21×10^{-1}	2.8×10^2	20
90	4.4×10^{-2}	1.4×10^{-3}	4.41×10^{-2}	3.8×10^3	2.0×10^2
100	4.6×10^{-2}	1.9×10^{-4}	4.6×10^{-2}	2.6×10^4	3.0×10^3

Table 5.2. Balance of ionization in the polar lower ionosphere during the appearance of geomagnetic storms and the AA events.

	Geomagnetic storms with sudden commencements S_c		Moderate (3 dB) auroral absorption	
H (km)	Rate of ionization (el/cm^{3}s)	Electron density (el/cm^3)	Rate of ionization (el/cm^{3}s)	Electron density (el/cm^3)
60	1.10^1	6.10^2	–	–
70	3.10^2	4.10^3	1.10	2.10^2
80	3.10^3	5.10^4	4.10	8.10^3
90	4.10^4	7.10^4	5.10^2	9.10^4
100	2.10^4	4.10^5	2.10^4	4.10^5
110	7.10^3	2.10^5	1.10^4	2.10^5

type of absorption is magnetospheric electrons with energy $E_e = 20$–40 kiloelectron Volts (keV) or more, which precipitate into the ionosphere from the night side of the magnetosphere. Balance of ionization in the polar lower ionosphere during the appearance of geomagnetic storms and AA events is shown in Table 5.2. With the start of the riometer observations it was possible to make quantitative evaluations of AA and determination of the synoptic maps of its spatial–temporal distribution in the high-latitude ionosphere. Significant contribution in this type of ionospheric research was made by the analysis of observations made at the 'NP' ice drifting stations, which allowed one to determine accurately the high-latitude boundary of the AA zone. The scientists of the AARI headed by V. M. Driatskiy made important contributions to the studies of this type of abnormal absorption. The summary of these studies are given in Driatskiy (1974) and Zhulina *et al.* (1983).

The balance of ionization of the polar lower ionosphere during AA events differs greatly from that during undisturbed periods.

It is clear that electron density in the polar lower ionosphere increases greatly during geomagnetic storms and AA events as compared with undisturbed periods. This increase is especially notable at altitudes of 70–80 km. It is worthy to indicate that intensity of absorption in some AA events can be as great as 10 dB and radio communication under these circumstances would be impossible due to total absorption of radio wave energy in the layer of abnormal ionization.

The spatial–temporal distribution of the auroral absorption zone in the Russian sector of the Arctic as functions of the geomagnetic latitude, the cycle of solar activity and daytime is given at Figure 5.5 (Zhulina *et al.*, 1983). Probability of appearance of AA events with an intensity of more than 0.5 dB for periods 04 (a), 12 (b), and 20 (c) of local geomagnetic times for spring months (maximum of seasonal variations) in different epochs of solar activity are shown in this figure. The polar boundary of the AA zone was determined mainly from observations taken at different 'NP' stations in 1964–1970. Figure 5.5 data show that a maximum of the AA zone is located in a comparatively narrow latitudinal belt (64°–68°). The position of this zone slightly changes as a function of daytime, season, cycle of the solar

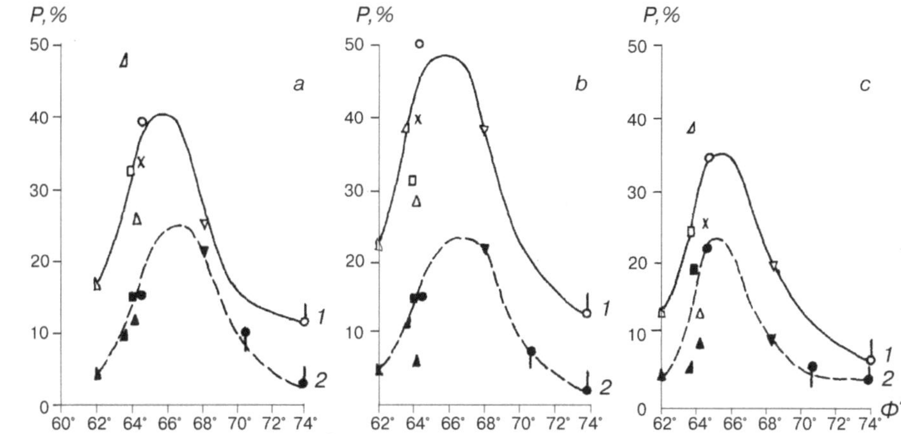

Figure 5.5. Probability (in percent) of the appearance of an AA event with intensity $A > 0.5\,\text{dB}$ in the Arctic in spring as a function of geomagnetic latitude. Curve 1 – epoch of maximum solar activity, curve 2 – epoch of minimum solar activity.

activity and geomagnetic activity level. Average duration of a sole AA event is equal to about 30 minutes. However, there are periods when the AA events appear one after another within several hours.

Another type of abnormal absorption of radio waves in the polar lower ionosphere is absorption in the polar cap (PCA). These events appear after solar flares, which produced an injection of solar cosmic rays into interplanetary space. The solar cosmic rays consist of protons with energy from one to thousands of Megaelectron-Volts (MeV) as well as from nuclei of Helium and other heavy elements. The sequence of events during PCA cases is as follows: (a) an optical flare on the visible limb of the Sun produces a sudden ionospheric disturbance (SID), which is a result of solar X ray radiation. After this, a noise storm (of IY type) in radio emissions of the Sun appears. These events are precursors of the start of the PCA events and they are distinguished easily by ground-based observations. (b) Several hours after the optical flare, intense absorption of radio waves begins in a wide range of frequencies (from tens of kHz up to hundreds of MHz) over the entire polar regions of both hemispheres (geomagnetic latitudes above ~60°). (c) The amplitude of absorption of radio waves damps down to its background value during several days after the start of the event depending of degree of the Sun illumination of the ionosphere. As a rule a geomagnetic storm with sudden commencement (S_c) starts after 1–2 days of the event start. During a storm, the equatorial boundary of the PCA zone expands down to more low latitudes and the duration of the event becomes longer.

Balance of ionization in the polar lower ionosphere during two PCA events of different intensity is shown in Table 5.3. The chosen PCA events are: (a) the event of 4 August, 1972 (amplitude of absorption more than 12 dB, total solar illumination of the Arctic ionosphere); (b) the event of 2 November, 1969 (amplitude of absorption ~7 dB, condition of the polar night in the Arctic).

Table 5.3. Balance of ionization in the polar lower ionosphere during two PCA events of different intensity.

	Moderate PCA event of 2 November, 1969 ($A = 7$ dB)		Intense PCA event of 4 August, 1972 ($A > 12$ dB)	
H (km)	Rate of ionization (el/cm^{3}s)	Electron density (el/cm^3)	Rate of ionization (el/cm^{3}s)	Electron density (el/cm^3)
40	6×10^2	1×10^2	4×10^4	1.5×10^2
50	8×10^2	1×10^3	3×10^4	8×10^3
60	9×10^2	1×10^4	2×10^4	2×10^4
70	6×10^2	2×10^4	8×10^3	5×10^4
80	2.5×10^2	2.5×10^4	2.5×10^3	6×10^4
90	6×10^1	6×10^4	1×10^3	7×10^4
100	1.5×10^1	1×10^5	2×10^2	1×10^5

Comparing the data of Tables 5.2 and 5.3 one can see that the auroral electrons produced abnormal ionization at higher altitudes than solar cosmic rays. Intensity of the absorption during the PCA events is a magnitude higher than the auroral absorption. Furthermore the AA events are preferably local disturbances while the PCA events cover the entirety of the polar regions of both hemispheres. The special calculations show that the ionospheric effects produced by the intense PCA events could be compared with the correspondent effects produced by the powerful nuclear blasts in the Earth's atmosphere.

A total number of recorded PCA events demonstrates clear cyclic variations: probability of appearance of the PCA events is maximum in years of high solar activity while during epochs of minimum solar activity this probability decreases significantly. There are several years with no recorded PCA events (1964, 1994). Seasonal distribution of PCA event appearance is more or less uniform, however, the month of December demonstrates minimal probability of the PCA event appearance. Amplitude of the ionospheric absorption during the PCA events is maximal at the local noon and minimal at the local midnight. The ratio of these two amplitudes of absorption is equal to $\sim$4–6. This peculiarity of diurnal variations of ionospheric absorption during PCA events naturally affects the condition of radio wave propagation: these conditions are much better during night-time compared with daytime.

A special peculiarity of the PCA events is the effect of the so-called midday recovery, when the amplitude of absorption decreases significantly at the hours near local geomagnetic noon. This effect, observed in the low-latitudinal part of the PCA precipitation zone, could improve conditions of radio wave propagation events during the polar day when the polar lower ionosphere is fully under the Sun's illumination.

Figure 5.6 shows the typical variations of ionospheric absorption together with changes of the fluxes of solar protons during the PCA event of 11–18 April, 1969 (Zhulina *et al.*, 1983). Riometer measurements at a frequency of 32 MHz were performed at the ice drifting station 'NP-16' ($\varphi = 81.7°$N; $\lambda = 177.2°$E, geomagnetic

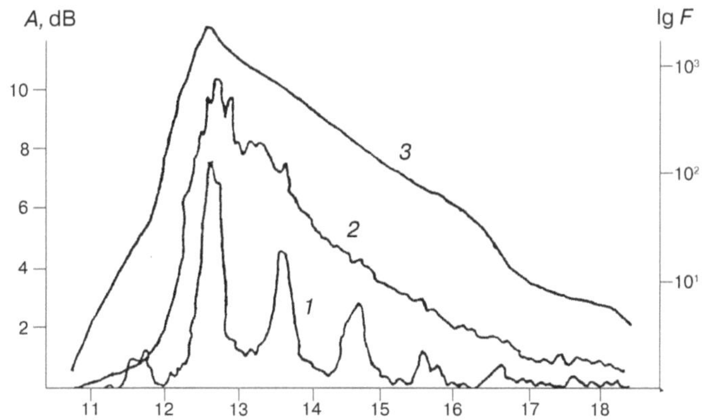

Figure 5.6. Temporal variations of intensity of the polar cap absorption event in April, 1969 from riometer observations at the Antarctic station Vostok (curve 1) and 'NP-16' (curve 2). Curve 3 shows simultaneous variations of the solar proton flux with energy $E_p > 10\,\text{MeV}$ recorded by the satellite 'Explorer-10'.

latitude 77.0°N) and at the Antarctic station Vostok ($\varphi = 78.4°\text{S}$; $\lambda = 106.8°\text{E}$, geomagnetic latitude 84.3°S). Intensity of the solar protons with an energy $E_p > 10\,\text{MeV}$ is given by measurements made at the satellite 'Explorer-34'. Ice drifting station 'NP-16' was located in the territory of the polar day and consequently variations of riometer absorption at this station (Figure 5.6, curve 2) were absolutely identical to variations of the solar proton fluxes (Figure 5.6, curve 3). The Vostok station was located in the region with changing solar illumination where the height of the Sun above the horizon changed between positive values at local noon ($h = 3°$) and negative values ($h = -20°$) at local midnight. Therefore, diurnal variations of absorption at the Vostok station demonstrates the 'day/night' effect when the ratio of the daytime absorption to the night-time absorption was equal to ~5. It is worthy to note that the noon values of absorption were very close at both stations located in opposite hemispheres.

The very high degree of correlation between intensity of the solar proton fluxes and values of riometer absorption at the near-Pole stations gives an excellent opportunity to make fruitful exploration of the physics of the solar terrestrial coupling. The results of these studies made by scientists of the AARI can be found in recently published scientific books (Driatskiy, 1974; Troshichev (ed.), 1986).

Riometer observations performed at the ultra-short radio wavelength (radio waves less than 10 meters) used as radio emissions from interplanetary space as a source. Besides that, the experts of the AARI made, at the ice drifting stations, very interesting measurements from the remote television transmitters operating at the same frequencies of the ultra-short wave range. For example, such observations were made at stations 'NP-6' and 'NP-7'. In October–November, 1957, when the station 'NP-7' was located in the region of 86.7°N, 180.0°E signals from the Russian television stations in Vladivostok and Khabarovsk were received regularly by the engineers at this station. Sometimes signals from several Japanese television stations were also received. In December 1957, the signals of several European television stations were received sporadically. The maximal distance between the transmitting and receiving points in these cases was more than 4,000 km, which exceeds the maximal distance for usual ultra-short wave communication.

The best conditions for receiving remote television signals were during near-noon hours (local time) (i.e., during the maximum ionization in the $F2$ layer of the local ionosphere). The most favourite condition for receiving television signals was the presence of multiplicative reflections from the $F2$ layer with its altitudes above 250 km and crucial frequency f_0F2 more than 8 MHz. These conditions were typical in 1957, which was the year of maximum solar activity. During magnetically disturbed periods the receiving of the television signals was impossible.

Experience obtained during observations of the remote television signals made at the ice drifting stations does not lose its scientific importance even today.

5.3.3 Oblique sounding of the ionosphere

Radio communication at short radio waves (HF communication) is under strong impact from the various high-latitude ionospheric disturbances, which means it has a decreased reliability. Specific features of the high-latitude ionosphere are the great variability and great range of changes of its parameters. Due to this it is very difficult to develop a reliable method of forecasting the conditions of high-latitude HF radio communication. Therefore, the problem of organization and planning HF radio communication in the Arctic becomes a very complicated task.

One of the most effective methods of improving this situation is by establishing a wide network of oblique soundings of the polar ionosphere. An ionogram of oblique ionospheric sounding differs significantly from its counterpart of vertical sounding (Figure 5.1). The former contains the following information: maximum usable frequency (MUF), lower usable frequency (LUF), modal structure of the HF signal, time delays between the different modes of the signal, abnormal modes of propagation, etc. A radar of oblique sounding of the ionosphere provides the possibility to observe the current condition of the ionospheric radio channel in a regime of real time and the tendency of its change, which is extremely important for reliable short-term forecasting of its conditions. These features of oblique ionospheric sounding are extremely important during the strong high-latitude ionospheric disturbances. Another advantage of this method is the possibility of receiving the signals of a sole transmitter by several spatially displaced receivers, which allows one to study conditions of HF radio communication in the vast territories where permanent points of ionospheric observations are absent.

The scientists of the AARI were pioneers in Russia of developing the techniques of oblique ionospheric sounding and its practical use. F. Ya. Zaborshchikov developed the original 'Ionospheric Probe' at the end of 1950s – designed to test conditions of HF radio wave propagation in the Arctic by means of oblique incidence sounding. Unfortunately, the level of technical industry at that time was comparatively low, which prevented use of this invention for practical purposes. In 1968 experts of the AARI developed the very first (in USSR) industrial radar of

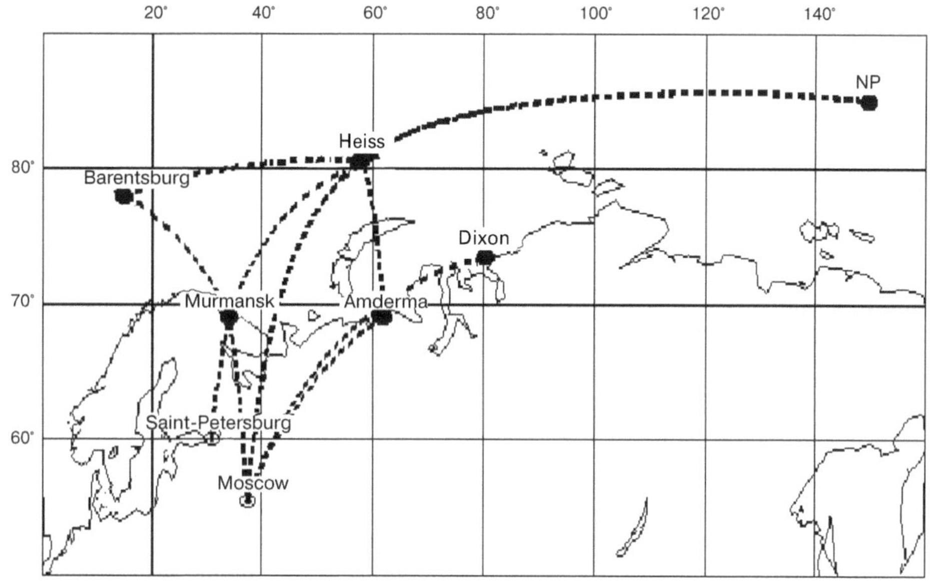

Figure 5.7. Map demonstrating the locations of the Russian oblique sounding radio paths in the Arctic.

oblique incidence sounding – operating in the frequency range 3.5–27.5 MHz. A description of this radar, its technical parameters and the results of its first practical use are published in Smirnov (1972a). This achievement was an important step forward in the exploration of the polar ionosphere in the USSR and in the practical improvement of high-latitude radio communications. Most of the works connected with research, development, field testing and industrial production of this radar was made by the experts of the AARI headed by V. M. Driatskiy and V. B. Smirnov.

The next step after the appearance of the first units of radar of oblique sounding was their practical use at different radio paths in the Russian Arctic. This work was made under the control of the experts of the AARI. An important part of this program was the establishment of a receiver of oblique sounding at the ice drifting station 'NP-22' in 1974. The first radio path for studies in this program was the link between Moscow and 'NP-22'. In 1977 another link, this time between Heiss Island and 'NP-22' was put into operation. In 1978 the link between Murmansk and 'NP-28' started to operate. Figure 5.7 shows the map with a disposition of acting radio paths of oblique sounding in the Arctic. Since the positions of the ice drifting stations have changed with time their locations were indicated at definite time intervals. The transmitting points of the system are indicated on the map by arrows. This program of study of HF radio wave propagation in the central Arctic continued up to the beginning of the 1990s, when all operations at the ice drifting stations were stopped. So, the AARI keeps, in its archives, many years of unique data of ionosphere oblique sounding for the central Arctic.

Table 5.4. Geographic characteristics of the oblique incidence sounding radio paths during operation of the ice drifting station 'NP-28' in January, 1988.

	Geographic coordinates			
Oblique sounding radio path	Transmitter	Receiver	Midpoint of radio path	Length of radio path
Moscow–'NP-28'	55.6°N, 38.9°E	87°N, 175°E	74°N, 61°E	4,650 km
Heiss Island–'NP-28'	80.6°N, 58°E	87°N, 175°E	82°N, 117°E	1,900 km
Murmansk–'NP-28'	68.9°N, 33.1°E	87°N, 175°E	81°N, 70°E	3,350 km

Table 5.4 demonstrates the geographic characteristics of the oblique incidence sounding radio paths during operation of the ice drifting station 'NP-28' in 1978.

The radio paths indicated in Table 5.4 were located in different geophysical zones. The path Moscow–'NP-28' was a typically transauroral link since it crossed the auroral oval and its terminal points were located in the middle latitudes (Moscow) and in the polar cap (NP-28). Reliability of radio wave propagation at this path was controlled by ionospheric conditions at the midpoint of the path located at the centre of the auroral oval. The path Heiss Island–'NP-28' is entirely located inside the polar cap region, which means that it will experience the influence of typical auroral disturbances at a much lesser extent. However, it will be under strong influence of the PCA events (Smirnov, 1972a). Another feature of this link is a comparatively low level of industrial radio noise. This factor leads to a significant increase of the ratio 'signal-to-noise' at the receiver input, which leads to better communication. However, the strong snow winds, which are the permanent feature at polar latitudes, increase electrostatic radio noise and deteriorate the quality of the receiving signal.

Radio path Murmansk–'NP-28' is intermediate in relation to the previous paths. Its terminate points are located in the auroral oval (Murmansk) and in the polar cap ('NP-28'). Midpoint of this path is located inside the polar cap but a significant part of this path is located in the auroral oval due to which this path is under the impact of auroral disturbances.

There are some special aspects of the organization of oblique sounding signals at the ice drifting station. The ice floe constantly changed direction and speed of drift. So, the distance between receiver and transmitter also changed as well as the direction to the transmitter. Construction of the antenna system of the receiving point must take into account these factors. If the path is long enough these factors influence the quality of the receiving signal insignificantly. In order to compensate losses of the signal strength due to a change of the ice floe drift direction four mutually-prependicular V-shaped antennas were erected at receiving points, and an appropriate antenna was switched on to receiver in dependence of the drift direction. So, it was possible to find the optimal direction to a transmitting point (Lukashkin *et al.*, 1978).

The standard regime of transmitting signals of oblique sounding was once an hour for the whole day. Registration of the signal at the receiving point was made on the film. Current information about ionospheric conditions is transmitted regularly to centres of radio service.

Briefly, let us consider the main scientific results obtained during the fulfillment of the program of oblique sounding of the ionosphere in the central Arctic. This method allows one to get a real physical picture of the ways radio waves propagate in a given frequency range through the ionosphere. At the long-distance radio path (Moscow–'NP-22') the combined modes of M-type were observed when reflection took place from the $F2$ layer, after that – from the Es layer and once again from the $F2$ layer. Sometimes the modes of N-types were observed, when reflection took place consequently from the $F2$ layer and Es layer. These modes were typical for summer months. It was found that existing methods of calculations could not reproduce these complicated modes of radio wave propagation (Driatskiy *et al.*, 1978b; Vovk *et al.*, 1978; Vovk and Shirochkov, 1978). In wintertime the mode structure of the signal is comparatively simple and consists of double reflections from the $F2$ layer. In one scientific paper (Vovk *et al.*, 1978) variations of the multiray propagation of radio waves (time delays of the different modes of the signals) are investigated for the radio path Moscow–'NP-22'. It was shown that the delays are a minimum in wintertime while in summer they are the greatest especially in the morning and noon hours. The most important characteristics of the HF radio channel are the boundaries of the frequency range (MUF and LUF), which depend on ionospheric conditions and are a function of many geophysical factors. This parameter was the main subject of study by means of oblique sounding made by the scientists of the AARI.

Figure 5.8 shows the diurnal variations of this parameter (MUF and LUF) for the radio path Moscow–'NP-22' for two months (January and July) of 1975, which was a year of minimum of solar activity. Thin lines correspond to everyday values of each parameter while the thick lines correspond to the median values of these parameters. One can see that during the summer months of the year of minimum solar activity the diurnal variations of the frequency range are small and the range of usable frequency is rather wide over the whole day. In the wintertime the frequency range was rather narrow at the noon period while during evening and night-time only casual opportunities for the propagation of radio waves existed.

Figure 5.9 shows the diurnal variations of the frequency range (MUF and LUF) for radio path Heiss Island–'NP-22' for December, 1979 (maximum solar activity). One can see that the Es layer plays important role in the propagation of radio waves at this path (Figure 5.9, upper panel). The diurnal variations of the MUF and LUF parameters are practically negligible in the dark ionosphere.

Absorption of radio waves in the lower ionosphere is one of the main factors determining conditions of propagation of radio waves at high latitudes. The lowest usable frequencies of HF radio waves depend on the magnitude of absorption of radio waves in the lower ionosphere. The main types of absorption of radio waves at high latitudes are AA and PCA. Analysis of the data of the oblique soundings obtained at the 'NP' ice drifting stations showed that the radio paths Moscow–'NP' and Heiss Island–'NP' are strongly influenced by both types of abnormal absorption (Driatskiy *et al.*, 1978; Vovk *et al.*, 1978). During strong PCA events propagation of the signal at both these radio paths was either absent completely or

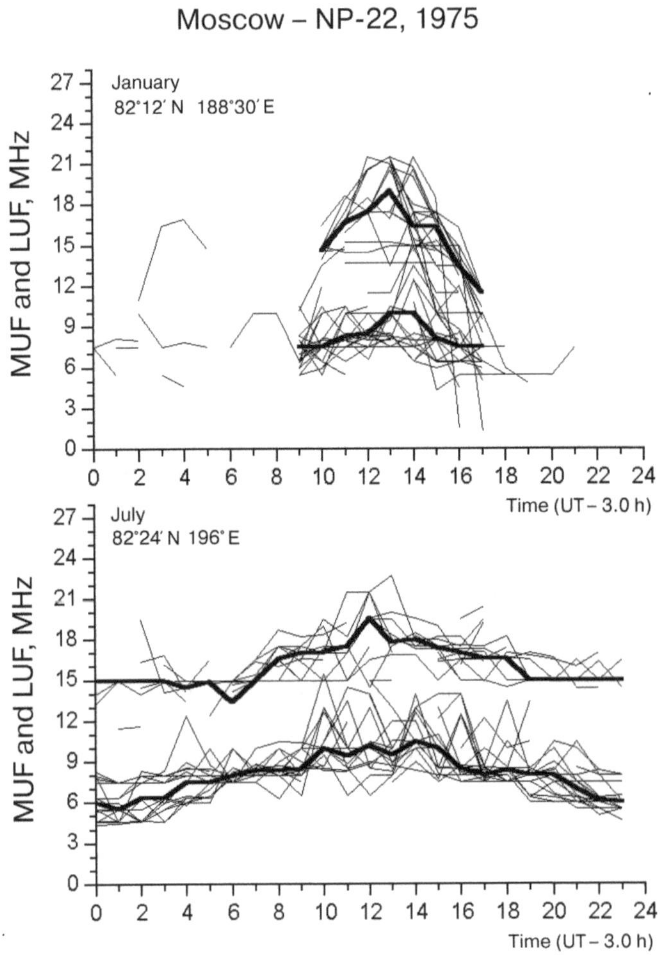

Figure 5.8. Temporal variations of the operating frequency ranges at the oblique incidence radio path Moscow–'NP-22' in the summer of 1975 (a) and winter of 1975 (b).

was observed sporadically for several hours only. The data of oblique soundings obtained at the 'NP' ice drifting stations were used widely for development of the methods of short-term forecasting of HF communication (Khodzha-Achmedov and Kashurko, 1978; Vovk and Shirochkov, 1988; Khodzha-Achmedov and Lukashkin, 1983; Driatskiy *et al.*, 1978b). It was shown that the accuracy of the short-term forecasting of HF communication depends on the position of the midpoint of the HF radio path relative to the position of the main structure of the high-latitude ionosphere. Error of the forecast increases if the midpoint of the radio path is located inside the main ionospheric trough and decreases if this point is located inside the auroral oval.

The data of oblique sounding obtained at the 'NP' ice drifting stations were used

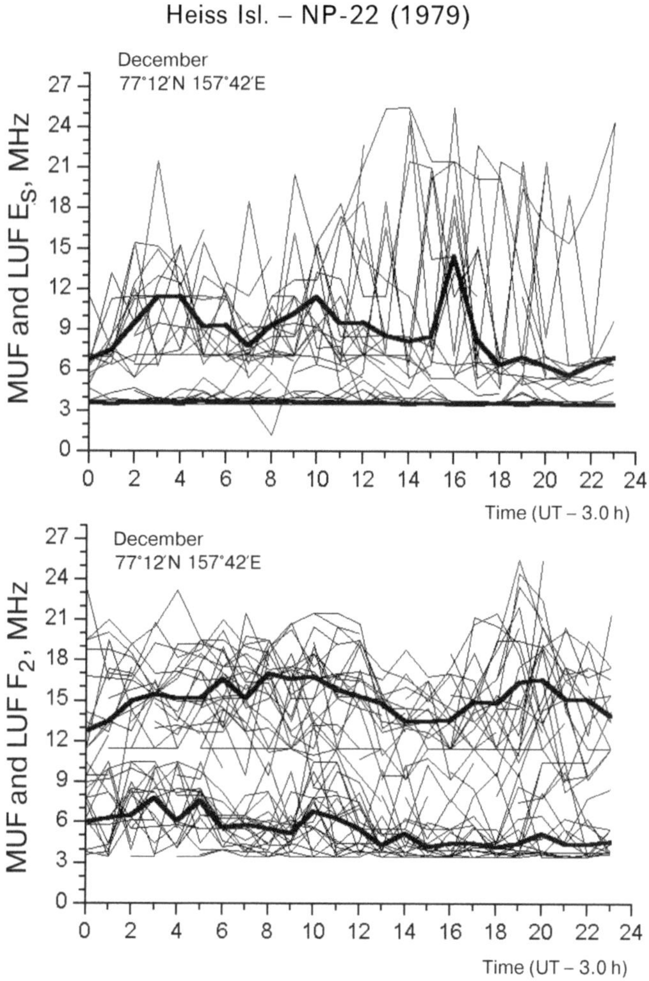

Figure 5.9. Temporal variations of the operating frequency ranges at the oblique incidence radio path Heiss Island–'NP-22' in December, 1979. (*upper panel*) Variations of the LUF and MUF for the *Es* layer. (*lower panel*) Variations of the LUF and MUF for the *F2* layer.

for studies of the features of the polar cusp and dynamics of ionospheric substorms (Egorova and Lukashkin, 1985; Egorova and Lukashkin, 1991; Egorova, 1988).

The data of the oblique soundings obtained at the ice drifting stations contain very valuable scientific information, which has not been studied carefully up to now. The experts of the AARI use this information for validation and development of high-latitude HF radio channel models (Khodzha-Achmedov and Ignatov, 1991; Blagoveshchensky *et al.*, 1989; Blagoveshchensky *et al.*, 1991). Figure 5.10 shows the results of comparison of experimental and modeled values of the MUF at radio path Moscow–'NP-22' for October, 1975. The conclusion can be made that taking

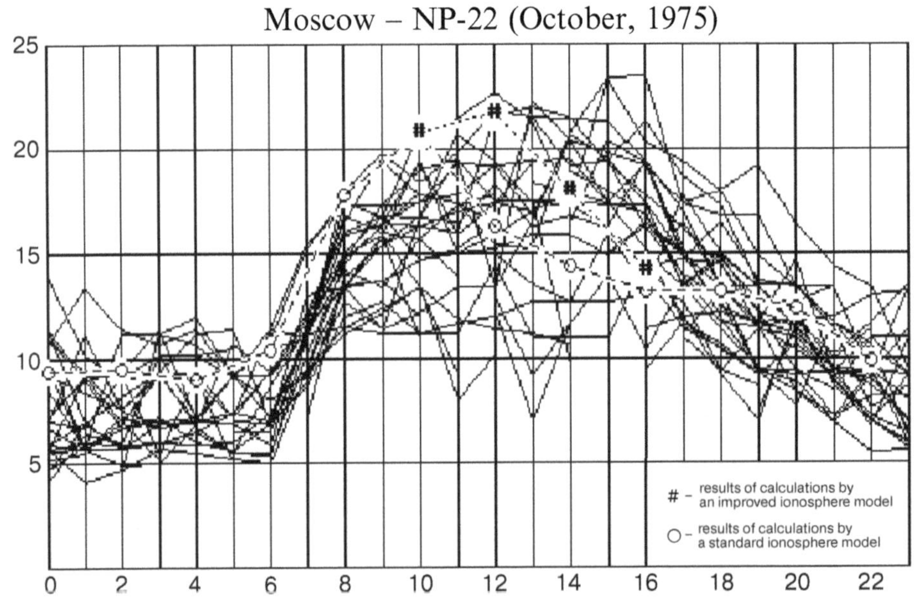

Figure 5.10. Comparison of the experimental MUF $F2$ values obtained at radio path Moscow–'NP-22' in October, 1975 with the same parameter obtained by model calculations.

into account additional gradients of electron density allows one to describe satisfactorily diurnal variation of the operating frequencies at this radio path.

Figure 5.11 shows the results of calculations of the trajectories of the HF radio signal at the radio path Moscow–'NP-22' for 15 October, 1975. Calculations were made for the operating frequency 17 MHz at 12 hours local time and for sunspot number $W = 90$.

5.3.4 Observations of very low frequency signals

Various ionospheric observations performed at the 'NP' ice drifting stations included observations of the signals of remote very low frequency (VLF) transmitters operating at frequencies 1–30 kHz. Attention to studies of the propagation of VLF radio waves was stipulated by extensive development of the various systems of radio navigation in the middle of the last century. Very high stability of phase of the VLF signals even during strong geophysical disturbances stimulated very extensive use of VLF radio waves for long-distance communication, transmitting of the standard frequencies as well as the signals of the precise time, development of the accurate radio navigation system and solving the other problems including detection of the atmospheric nuclear blasts. The vast territory of the USSR naturally stimulated the importance of using these radio waves for various purposes.

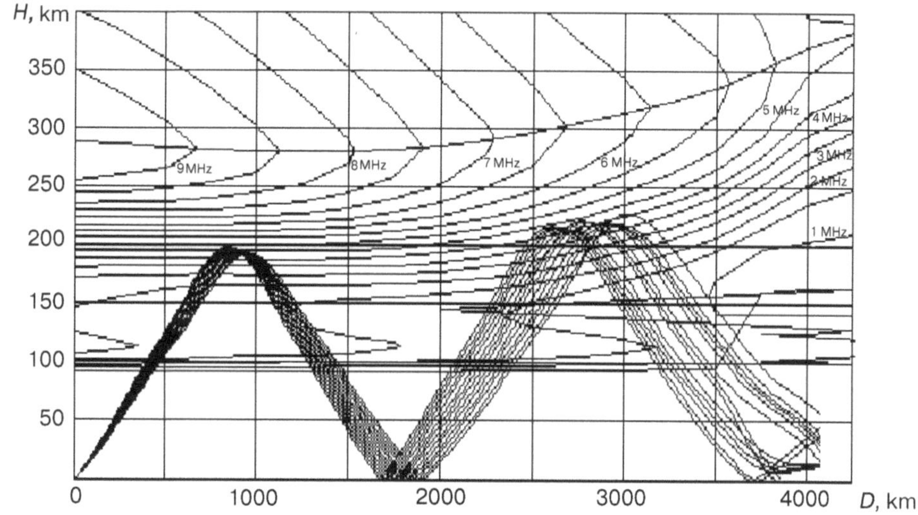

Figure 5.11. Results of the model calculations of the HF radio wave trajectories for radio path Moscow–'NP-22' at 12 hours Moscow time on 15 October, 1975. Operating frequency was 17 MHz.

VLF radio waves propagate in the waveguide 'lower ionosphere–ground surface' and conditions of propagation of these radio waves depend on electroconductivity of both ionosphere and ground surface. These factors play a decisive role in the spatial distribution of the amplitude and phase of the VLF radio signals. If these signals propagate above the ocean water (almost ideal conductor) a minimal decrease of the VLF signal takes place. In the case of propagation above such kinds of ground surface like permafrost, tundra, snow cover with low temperature and grain structure, significant absorption of the VLF signal takes place.

The measurements of signals of remote VLF transmitters at slowly moving platforms such as the 'NP' ice drifting station is a very interesting and important experiment due to the almost unknown situation for propagation of these radio waves in the central Arctic.

During operation of 'NP-13' (August, 1965–January, 1966; May, 1966–March, 1967) the permanent measurements of the amplitude of the VLF transmitter of station NPG/NLK (Seattle, USA) were made by experts of the AARI (Fedjakina, 1980). The observations were made at a frequency of 18.6 kHz. During this time 'NP-13' was located at geographic latitudes of more than 80°N. It is interesting to note that during the second period the station was very close to the point of the North Geographic Pole.

The data obtained during observations of the VLF signals made at 'NP-13' gave the possibility of studying the impact of the different geophysical disturbances on the conditions of propagation of VLF radio waves. Unfortunately the technical problems did not allow the registration of the phase variations of the VLF signals.

Comparison of the amplitude variations of the VLF signal averaged for five magnetically quiet days, with their counterparts averaged for five magnetically

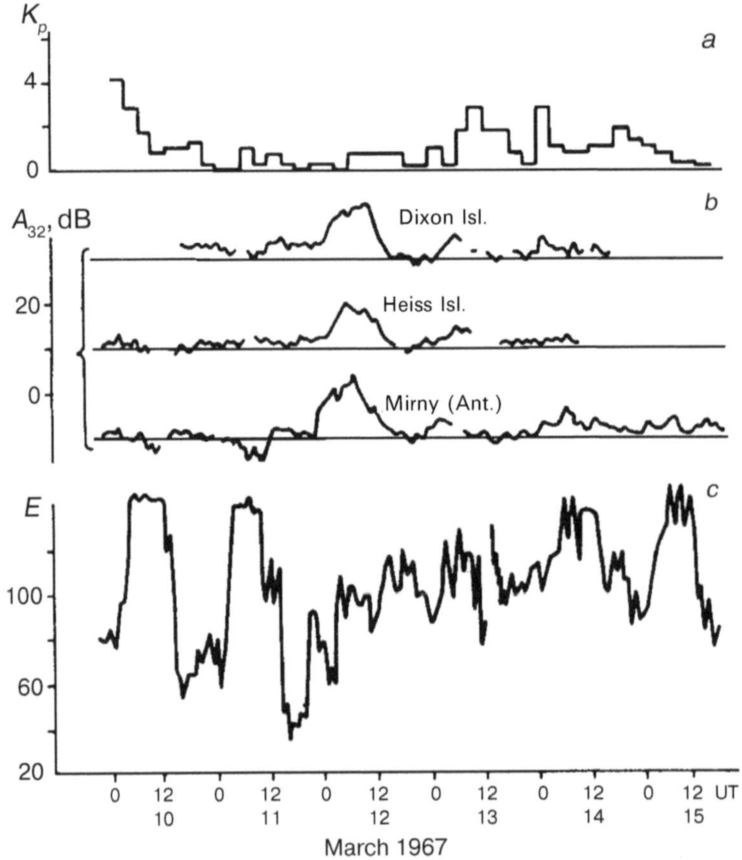

Figure 5.12. Temporal variations of the planetary index of geomagnetic activity K_p (a), values of riometer absorption recorded at 'NP-13' (b) and amplitude of the VLF signal transmitted by NPG/NLK station at a frequency of 18.6 kHz and measured at 'NP-13' (c) during the PCA event on 15 March, 1967.

disturbed days, does not reveal any notable difference between these two data sets. It is no wonder since 'NP-13' was located inside the polar cap region during the whole cycle of the observations – with very small levels of auroral disturbances in this area. On the other hand the PCA events described in the previous paragraphs impact significantly upon the variations of the VLF signal amplitudes. Figure 5.12 shows the results of the VLF signal amplitude measurements performed on 12–14 March, 1967 during the strong PCA event. The data of the measurements made at 'NP-13' during 10–17 March are shown. The upper panel of Figure 5.12 shows temporal variations of the planetary index of geomagnetic activity K_p expressed in special units. The middle panel shows values of riometer absorption in dB recorded at the Arctic stations Dixon (73.55°N; 80.57°E; curve 1), Heiss Island (80.62°N; 58.05°E; curve 2) and Antarctic station Mirny (66.55°S; 93.02°E; curve 3). The lower panel shows variations of the amplitude of VLF signal expressed in some artificial units. The level of the VLF signal for 10 March demonstrates typical undisturbed variations when high night-time amplitudes of the signals are smoothly transformed into comparatively low daytime values. This type of regular variation changed abruptly on 11 March at 10 hours when the level of the signal decreased greatly. It is worthy to note that riometer absorption at all stations demonstrated a later start for the ionospheric disturbances. Variations of the VLF signal returned to its undisturbed values on 14 March only, later than when the riometer absorption recovered to its quiet level. A conclusion can be made that registration of the VLF signal level is a more sensitive detector of the penetration of the solar proton fluxes into lower ionosphere than standard riometer operating at frequency 32 MHz. In fact the real relation between spatial–temporal variations of the riometer absorption and the changes of the phase of the VLF signal during the PCA events is much more complex and it was studied later on at the base of the Russian observations in the Arctic (Shirochkov *et al.*, 1998). The further Russian research of the propagation of the VLF radio waves in the polar ionosphere demonstrated that this type of ionospheric measurements is very effective ground-based method for study of the effects produced by precipitation of the energetic particles of different types.

5.3.5 Explorations of ionospheric structure by means of signals from the Earth's satellites

The launch of the World's first artificial satellite by Soviet scientists in 1957 began the start of studies of near-Earth space by a principally new technical method – satellite observations. The study of the ionosphere was also gained from using satellite measurements. One of them was an interferometer beacon method – a radio transmitter on the satellite radiates a spectrum of coherent frequencies. Receiving units on the ground surface registers phase differences of the coherent radio waves transmitted by satellite passing through the Earth's ionosphere. This information gives the possibility of calculating values of electron density in the Earth's ionosphere along a trajectory of a transmitting radio wave by the use of special numerical methods. Additionally, such important parameters as refraction of radio waves, phase and group lag of the signal, the Doppler shift of the frequency, and angle and velocity of the polarization rotation due to influence of the geomagnetic field can be calculated. This information is vital to expanding our knowledge about ionospheric conditions, which is necessary in order to establish a reliable radio communication with satellites.

One of the main parameters recorded by the standard observations of the signals from the satellites is the depth of amplitude decrease in these signals. This parameter is expressed by relative fluctuations of the amplitude of power of electrical field created by the radio waves. This vector parameter is called the index of scintillations S_4 and is expressed by:

$$S_4 = 1/E^2 [\overrightarrow{(E^2 - E^2)^2}]^{0.5}$$

where E is the strength of the electric field of the radio wave.

Table 5.5. The parameters of the satellite orbits, the periods of the observation and ice drifting station coordinates.

Satellite	Apogee (km)	Perigee (km)	Orbit period (min)	Orbit inclination (degree)	Time of observations	Latitude	Longitude
Cosmos-321	507	280	92	71	February–March, 1970	83°N	203°E
Cosmos-356	600	240	92.6	82	August–September, 1970	78°N	148°E
Cosmos-381	1030	970	105	74	December, 1970–January, 1971	80.5°N	146.4°E
Transit	10,916–11,694	925–1078	105–107	89.6–90.2	April 1981	85°N	158°E

Intensity of this parameter demonstrates definite diurnal variations with maximum in the morning and evening hours (local time). Amplitude of this parameter increases with enhancement of geomagnetic activity (amplitude S_4 is proportional to *Kp* index value). There are no significant solar cycle variations of this parameter. The scintillations of the satellite radio signal are produced by irregularities of the ionosphere of size 3–20 km as well as by travelling ionospheric disturbances caused by many sources including meteorological disturbances.

The statistic characteristics of the satellite signals at the radio paths of satellite–ground surface were measured at the ice drifting stations 'NP-16, 19, 22'. The signals of the following satellites were recorded: 'Cosmos-321', 'Cosmos-356', 'Cosmos-381' (operating frequency 20 MHz) and satellite 'Transit' at frequency 150 MHz (Solodovnikov *et al.*, 1988). The parameters of the satellite orbits, periods of the observations and ice drifting station coordinates are given in Table 5.5.

The results of the measurements indicated that the high level of signal scintillations is a permanent feature of the high-latitude data. Values of the S_4 index for frequency 20 MHz are more than 0.6, as a rule. In the auroral zone values of the S_4 index exceed the correspondent values of this parameters in the polar cap. Variability of this parameter in the auroral oval is much greater than in the polar cap with maximum of changes in August–September.

Usual development of the S_4 index variations during elementary magnetospheric disturbance – substorm as recorded at the 'NP-22' ice drifting station in April, 1981, could be described as following. During a preliminary phase of substorm, several small disturbances of the satellite signal amplitude were observed with values of S_4 index equal to 0.3–0.4. Spectral density of amplitude scintillations actually does not depend on frequency F and is expressed as a uniform spectral noise. During the break-up phase of the substorm, density of fluctuations increases several times, intense packages of waves appear, which are evidence of excitements of various instabilities in the ionospheric plasma. These processes decrease slowly (up to several days) remaining very active even under low levels of geomagnetic activity. The character of variations of satellite signal fluctuations in the Arctic polar cap differ greatly from these events in the auroral and subauroral regions of the Arctic, which suggests a different nature of the physical processes controlling ionospheric conditions in these parts of the Arctic.

Such observations performed in the central Arctic give the possibility of better understanding the complex electrodynamic processes typical for the polar ionosphere during magnetospheric disturbances and to correctly organize the telemetric service of the different satellite complexes.

5.4 GEOPHYSICAL OBSERVATIONS IN THE CENTRAL ARCTIC AS A PART OF THE SYSTEMATIC STUDIES OF SOLAR–TERRESTRIAL PHYSICS

The very first systematic studies of the intense high-latitude geophysical disturbances revealed the solar origin of these disturbances (Ohl, 1969). The powerful energy of such events like the global geomagnetic storm and the solar proton flare (PCA) allowed one to suggest that these geophysical disturbances could influence the character of various meteorological processes in the polar atmosphere. The data of riometer observations showed that the maximum of energy of the most powerful geophysical disturbances dissipates at the altitudes of the mesosphere and stratosphere, sometimes descending even to tropospheric altitudes. Shchuka (1969) studied variations of baric pressure at the ground surface during more than 40 PCA events during the period 1948–1965. Duration of the solar proton fluxes precipitation was determined from the data of the ionospheric observation in the Arctic including measurements made at the 'NP' ice drifting stations. It was shown that the main atmospheric effect of the precipitation of the solar proton fluxes was a drop in the ground surface baric pressure. In several cases this drop of pressure was as great as 25–30 hPa. The region of maximum proton fluxe impact on the baric fields was located in the region of north Canada and the north Greenland Sea (i.e., in the region of the northern polar cap where intensity of the precipitation of the solar protons is maximal). The latest studies of the character of the atmospheric processes during periods of PCA events showed that the times of the spring transformation of atmospheric circulation in the Antarctic stratosphere at levels of 50–100 hPa begin 10–15 days earlier than in the quite periods (Shirochkov and Maurits, 1986). Interesting results were obtained during the scientific expedition on board the atomic ice-breaker *Siberia* to the North Geographic Pole in May–June, 1987. In this expedition the ice-breaker travelled along the geographic meridian 40°E from Murmansk and crossed the regions of the auroral oval and the polar cap. Permanent measurements of the total ozone content (TOC) during this expedition demonstrated a principally different character of the diurnal variations of the TOC in the regions of the auroral oval of the polar cap (Shirochkov and Nagurny, 1992). It is well known that the intensity and variability of the auroral electron precipitation are extremely high in the auroral oval while in the polar cap they are almost negligible.

It was found that the TOC variations in the auroral oval demonstrate great diurnal variations in the auroral oval increasing with an increase of geomagnetic activity level. On the contrary, in the polar cap area these variations were negligible and the total level of TOC was much lower in the polar cap compared with that in the auroral oval. These unique results are direct evidence of the strong influence of the magnetospheric–ionospheric disturbances on the dynamic of the ozone layer in the high latitudes.

In more recent years scientists of the AARI made more intense and large-scale studies of the impact of heliogeophysical disturbances on the hydrometeorological and biological systems in the polar region of the Earth.

6

Conclusion

This book presents evidence on the characteristics of the ocean column, its ice cover, surface atmosphere and ionosphere in the Arctic. The earlier generalized evidence on these parts of the natural environment at high latitudes was presented in the form of cartographic material in the *Atlas of the Oceans. The Arctic Ocean* (1980) and in the *Atlas of the Arctic* (1985). For their preparation and compilation, materials of observations at drifting stations up to 1980 were used.

The authors attempt in this present edition to describe in greater detail both the characteristics proper and the parameters of their variability on different timescales – from day-to-day to multiyear. For this purpose, data of observations at all drifting stations were used: beginning from 'NP-1' in 1937–1938, then 'NP-2' in 1950–1951, and later from 'NP-3' in 1954 to 'NP-31', which ended the continuous cycle of observations in 1991.

The evidence published in the book has, to a great extent, scientific application and, as the authors hope, introduces into scientific circulation a new systematized material on the state of oceanic water, ice cover, near-water/near-ice layer of the atmosphere and near-space characteristics of the recent past.

Section 4.7 devoted to the description of the biotic component of water of the Arctic Ocean is slightly apart from the general context of the information presented in the book. Although these observations were not as systematic as the observations of abiotic characteristics, nevertheless we considered it necessary to include the generalized results of biological studies at the drifting stations into this book. In our opinion this makes them available to a wider range of researchers, involved in issues/problems related to the control of reproduction of the Earth's bio-resources, who investigate their dependencies on environmental parameters.

The authors intentionally ignored or considered to a minor extent the issues of interaction of the natural environmental compartments and their mutual influence. The authors have a clear understanding of a strong relationship and often of a correlation of the processes occurring in these media. Forecasting of meteorological and ice conditions and conditions of radio wave propagation in some or other areas of the Arctic in different seasons of the year is performed exactly on the basis of such established interrelations. The results obtained on the basis of multiyear observations at the 'NP' drifting stations, and published in hundreds of scientific articles and monographs and frequently referred to in bibliography, allowed us to clearly define the role and place of the Arctic region in the global climatic system. The observations at the drifting stations allowed us to detect and describe the systems of atmospheric and oceanic circulation, penetration and spreading of Atlantic water in the Arctic Ocean and to discover the anticyclonic ice circulation in the eastern Arctic. A theory of the ice drift related to the peculiarities of atmospheric circulation was constructed and substantiated.

A distinguishing feature of the Arctic climatic system is that the interannual variability of meteorological parameters in the Arctic is much greater than their interannual variability at temperate latitudes. This is primarily determined by the high-latitudinal location of this region and the related specific features of incoming solar radiation to the surface – existence of the polar day and the polar night. A decisive influence on the radiation regime of the Arctic basin is produced by the presence, over much of the year or during the entire year, of snow-covered ice with a high albedo. The temperature regime of the Arctic basin forms, to a great extent, under the influence of advection of warmer air masses from temperate latitudes; the energy exchange intensity in the ocean–ice–atmosphere system plays a significant role in its formation. These and a number of other factors determine the climate parameters and their variability in the Arctic region.

To date, the role of each of the aforementioned climate-forming factors in the formation of climatic conditions in this Earth's region is more or less clear. However, to take into account their interactions and combined influence on the state of the climatic system, including distant implications, is still a weak area. The estimates of the role of each of the factors and their combined impact on climate will be more reliable if based on the analysis of already available and obtained data. The role of new information is extremely significant.

During the 12 years from 1991, after the end of operation of 'NP-31' until 2003 when observations at NP-32 were resumed, there was no reliable information on the meteorological, ice and oceanographic characteristics in the central Arctic basin. The parameters obtained during the drift of 'NP-32' and measured now at 'NP-33' showed a drastic difference between their modern and regime values obtained over the period 1950–1991.

Resumed operation of the drifting stations, expansion of the volume and types of observations in the Arctic in general envisaged by national and international projects in the framework of holding the International Polar Year 2007/2008, will allow us to obtain new information on the processes occurring in different parts of the Arctic climatic system and to more reliably assess their role in climate change at high latitudes and the contribution of these changes to the global climate changes.

References

Afanasyev, I. F. (1978). Research into the benthic fauna of the central part of the Arctic Basin. *Okeanologiya [Oceanology]*, **18**(5), 950–951 [in Russian].

Afanasyev, I. F. and Filatova, Z. A. (1980). The problem of the investigation of the deep sea benthic fauna in the Canadian Basin of the Arctic Ocean. *Biology of the Central Arctic Basin*, 219–229 [in Russian].

Aleksandrov, Ye. I., Bryazgin, N. N., and Dementyev, A. A. (1999a). *Meteorological Regime of the Arctic Basin: Results from the Drift Station* (Volume I) (344 pp.). Gidrometeoizdat, St. Petersburg [in Russian].

Aleksandrov, Ye. I., Bryazgin, N. N., Dementyev, A. A., Marshunova, M. S., Radionov, V. F., and Svyashchennikov, P. N. (1999b). The meteorological and radiation regime of the Arctic Basin: Results from the 'North Pole' drifting station. *Problemy Arktiki i Antarktiki [Problems of the Arctic and Antarctic]*, **71**, 61–77 [in Russian].

Aleksandrov, Ye. I., Bryazgin, N. N., Dementyev, A. A. and Radionov, V. F. (2004). *Meteorological Regime of the Arctic Basin: Results from the Drift Station, Vol. II: Climate of the Near-ice layer of the Atmosphere* (144 pp.). Gidrometeoizdat, St. Petersburg [in Russian].

Alekseyev, G. V. (1976). On the interdependence between the ice cover and atmosphere in the Central Arctic. *Proceedings of the AARI*, **332**, 109–113 [in Russian].

Alekseyev, G. V. and Buzuev, A. Ya. (1973). Lateral melting of ice in fractures. *Proceedings of the AARI*, **307**, 169–178 [in Russian].

Alekseyev, G. V., Myakoshin, O. I., and Smirnov, N. P. (1997). The variability of ice transport through the Fram Strait. *Meteorologiya i Gidrologiya [Meteorology and Hydrology]*, **9**, 52–57 [in Russian].

Alekseyev, G. V., Bulatov, L. V., Zakharov, V. F., and Ivanov, V. V. (1998). Atlantic water thermal expansion in the Arctic Basin. *Meteorologiya i Gidrologiya [Meteorology and Hydrology]*, **7**, 69–78 [in Russian].

AMAP (2003) *AMAP Assessment 2003: The Influence of Global Change on Contaminant Pathways to, within, and from the Arctic*. Arctic Monitoring Assessment Program, Oslo, 65pp.

Anapolskaya, A. Ye. and Gandin, L. R. (1958). Methodology for determining the wind speed for calculating the wind loads on structures. *Meteorologiya i Gidrologiya [Meteorology and Hydrology]*, **10**, 9–17 [in Russian].

Andreev, M. Yu., Kudryashov, V. M., Lebedev, G. A., and Petnicov, V. G. (1995). Long-range sound propagation of explosive signals in the Arctic. *Proceedings of 15th International Acoustic Congress, Trondheim, Norway* (Vol. 4, pp. 513–516).

Angstrom, A. (1933). On the total radiation from sun and sky at Sveanor. *Geogr. Ann. Arg.*, **XV**(2/3).

Appel, I. L. and Gudkovich, Z. M. (1979). The catoptric capability of ice cover during the period of melting in the southeastern Laptev Sea. *Poleks-Sever-76* (Vol. II, pp. 27–31). Gidrometeoizdat, Leningrad [in Russian].

Appel, I. L. and Gudkovich, Z. M. (1984). Studies of possible changes of mean salinity of the Kara Sea upper layer caused by the stable anomalies of river run-off. *Problemy Arktiki i Antarktiki [Problems of the Arctic and Antarctic]*, **58**, 5–14 [in Russian].

Appel, I. L. and Gudkovich, Z. M. (1992). *Numerical Modeling and Forecasting the Arctic Seas Ice Cover Evolution in Summer* (144 pp.). Gidrometeoizdat, St. Petersburg [in Russian].

Appel, I. L., Gudkovich, Z. M., and Frolov, I. Ye. (1976). The forces of pressure and internal interaction within the ice cover at the coastward drift. *Proceedings of the AARI*, **320**, 153–161 [in Russian].

Appel, I. L., Gudkovich, Z. M., Nikolayev, S. Yu., and Pozdnyshev, S. P. (1994). The results of the Arctic Basin ice dynamics modeling for long periods of time. *Problemy Arktiki i Antarktiki [Problems of the Arctic and Antarctic]*, **67–68**, 108–120 [in Russian].

Arctic Ocean Snow and Meteorological Observations from Drifting Stations: 1937, 1950–1991 (1996). National Snow and Ice Data Center, Boulder, CO (Version 1.0, CD-ROM).

Arkhipov, A. V., Gudkovich, Z. M. and Klyachkin, S. V. (2001). Modeling the ice cover deformation caused by the unevenness of ice thickness and ridge concentration. *Proceedings of the AARI*, **443**, 65–75 [in Russian].

Ashik, I. M. (1997). Numerical calculations and forecasts of level fluctuations, currents and ice drift in the offshore Western Arctic seas. *Navigatsiya i Gidrografiya [Navigation and Hydrography]* **4**, 85–94 [in Russian].

Atlas of the Arctic (1985). Arctic and Antarctic Research Institute, St. Petersburg and Central Directorate for Geodesy and Cartography, Moscow (204 pp.) [in Russian].

Atlas of the Oceans. The Arctic Ocean. (1980). VMF (the USSR Ministry of Defense publishers) (184 pp.) [in Russian].

Bagrov, N. A. (1975). On the circulation center. *Meteorologiya i Gidrologiya [Meteorology and Hydrology]*, **2**, 3–11 [in Russian].

Baranova, Z. I. (1989). New species of holothurians *Elpidia* from the Arctic Ocean. In: D. L. Kalyo (ed.), *Problems in the Study of Fossil and Modern Echinodermata* (pp. 218–222). Tallinn.

Belov, M. I. (1956). *Arctic Navigation from Ancient Times to the Middle of the 19th Century* (Vol. 1, 592 pp.). Morskoy Transport, Moscow [in Russian].

Belov, M. I. (1959). *The History of the Discovery and Development of the North Sea Route* (Vol. III, 511 pp.). Morskoy Transport, Leningrad [in Russian].

Belov, M. I. (1969). *The History of the Discovery and Development of the North Sea Route* (Vol. IV, 616 pp.). Gidrometeoizdat, Leningrad [in Russian].

Belov, M. I. and Vedernikov, A. V. (1968). News about the drift of the station 'North Pole-7'. *Morskoy Flot*, **4**, 34–35 [in Russian].

Belyaeva, T. V. (1980). Phytoplankton within the drift area of 'Severniy Polus-22'. *Biology of the Central Arctic Basin* (pp. 133–142). Nauka, Moscow. [in Russian].

Belyakov, L. N. (1974). The hydromechanical peculiarities of the under-ice layer (from experimental data). *Problemy Arktiki i Antarktiki [Problems of the Arctic and Antarctic]*, **43–44**, 147–152 [in Russian].

Belyakov, L. N. and Grishchenko, V. D. (1980). Currents induced by ice drift. In: L. A. Timokhov (ed.), *Vertical Structure and Dynamics of the Ocean's Under-ice layer* (pp. 91–96). Gidrometeoizdat, Leningrad [in Russian].

Blagoveshchensky, D. V., Egorova, L. V., and Lukashkin, V. M. (1989). Signatures of the polar cusp at high-latitude radio paths. *Investigations of the Subauroral Ionosphere* (pp. 193–197). Nauka, Moscow [in Russian]

Blagoveshchensky, D. V., Egorova, L. V., Lukashkin, V. M., Michailova, L. N., and Shumilov, I. A. (1991). Dynamics of auroral substorms from the data of the network of oblique sounding radio paths. *Proceedings of the AARI*, **427**, 160–171 [in Russian]

Blinov, N. I. and Vorobiev, V. N. (1976). Variability of temperature and depths of occurrence of the Pacific water to the north of the Chukchi Sea. *Proceedings of the AARI*, **319**, 73–80 [in Russian].

Bogorodskiy, V. V. and Gaitskhoki, B. Ya. (1979). Optic classification of natural ice. *Optical Methods of Investigations of the Oceans and Internal Water Bodies* (pp. 341–347). Nauka, Novosibirsk [in Russian].

Bogorodskiy, V. V. and Gavrilo, V. P. (1976). Possibilities and perspectives of use of acoustic methods in the study of the ice cover in rivers. *Proceedings of Coordinating Workshops in Hydrotechnics*, **111**, 189–193 [in Russian].

Bogorodskiy, V. V. and Gavrilo, V. P. (1980). *Ice: Physical Properties and Up-to-date Methods of Glaciology* (384 pp.). Gidrometeoizdat, Leningrad [in Russian].

Bogorodskiy, V. V. and Gusev, A. V. (1973). Attenuation of the acoustic waves in ice in the frequency range of 200–1,100 kHz. *Acoustic Journal*, **19**(2), 133–139 [in Russian].

Bogorodskiy, V. V. and Lebedev, G. A. (1978). Acoustic signals of narwals. *Acoustic Journal*, **24**(6), 826–834 [in Russian].

Bogorodskiy, V. V. and Sukhorukov, K. K. (1983). The physical conditions of sea ice cover thermal destruction on the lower surface. *Meteorologiya i Gidrologiya [Meteorology and Hydrology]*, **7**, 67–73 [in Russian].

Bogorodskiy, V. V. and Volodin, E. S. (1982). Weakening of the beam lights in layers of snow and sea ice. *Proceedings of the USSR Academy of Sciences*, **263**(3), 591–594 [in Russian].

Bogorodskiy, V. V., Gavrilo, V. P., and Gusev, A. V. (1970). On the nonlinear effects of ice destruction in fluids. *Proceedings of the AARI*, **295**, 159–165 [in Russian].

Bogorodskiy, V. V., Gavrilo, V. P., Gusev, A. V., Gudkovich, Z. M., and Poliakov, A. P. (1972). Stress state of the ice cover caused by action of a temperature wave and its relationship with acoustic noise in the under-ice layer of the ocean. *Proceedings of International Symposium 'Ice and Its Action on Hydrotechnical Structures'* (pp. 31–36). Energy, Leningrad [in Russian].

Bogorodskiy, V. V., Gavrilo, V. P., and Gusev, A. V. (1974). Acoustic effects of ice friction. *Proceedings of the AARI*, **324**, 97–103 [in Russian].

Bogorodskiy, V. V., Smirnov, G. E., and Smirnov, S. A. (1975a). Adsorption and scattering of soundwaves by sea ice. *Proceedings of the AARI*, **326**, 128–134 [in Russian].

Bogorodskiy, V. V., Gavrilo, V. P., and Poliakov, A. P. (1975b). Radio hydroacoustic method of investigations of mesoscale characteristics of sea ice dynamics. *Proceedings of the AARI*, **326**, 219–228 [in Russian].

Bogorodskiy, V. V., Gavrilo, V. P., Gusev, A. V., Zubkov, L. I., and Savchenko, V. G. (1977). High-frequency internal waves in the Arctic Basin. *Proceedings of the AARI*, **342**, 75–87 [in Russian].

Bogorodskiy, V. V., Baranov, G. I., and Gusev, A. V. (1980). Spatiotemporal variability of hydrophysical fields in the Arctic Basin. *Proceedings of the USSR Academy of Sciences*, **253**(4), 921–924 [in Russian].

Bogorodskiy, V. V., Gavrilo, V. P., and Smirnov, V. N. (1982a) Slow waves in the ice cover in the Arctic Ocean. In: *Impact of Large Scale Internal Waves on the Sea Surface* (pp. 158–161). Institute of Applied Physics of the USSR Academy of Sciences, Gorkiy [in Russian].

Bogorodskiy, V. V., Gusev, A. V., and Zubkov, L. I. (1982b). Mesoscale eddy structure of waters in the Arctic Basin. *Proceedings of the USSR Academy of Sciences*, **262**(5), 1250–1252 [in Russian].

Bogorov, V. G. (1946a). Zooplankton of the Polar Basin. *Scientific Results from the Airborne Expedition aboard the 'USSR IL-169' to Inaccessible Polar Areas*. Glavsevmorput, Moscow [in Russian].

Bogorov, V. G. (1946b). Zooplankton collections from an icebreaker. *The Drifting Expedition on the Icebreaker G. Sedov in 1937–1940* (pp. 336–370). Glavsevmorput, Moscow [in Russian].

Borodachev, V. Ye. and Frolov, I. Ye. (1997). *Typology of Ice Distribution in the Seas of the Russian Arctic* (155 pp.). Gidrometeoizdat, St. Petersburg [in Russian].

Borodachev, V. Ye., Volkov, N. A., and Grischenko, V. D. (1981). Peculiarities of the Arctic ice cover spatial structure and seasonal variability. *Proceedings of the AARI*, **372**, 35–43 [in Russian].

Bolotinskaya, M. Sh. and Ivanov, V. V. (1999). Synoptic and statistical characteristics of the variability of macroprocess development natural stages on a variable temporal scale. *Proceedings of the AARI*, **441**, 6–18 [in Russian].

Bourke, R. H. and Garret, R. P. (1987). Sea ice thickness distribution in the Arctic Ocean. *Cold Regions Science and Technology*, **13**(3), 259–280.

Brestkin, S. V. and Yulin, A. V. (eds.) (1998). *Natural Conditions along Selected Routes* (INSROP Working Paper No. 121, 36 pp.). Fridtjof Nansen Institute (INSROP), Oslo.

Brodskiy, K. A. (1956). Life in the water column of the Polar Basin. *Priroda*, **5** [in Russian].

Brodskiy, K. A. and Pavshtiks, E. A. (1976). Plankton of the central Arctic Basin. *Voprosy Geografii*, **101**, 148–167 [in Russian].

Brown, A. M. and Grary, A. P. (1958). The movement of the ice in the Arctic Ocean. In: *Arctic Sea Ice* (pp. 191–207). National Academy of Sciences, Washington, DC.

Bryazgin, N. N. (1966). Collected precipitation loss caused by evaporation from a precipitation-measuring vessel under Arctic conditions. *Proceedings of the MGO*, **195**, 163–166 [in Russian].

Bryazgin, N. N. (1976). Average precipitation in the Arctic taking into account the errors of precipitation gauges. *Proceedings of the AARI*, **323**, 40–74 [in Russian].

Bryazgin, N. N. (1983). Amount and frequency of occurrence of atmospheric precipitation in the Arctic Ocean. *Proceedings of the AARI*, **381**, 77–87 [in Russian].

Bryazgin N. N. (1996). Method of measurement and correction of solid precipitation in the Russian Arctic. *Proceeding of Workshop on the ACSYS Solid Precipitation Climatology Project* (WCRP-93, pp. 30–32, WMO/TD No. 739). World Meteorological Organization, Geneva.

Bryazgin, N. N. (1998). Precipitation characteristics in the Russian Arctic influencing atmospheric pollution removal. *Proceedings of the AARI*, **439**, 135–148 [in Russian].

Bryazgin, N. N. and Voskresensky, A. I. (2000). Atmospheric precipitation in the Russian Arctic. *Proceedings of the AARI*, **440**, 102–114 [in Russian].

Bryazgin, N. N., Kuznetsova, L. P., Maksimov, G. A., Nechaev, I. N., and Shver, Ts. A. (1967). The precipitation loss as a result of moistening the precipitation gauge. *Meteorologiya i Gidrologiya [Meteorology and Hydrology]*, **3**, 106–109 [in Russian].

Budyko, M. I. (1966). The possibilities of changing the climate by the impact on polar ice. *Modern Problems of Climatology* (227 pp.). Gidrometeoizdat, Leningrad [in Russian].

Budyko, M. I. (1969). *Polar Ice and the Climate* (36 pp.). Gidrometeoizdat, Leningrad [in Russian].

Burova, L. P. (2000). Water vapor in the Arctic atmosphere. *Proceedings of the AARI*, **440**, 12–28 [in Russian].

Burova, L. P. and Voskresensky, A. I. (1962). The meteorological conditions of icing in the clouds of As and Ac types. *Proceedings of the AARI*, **239**, 95–103 [in Russian].

Bushuev, A. V. and Loschilov, V. S. (1967). Accuracy of airborne observations and charting the sea ice. *Proceedings of the AARI*, **257**, 84–92 [in Russian].

Bushuev, A. V., Volkov, N. A., Gudkovich, Z. M., and Loschilov, V. S. (1967). Results of expedition studies of the ice cover drift and dynamics of the Arctic Basin in spring. *Proceedings of the AARI*, **257**, 26–45 [in Russian].

Buynitskiy, V. Ch. (1946). Diary of the observations of mammals and birds. *The Drifting Expedition on the Icebreaker G. Sedov in 1937–1940* (pp. 5–13) Glavsevmorput, Moscow [in Russian].

Buynitskiy, V. Kh. (1951). Formation and drift of the ice cover in the Arctic Basin. *The Drifting Expedition on the Icebreaker G. Sedov in 1937–1940* (Vol. 4, pp. 74–151). Glavsevmorput, Moscow [in Russian].

Buzuev, A. Ya. (1975). Statistical estimate of ice cover spatial distribution. *Proceedings of the AARI*, **326**, 187–192 [in Russian].

Buzuev, A. Ya. and Dubovtsev, V. F. (2002). Generalization of the ice cover morphometric characteristics for estimating climatic changes in the Arctic Basin and seas of the Siberian Shelf. *Scientific Conference of C.I.S.* (pp. 222–223). Gidrometeoizdat, St. Petersburg [in Russian].

Buzuev, A. Ya. and Ryvlin, A. Ya. (1976). The non-uniformity of ice cover characteristics distribution for estimating the ice-passing capacity of vessels. *Meteorologiya i Gidrologiya [Meteorology and Hydrology]*, **4**, 68–77 [in Russian].

Buzuev, A. Ya. and Spichkin, V. A. (1977). Role of water puddles in forming the spatial non-uniformity of the multiyear ice thickness in winter. *Problemy Arktiki i Antarktiki [Problems of the Arctic and Antarctic]*, **49**, 53–58 [in Russian].

Buzuev, A. Ya., Romanov, I. P., and Fedyakov, V. Ye. (1979a). Variability of snow distribution on the ice cover of the Arctic Ocean. *Meteorologiya i Gidrologiya [Meteorology and Hydrology]*, **9**, 78–85 [in Russian].

Buzuev, A. Ya., Gorbunov, Yu. A., Gudkovich, Z. M., and Losev, S. M. (1979b). Estimate of water area on ice by air photography. *Poleks-Sever-76* (Vol. II, pp. 92–106). Gidrometeoizdat, Leningrad [in Russian].

Campbell W. J. (1964). *On the Steady-state Flow of Sea Ice* (O.N.R. Sci. Rep. No. 307/352, 167 pp.). University of Washington, Seattle.

Cherepanov, N. V. (1976). The classification of ice of natural water bodies. *Proceedings of the AARI*, **331**, 77–99 [in Russian].

Chernigovsky, N. T. (1946). Actinometric observations. *Results from the Airborne Expedition aboard the 'USSR IL-169' to Inaccessible Polar Areas* (pp. 155–161). Glavsevmorput, Moscow [in Russian].

Chernigovsky, N. T. and Marshunova, M. S. (1965). *The Climate of the Soviet Arctic: Radiation Regime* (198 pp.). Gidrometeoizdat, Leningrad [in Russian].

Coachman, L. K. and Barnes, S. A. (1961). The contribution of Bering Sea water to the Arctic Ocean. *Arctic*, **14**, 147–161.

Coachman, L. K. and Barnes, S. A. (1962). Surface water in the Eurasian basin of the Arctic Ocean. *Arctic*, **15**(4), 251–277.

Coachman, L. K. and Barnes, S. A. (1963). The movement of Atlantic water in the Arctic Ocean. *Arctic*, **16**(1), 9–16.

Coon, M. D. *et al.* (1974). Modelling the pack ice as an elastic-plastic material. *AIDJEX Bulletin*, **24**, 1–105.

Crairy, A. P. (1958). The scientific study of the Arctic ice island T-3. In: K. Rodal (ed.). *The North* (pp. 209–247). Geografgiz, Moscow [in Russian].

Crowley, G. (1996). Critical review on ionospheric patches and blobs. In: W. R. Stone (ed.), *The Review of Radio Science 1993–1996* (pp. 619–648). Oxford University Press, Oxford, UK.

Dmitriev, F. A. and Pivovarov, S. V. (1981). Distribution of hydrocarbons in fresh snow and ice at the station 'North Pole-22' (from observations in 1977–1978). *Meteorologiya i Gidrologiya [Meteorology and Hydrology]*, **5**, 65–69 [in Russian].

Dolgin, I. M. (ed.) (1971). *The Meteorological Regime of the Foreign Arctic* (227 pp.). Gidrometeoizdat, Leningrad [in Russian].

Doronin, Yu. P. (1959). On thermal transformation of the atmospheric lower layer in the Arctic. *Proceedings of the AARI*, **226**, 76–98 [in Russian].

Doronin, Yu. P. (1968). On the problem of Arctic ice annihilation. *Problemy Arktiki i Antarktiki [Problems of the Arctic and Antarctic]*, **28**, 21–28 [in Russian].

Doronin, Yu. P. (1969). *Thermal Interaction between the Atmosphere and Hydrosphere in the Arctic* (300 pp.). Gidrometeoizdat, Leningrad [in Russian].

Doronin, Yu. P. (ed.) (1980). *Ocean Dynamics* (304 pp.). Gidrometeoizdat, Leningrad [in Russian].

Doronin, Yu. P. and Kheysin, D. Ye. (1975). *Sea Ice* (318 pp.). Gidrometeoizdat, Leningrad [in Russian].

Doronin, Yu. P. and Kubyshkin, N.V. (2001). *Sea Ice Growth and Melting* (44 pp.). Gidrometeoizdat, St. Petersburg [in Russian].

Dorsey, H. G. (1951). Arctic meteorology. In: T. F. Malone (ed.), *Compendium of Meteorology* (pp. 942–951). American Meteorological Society, Boston.

Driatsky, V. M. (1956). Studies of the ionosphere and radiowave propagation in the Arctic. *Proceedings of the AARI*, **84**, 5–19 [in Russian]

Driatsky, V. M. (1974). *The Nature of Abnormal Absorption of Cosmic Radio Emission in the Lower Ionosphere* (224 pp.). Gidrometeoizdat, Leningrad [in Russian]

Driatsky, V. M. and Besprozvannaya, A. S. (1958). Ionospheric conditions in the circumpolar region. *Annales de Géophysique*, **14**, 438–455.

Driatsky, V. M., Smirnov, V. B., and Kosterin, I. N. (1978a). Experimental study of HF radio wave propagation in Arctic radio paths. *Proceedings of the AARI*, **351**, 5–9 [in Russian]

Driatsky, V. M., Smirnov, V. B., Kosterin, I. N., and Lukashkin, V. M. (1978b). Impact of radiowave absorption on LUF values at polar HF radio paths. *Proceedings of the AARI*, **351**, 78–88 [in Russian]

Drogaitsev, D. A. (1949). Interaction between the Arctic seas and the atmosphere. *Proceedings of the CIP [Central Institute of Forecasting]*, **13**, 147 [in Russian].

Drogaitsev, D. A. (1956). *Zones of Ice Compression and Divergence in the Field of Atmospheric pressure* (Geophysics Series No. 11, pp. 1332–1337). Academy of Sciences of the USSR, Moscow [in Russian].

Dvorkin, Ye. N. (1970). The tides. *Soviet Arctic: The Seas of the Arctic Ocean* (pp. 191–197). Nauka, Moscow [in Russian].

Dzerdzeyevsky, B. L. (1954). Atmospheric circulation in the Central Polar Basin. *Proceedings of the Drifting Station 'North Pole-2'*, **2**, 64–200 [in Russian].

Egorova, L. V. (1988). On short-term forecasting of the MUF at auroral HF radio paths. *Applied Problems of Electrodynamics* (pp. 54–59). LIAP, Leningrad [in Russian]

Egorova, L. V. and Lukashkin, V. M. (1985). Numerical short-term forecasting of HF radiowave propagation in polar radio paths. *Forecasting of Ionosphere and Radiowave Propagation* (pp. 54–59). Nauka, Moscow [in Russian]

Egorova, L. V. and Lukashkin, V. M. (1991). Short-term forecasting of the MUF at high-latitude HF radio paths under different geophysical conditions. *Proceedings of the AARI*, **427**, 85–90 [in Russian]

Fedjakina, N. I. (1980). *Propagation of VLF Radiowaves at Transpolar Radio Paths* (104 pp). Nauka, Novosibirsk [in Russian]

Felsenbaum, A. I. (1957). On ice pressure and diverging in the Arctic Basin. *Doklady Akademii Nauk SSSR [Reports of the USSR Academy of Science]*, **116**(2), 217–220 [in Russian].

Felsenbaum, A. I. (1959). The mean climatic ice drift in the Central Arctic Basin. *Doklady Akademii Nauk SSSR [Reports of the USSR Academy of Science]*, **126**(1), 66–69 [in Russian].

Fetterer, F. and Radionov, V. (eds) (2000) The Arctic Meteorology and Climate Atlas. *Arctic Climatology Project. Environmental Working Group*. National Snow and Ice Data Center, Boulder, CO (CD-Rom).

Frolov, I. Ye. (1981). The numerical model of the autumn–winter ice phenomena. *Proceedings of the AARI*, **372**, 73–81 [in Russian].

Frolov, I. Ye. (1987). On the question of ice drift in the marginal zone. *Proceedings of the AARI*, **402**, 164–170 [in Russian].

Frolov, I. Ye., Kolesov, S. A., and Makshtas, A. P. (1997). Mathematical modeling of sea ice cover. In: I. Ye. Frolov and V. P. Gavrilo (eds.), *Sea Ice* (pp. 227–274). Gidrometeoizdat, St. Petersburg [in Russian].

Gaitskhoki, B. Ya. (1970). Photometric model of snow and ice cover. *Proceedings of the AARI*, **295**, 55–59 [in Russian].

Gaitskhoki, B. Ya. (1974). Some results of investigations of natural illumination in the Central Arctic. *Proceedings of the AARI*, **324**, 172–181 [in Russian].

Gaitskhoki, B. Ya. (1975). Optical characteristics of some species of natural ice. *Proceedings of the AARI*, **326**, 55–59 [in Russian].

Gaitskhoki, B. Ya. and Ivashnev, A. G. (1980). Objective classification of the curves of vertical distribution of the weakening index of water illumination. *Proceedings of the AARI*, **374**, 142–148 [in Russian].

Gaitskhoki, B. Ya., Grigoriev, A. I., and Grishchenko, V. D. (1978). Some results of measurements of radiant energy under the ice near the edge of an ice island. *Proceedings of the AARI*, **359**, 99–103 [in Russian].

Galkin, R. M. (1970). On the relation between geomagnetic field and the relief forms of the Arctic Ocean bottom. *Proceedings of the AARI*, **288**, 80–84 [in Russian]

Galt, I. A. (1967). *Current Measurements in the Canada Basin of the Arctic Ocean, Summer, 1965* (Technical Report No. 4, pp. 17–25). University of Washington, Department of Oceanography, Seattle.

Gavrilo, V. P. and Gusev, A. V. (1975). Use of acoustical methods of snow and ice investigations. *Proceedings of the AARI*, **326**, 121–127 [in Russian].

Gavrilo, V. P., Gusev, A. V., and Poliakov, A. P. (1970). On the possibility of acoustic registration of the critical stress state in the ice cover. *Proceedings of the AARI*, **295**, 166–173 [in Russian].

Gavrilo, V. P., Gusev, A. V., Zaretskiy, Yu. K., and Fish, A. M. (1978). Acoustic emission as an index of ice deformation and destruction. *Proceedings of the AARI*, **359**, 118–126 [in Russian].

Gavrilo, V. P., Gusev, A. V., and Nikitin, V. A. (1980). Seasonal variations of the elastic properties of sea ice cover. *Proceedings of the AARI*, **374**, 37–42 [in Russian].

Gevorkyan, M. G. (1941). The theory of the ice floes drift. *Problemy Arktiki [Problems of the Arctic]*, **4**, 5–31 [in Russian].

Girs, A. A. (1974). *The Macrocirculation Method for Long-term Meteorological Forecasts* (480 pp.). Gidrometeoizdat, Leningrad [in Russian].

Glagoleva, M. G. (1950). Ice drift in coastal areas. *Proceedings of the CIP [Central Institute of Forecastings]*, **17**, 159–166 [in Russian].

Golovin, P. N., Kochetov, S. V., and Timokhov, L. A. (1993). Specific features of the thermohaline structure of fractures in summer in Arctic ice. *Oceanology*, **33**(6), 833–838 [in Russian].

Golovin, P. N., Kochetov, S. V., and Timokhov, L. A. (1995). Freshening of the under-ice layer by ice melting. *Oceanology*, **35**(4), 525–530 [in Russian].

Gorbunov, G. P. (1946). Bottom fauna of new Siberian shallow waters and central part of the Arctic Ocean. *The Drifting Expedition on the Icebreaker G. Sedov in 1937–1940* (pp. 31–138). Glavsevmorput, Moscow [in Russian].

Gorbunov, Yu. A. (1956). Influence of the coastline on ice drift in the East Siberian Sea. *Problemy Arktiki [Problems of the Arctic]*, **2**, 107 [in Russian].

Gorbunov, Yu. A. and Timokhov, L. A. (1968). Studying of ice dynamics. *Reports of the USSR Academy of Science, Physics of Atmosphere and Ocean*, **4**(10), 1086–1091 [in Russian].

Gorbunov, Yu. A., Karelin, I. D., and Losev, S. M. (1986). The reasons for ice cover entirety disturbance in winter. *Problemy Arktiki i Antarktiki [Problems of the Arctic and Antarctic]*, **62**, 110–116 [in Russian].

Gorbunov, Yu. A., Gudkovich, Z. M., and Losev, S. M. (1998). Study into the ice cover dynamics in the Arctic Basin from data of automated buoys. *Mini-Conference, 'The Arctic Buoy Program', 3–5 August 1998, Seattle, Washington* (abstract).

Gorbushina, G. N., Driatsky, V. M., and Zhulina, E. M. (1969). *Instruction for Numerical Calculation of HF Radio Paths in High Latitudes.* (93 pp.), Nauka, Moscow [in Russian].

Gordienko, P. A. (1958). Ice drift in the central part of the Arctic Ocean. *Problemy Severa [Problems of the North]*, **1**, 3–29 [in Russian].

Gordienko, P. A., Buzuyev, A. Ya., and Sergeyev, G. N. (1967). Studying ice cover as a navigation environment. *Problemy Arktiki i Antarktiki [Problems of the Arctic and Antarctic]* **27**, 93–104 [in Russian].

Grishchenko, V. D. (1981). Some peculiarities of melting and growth in the underwater part of ice in the Arctic Basin. *Proceedings of the AARI*, **372**, 123–128 [in Russian].

Grishchenko, V. D. and Kochetov, S. V. (1989). Microstructure of vertical distribution of temperature and salinity in the under-ice layer. In: L. A. Timokhov (ed.), *Vertical Structure and Dynamics of the Ocean Under-ice Layer* (pp. 61–68). Gidrometeoizdat, Leningrad [in Russian].

Gudkovich, Z. M. (1955a). On ice floe rotation. *Problemy Arktiki [Problems of the Arctic]*, **1**, 16–21 [in Russian].

Gudkovich, Z. M. (1955b). Results of preliminary analysis of deep sea observations of hydrological observations. *Observations of the Research Drifting Station in 1950–1951* (Vol. 1). Leningrad [in Russian].

Gudkovich, Z. M. (1957). Ice drift in the central part of the Arctic Basin (105 pp.). *Proceedings of the AARI*, **87** [in Russian].

Gudkovich, Z. M. (1961a). On the basic regularities of ice drift in the Central Polar Basin. *Conference 'The Atmosphere and Hydrosphere Interaction in the Northern Atlantic Ocean'* (Vols. 3–4, pp. 75–88). Gidrometeoizdat, Moscow [in Russian].

Gudkovich, Z. M. (1961b). Relationship of ice drift in the Arctic Basin with ice conditions in Arctic seas. *Proceedings of Oceanographic Commission of the USSR Academy of Sciences*, **II**, 13–20 [in Russian].

Gudkovich, Z. M. (1965). The correlation method of ice drift data processing. *Problemy Arktiki i Antarktiki [Problems of the Arctic and Antarctic]*, **21**, 56–59 [in Russian].

Gudkovich, Z. M. (1966). The anticyclone circulation of ice cover in the Arctic Basin and drift of the station 'North Pole-8'. *Problemy Arktiki i Antarktiki [Problems of the Arctic and Antarctic]*, **213**, 5–10 [in Russian].

Gudkovich, Z. M. (1979a). Ice concentration changes and ice pressure in the baric systems. *Proceedings of the AARI*, **364**, 75–82 [in Russian].

Gudkovich, Z. M. (1979b). The turbulent motion of ice cover in the Arctic Basin. *Proceedings of the AARI*, **364**, 34–39 [in Russian].

Gudkovich, Z. M. and Doronin, Yu. P. (2001). *Sea Ice Drift* (110 pp.). Gidrometeoizdat, St. Petersburg [in Russian].

Gudkovich, Z. M. and Klyachkin, S. V. (2001). Numerical model of ice cover evolution in autumn–winter period on the example of the Pechora Sea. *Proceedings of the AARI*, **443**, 57–64 [in Russian].

Gudkovich, Z. M. and Kovalev, Ye. G. (1967). On the influence of water circulation in the Arctic Basin on ice distribution in the eastern sector of the Soviet Arctic. *Proceedings of the AARI*, **116**, 7–20 [in Russian].

Gudkovich, Z. M. and Kovalev, Ye. G. (2002a). Fluctuations in ice cover extent of the Russian Arctic seas in the 20th century and estimate of possible changes in the 21st century. *Scientific-practical Conference 'Hydrometeorological Support of Economic Activities in Arctic and Freezing Seas'* (pp. 35–45). Gidrometeoizdat, St. Petersburg [in Russian].

Gudkovich, Z. M. and Kovalev, Ye. G. (2002b). Some mechanisms behind cyclic changes of the climate in the Arctic and Antarctic. *Okeanologiya [Oceanology]*, **42**(6), 815–821 [in Russian].

Gudkovich, Z. M. and Kudryashov, L. I. (1985). Some peculiarities of oceanographic conditions in the Kara Sea at the beginning of winter in 1979. *Proceedings of the AARI*, **400**, 156–161 [in Russian].

Gudkovich, Z. M. and Nikiforov, Ye. G. (1963). The stationary drift of the single ice floe. *Proceedings of the AARI*, **253**, 197–209 [in Russian].

Gudkovich, Z. M. and Nikiforov, Ye. G. (1965). Some important peculiarities of water density anomaly formation and its influence on the ice and hydrological conditions in the Arctic Basin and marginal seas. *Okeanologiya [Oceanology]*, **5**(2), 250–260 [in Russian].

Gudkovich, Z. M. and Nikolayeva, A. Ya. (1963). Ice drift in the Arctic Basin and its correlation with the ice cover extent of Soviet Arctic seas. *Proceedings of the AARI* **104**, 212 [in Russian].

Gudkovich, Z. M. and Pozdnyshev, S. P. (1995). Seasonal and spatial changes of the average speed of ice drift and gradient currents in the East Greenland ice stream. *Problemy Arktiki i Antarktiki [Problems of the Arctic and Antarctic]*, **69**, 116–123 [in Russian].

Gudkovich, Z. M. and Romanov, M. A. (1970). Method of calculation of the equivalent ice thickness in the Arctic seas in winter. *Proceedings of the AARI*, **292**, 4–48 [in Russian].

Gudkovich, Z. M. and Yevdokimov, S. N. (1971). Influence of long-term tides on ice drift in the Arctic Basin. *Proceedings of the AARI*, **303**, 98–107 [in Russian].

Gudkovich, Z. M. and Zakharov, V. F. (1998). Role of marginal dynamic processes in ice concentration changes in the Arctic seas in summer. *Meteorologiya i Gidrologiya [Meteorology and Hydrology]*, **3**, 65–71 [in Russian].

Gudkovich, Z. M., Melkonyan, G. I., and Nikiforov, Ye. G. (1963). Aerodynamic studies of ice floe models. *Proceedings of the AARI*, **253**, 219–231 [in Russian].

Gudkovich, Z. M., Kirillov, A. A., Kovalev, Ye. G., Smetannikova, A. V., and Spichkin, V. A. (1972). *Basis of the Methodology of Long-term Ice Forecasts for the Arctic Seas* (348 pp.). Gidrometeoizdat, Leningrad [in Russian].

Gudkovich, Z. M., Gladkov, M. G., and Lukyanchikov, S. N. (1979). Ice cover volume and ice thickness distribution in the southeastern Laptev Sea at the end of winter in 1976. *Poleks-Sever-76* (Vol. II, pp. 16–19). Gidrometeoizdat, Leningrad [in Russian].

Gudkovich, Z. M., Voynov, G. N., and Losev, S. M. (1989). Methodology for calculating the ice drift maximum speed. *Regime-forming Factors, Information Base and Methods of Its Analysis* (pp. 232–237). Gidrometeoizdat, Leningrad [in Russian].

Gudoshnikov, Yu. P., Kirpichionok, T. E., and Smagin V. M. (1987). *Chemical Characteristics of Sea Ice at Different Stages of Development according to WMO Nomenclature* (WDC 06.03.87, No. 616, 11 pp.). Arctic and Antarctic Research Institute, St. Petersburg [in Russian].

Guryanova, E. F. (1957). Zoogeography of the Arctic Basin. In: A. Ph. Treshnokov (ed.), *Observational Materials from the Scientific Research Drifting Stations 'Severniy Polus-3' and 'Severniy Polus-4' 1954/55* (Vol. 1, (pp. 343–362). Marine Transport, Leningrad [in Russian].

Guryanova, E. F. (1970). Peculiarities of Arctic Ocean fauna and their meaning for understanding the history of its development. *The Arctic Ocean and Its Coast in the Cenozoic* (pp. 126–161). Gydrometeoizdat, Leningrad [in Russian].

Guryanova, E. F. and Musina, A. A. (1960). The main features of oxygen and alkalinity in the Atlantic waters of Arctic seas. *Proceedings of the AARI* **218**, 125–158 [in Russian].

Gushchenkov, E. M. (1964). Distribution and metamorphization of Pacific waters in the Arctic Basin. *Oceanology*, **IV**(1), 36–42 [in Russian].

Guzenko, R. B. (2001). Modeling the oil pollution of Arctic sea water and ice, and estimating the influence of oil products on ice melting and growth (23 pp.). Doctoral thesis, University of St. Petersburg [in Russian].

Harder, M., Hilmer, M., and Lemke, P. (1998). Simulated sea ice transport through the Fram Strait. *ACSYS Arctic Forecast*, **3**, 1028–1114.

Hare, F. K. and Boville, B. W. (1965). Polar circulation. *Circulation in the Stratosphere, Mesosphere and Lower Thermosphere* (WMO Tech. Note 70, pp. 43–78). World Meteorological Organization, Geneva.

Hibler, W. D. (1977a). A viscous sea ice law as a stochastic average of plasticity. *Journal of Geophysical Research*, **82**, 3932–3938.

Hibler, W. D. (1977b). Modelling pack ice as a viscous-plastic continuum: Some preliminary results. *Symposium on Ice Processes and Models, 6–9 September, University of Washington, Seattle* (Vol. 2, pp. 46–55).

Hibler, W. D. (1979). A dynamic-thermodynamic sea ice model. *J. Phys. Ocean*, **9**, 815–846.

Hibler, W. D. (1980). Modeling a variable thickness sea ice cover. *Monthly Weather Review*, **108**, No. 12, 1949–1975.

Hibler, W. D., Weeks, W. F., Ackley, S., Kovacs, A., and Campbell, W. J. (1972). Mesoscale strain measurements on Beafort sea pack ice. *AIDJEX Bulletin*, **13**, 35–76.

Hunkins, K. (1962). Waves in the Arctic ocean. *Journal of Geophysical Research*, **67**(6), 2477–2489.

Hunkins, K. L. (1966). Ekman drift current measurements in the Arctic Ocean. *Deep Sea Research*, **13**, 607–620.

Hunkins, K. (1967). Inertial oscillations of Fletcher's Ice Island (T-3). *Journal of Geophysical Research*, **72**(4), 1165–1174.

Hydrochemical Atlas of the Arctic Ocean (2002). International Arctic Research Center, University of Alaska, Fairbanks and Arctic and Antarctic Research Institute of Roshydromet, St. Petersburg (CD-ROM edited by R. Colony and Ye. Nikiforov).

Ivanov, V. V. and Vinogradov, N. D. (1995). *Meteorological Forecasts* (INSROP Working Paper No. 10, pp. 7–23). Fridtjof Nansen Institute (INSROP), Oslo.

Ivanov, V. V. and Vinogradov, N. D. (1996). *Meteorological Forecasts* (INSROP Working Paper No. 36, pp. 7–20). Fridtjof Nansen Institute (INSROP), Oslo.

Ivanov, V. V. and Vinogradov, N. D. (1999). Role of meteorological observations from the 'North Pole' drifting stations and high-latitude expeditions for development and improvement of long-term meteorological forecasts. *Problemy Arktiki i Antarktiki [Problems of the Arctic and Antarctic]*, **71**, 20–28 [in Russian].

Ivchenko, V. O. and Kheysin, D. Ye. (1974). Determination of internal stress within the ice cover originating from ice drift. *Problemy Arktiki i Antarktiki [Problems of the Arctic and Antarctic]*, **43–44**, 84–91 [in Russian].

Johannessen, O. (1970). Note on some vertical profiles below ice floes in the Gulf of St. Lawrence and near the North Pole. *Journal of Geophysical Research*, **75**(15), 2857–2861.

Johannessen, O. M. *et al.* (2004). Arctic climate change: Observed and modeled temperature and sea ice variability. *Tellus*, **56A**, 1–18.

Johnson, F. S. (ed.). (1965). *Satellite Environment Handbook* (191 pp.). Stanford University Press, Stanford, CA.

Joint U.S.–Russian Atlas of the Arctic Ocean: Oceanography Atlas for the Summer Period (1998). Environmental Working Group, University of Colorado, Boulder, CO (CD-ROM edited by F. Tanis and L. Timokhov).

Joint U.S.–Russian Atlas of the Arctic Ocean: Oceanography Atlas for the Winter Period (1997). Environmental Working Group, University of Colorado, Boulder, CO (CD-ROM edited by F. Tanis and L. Timokhov).

Kamenskaya, O. E. (1980). Deep sea amphipods (Amphipoda, Gammaridea) from collections made during the expedition of the drifting station 'North Pole-22'. *Biology of the Central Arctic Basin* (pp. 241–251). Nauka, Moscow [in Russian].

Karelin, D. B. (1943). The influence of water salinity and currents on the ice growth. *Problemy Arktiki [Problems of the Arctic]*, **1**, 144–149 [in Russian].

Karklin, V. P. and Teytelbaum, K. A. (1987). The temporal structure of multiyear changes in ice cover extent. *Proceedings of the AARI*, **402**, 53–66 [in Russian].

Karklin, V. P., Yulin, A. V., Karelin, I. D., and Ivanov, V. V. (2001). Climatic fluctuations of ice cover extent of the Arctic seas of the Siberian Shelf. *Proceedings of the AARI*, **443**, 5–21 [in Russian].

Khanaychenko, N. K. (1946). Deformation fields of ice stress and displacement in the Arctic Basin. *Meteorologiya i Gidrologiya [Meteorology and Hydrology]*, **6**, 69–74 [in Russian].
Kheysin, D. Ye. (1967). *Ice Cover Dynamics* (215 pp.). Gidrometeoizdat, Leningrad [in Russian].
Kheysin, D. Ye. and Ivchenko, V. O. (1973). Numerical model of tidal ice drift taking into account the interaction of ice floes. *Reports of the USSR Academy of Science, Physics of Atmosphere and Ocean*, **9**(4), 420–429 [in Russian].
Khodzha-Achmedov, C. L. and Ignatov, V. S. (1991). Study of correlation of the LUF with other factors and the accuracy of their forecasting. *Proceedings of the AARI*, **427**, 71–79 [in Russian]
Khodzha-Achmedov, C. L. and Kashurko, M. S. (1978). Variations of the multiray propagation of HF radio waves. *Proceedings of the AARI*, **351**, 20–24 [in Russian]
Khodzha-Achmedov, C. L. and Lukashkin, V. M. (1983). Influence of the E_s layer and auroral absorption on the oblique propagation of HF radiowaves. *Proceedings of the AARI*, **390**, 118–124 [in Russian]
Khromov, S. P. and Mamonova, L. I. (1974). *Meteorological Glossary* (568 pp.). Gidrometeoizdat, Leningrad [in Russian].
Kobyakova, Z. I. (1957). A new species of *Bythocaris* from the Arctic Basin. In: A. Ph. Treshnikov (ed.), *Observational Materials from Scientific Research Drifting Stations 'Severniy Polus-3' and Severniy Polus-4' 1954/55* (Vol. 1, pp. 363–364). Marine Transport, Leningrad [in Russian].
Kochetov, S. V. (1973). Calculation of the annual cycle of the ice cover state at sea. *Proceedings of the AARI*, **307**, 17–27 [in Russian].
Koesner R. M. (1973). The mass balance of sea ice in the Arctic ocean. *Journal of Glaciology*, **12**, 173–186.
Kolchak, A. V. (1909). *Ice of the Kara and Siberian Seas* (190 pp.). St. Petersburg [in Russian].
Kolesov, S. A. (1990). Modeling the ice drift in the Arctic Basin. *Proceedings of the AARI*, **420**, 32–38 [in Russian].
Kolmogorov, A. N. (1941). Local structure of turbulence in the incompressible viscous liquid at very large Reynolds numbers. *Doklady Akademii Nauk SSSR [Reports of the USSR Academy of Science]*, **30**(4), 299–303 [in Russian].
Koltun, V. M. (1964). The benthos fauna research of the Greenland Sea and central part of the Arctic Basin. *Proceedings of the AARI*, **259**, 13–78 [in Russian].
Kondakov, N. N., Moskalev, L. I., and Nesis, K. N. (1980). *Benthoctopus sibiricus* Løyning: Endemic of the east Arctic. *Ecological Investigations of the Shelf* (pp. 42–56). Moscow [in Russian].
Kondratyev, K. Ya. (ed.) (1981). *Albedo and Angle Reflection Characteristics of Clouds and the Underlying Surface* (231 pp.). Gidrometeoizdat, Leningrad [in Russian].
Kondratyev, K. Ya. (2004). Changes in global climate: The unsolved problems. *Meteorologiya i Gidrologiya [Meteorology and Hydrology]*, **6**, 118–128 [in Russian].
Konstantinov, Yu. B. and Grachev, K. I. (2000). *The High-latitude Airborne Expeditions 'North' (1937, 1941–1993)* (176 pp.). Gidrometeoizdat, St. Petersburg [in Russian].
Koplan-Diks, I. S. (1968). *The basis of the Statistical Processing and Charting of Oceanographic Data* (129 pp.). Gidrometeoizdat, Leningrad [in Russian].
Korotkevich, Ye. S. (1972). *Polar Deserts* (420 pp.). Gidrometeoizdat, Leningrad [in Russian].
Kosobokova, K. N. (1980). Seasonal changes of the vertical distribution and age content of population of *Microcalanus pygmaeus*, *Oithona similes*, *Oncaea borealis* and *Oncaea notopus* in the central Arctic Basin. *Biology of the Central Arctic Basin* (pp. 167–182). Nauka, Moscow [in Russian].

Kowalik, Z. (1981). A study of the M2 tide in the ice-covered Arctic Ocean. *Modelling, Identification and Control*, **2**(4), 216–229.

Krenkel, E. T. (1973). *RAEM is My Personal Telegraph Code* (280 pp.). Sovetskaja Rossia, Moscow [in Russian]

Kriss, A. E. (1976). *Microbiological Oceanography* (269 pp.). Nauka, Moscow [in Russian].

Kulakov, I. Yu. and Timokhov, L. A. (1979). Ice pressure at the stage of compacting. *Proceedings of the AARI*, **364**, 129–137 [in Russian].

Krutskikh, B. A. (ed.) (1984). *International Symbols for Sea Ice Charts and the Nomenclature of Sea Ice* (56 pp.). Gidrometeoizdat, Leningrad [in Russian].

Krutskikh, B. A. (ed.) (1991). *The Climatic Regime of the Arctic at the Turn of the 20th and 21st Centuries* (200 pp.). Gidrometeoizdat, St. Petersburg [in Russian].

Kusakin, O. G. (1983). New species of Isopoda: Crustacea – The first from the abyss of the Canadian Hollow in the Arctic Basin. *Marine Biology*, **3**, 13–17 [in Russian].

Laktionov, A. F. and Shamontiev, V. V. (1957). Use of aviation in oceanographic investigations in the Arctic. *Problemy Arktiki [Problems of the Arctic]*, **2**, 19–31 [in Russian].

Lappo, S. D. (1958). On the rotary motion of drifting ice floes. *Problemy Severa [Problems of the North]*, **1**, 30–41 [in Russian].

La Seur, N. E. (1954). On the asymmetry of the middle-latitude circumpolar current. *Journal of Meteorology*, **11**(1).

Laykhtman, D. L. (1958). On ice floe drift. *Proceedings of the LHMI [Leningrad Hydrometeorological Institute]*, **7**, 129–137 [in Russian].

Lebedev, G. A. (1999). Hydrophysical and radiophysical investigations in the Arctic Basin. *Problemy Arktiki i Antarktiki [Problems of the Arctic and Antarctic]*, **71**, 78–105 [in Russian].

Lebedev, G. A. and Popov, I. K. (1973). Acoustic phenomena accompanying the process of interaction between snowstorm fluxes and the ice cover. *Proceedings of 8th All-Union Acoustic Conference* (pp. 19–22). Acoustic Institute, Moscow [in Russian].

Lebedev, G. A. and Popov, I. K. (1974). Nature and spectrum of acoustic signals caused by snowstorms in water bodies covered by ice. *Proceedings of the AARI*, **324**, 118–125 [in Russian].

Lebedev, G. A. and Sukhorukov, K. K. (2001). *Propagation of Electromagnetic and Acoustic Waves in Sea Ice* (81 pp.). Gydrometeoizdat, St. Petersburg [in Russian].

Lebedev, A. A. and Uralov, N. S. (1977). Ice balance in the Greenland Sea. *Proceedings of the AARI*, **341**, 43–52 [in Russian].

Legenkov, A. P. (1958). The theory of tidal diverging, compacting and pressure of ice cover in an open sea. *Problemy Arktiki [Problems of the Arctic]*, **5**, 5–18 [in Russian].

Legenkov, A. P. (1963). On tidal diverging, compacting and pressure of ice cover. *Proceedings of the AARI*, **248**, 52–61 [in Russian].

Legenkov, A. P. and Romanov, M. A. (1972). On ice shearing in the Arctic Basin. *Okeanologiya [Oceanology]*, **12**(4), 617–624 [in Russian].

Leontyeva, Ye. A. (1947). *Climate of the Soviet Sector of the Arctic* (500 pp.). Glavsevmorput, Leningrad [in Russian].

Linkov, Ye. M. (1958). Studies of the elastic properties of the ice cover in the Arctic. *Leningrad State University Bulletin, Physics and Chemistry*, **1**(16), 138–145 [in Russian].

Losev, S. M. and Gorbunov, Yu. A. (1979). Studying horizontal turbulent exchange at sea by air photography ice data. *Proceedings of the AARI*, **364**, 40–51 [in Russian].

Losev, S. M. and Gorbunov, Yu. A. (1995). The divergence of ice drift speed in the Arctic Basin. *Proceedings of the AARI*, **435**, 68–80 [in Russian].

Lukashkin, V. M., Smirnov, V. B., Vovk, V. Y., Vystavnoy, V. M., and Shumilov, I. A. (1978). Antenna systems for oblique sounding of the ionosphere at high-latitude radio paths. *Proceedings of the AARI*, **351**, 70–77 [in Russian].

Lukin, V. V. and Timokhov, L. A. (1989). Inversion layers under the ice cover. In: L. A. Timokhov (ed.), *Vertical Structure and Dynamics of the Ocean's Under-ice Layer* (pp. 82–90). Gidrometeoizdat, Leningrad [in Russian].

Makshtas, A. P. and Podgorny, P. A. (1995). Calculation of albedo for a meltpond. *Polar Research*, **14**(2).

Makshtas, A. P., Nazarenko, L. S., and Shutilin, S. V. (1988). Model of the sea ice cover of the Arctic Basin. In: V. I. Kuzin (ed.), *Mathematical Models for Studying Ocean Dynamics* (pp. 96–116). Nauka (Science), Novosibirsk [in Russian].

Makshtas, A. P., Andreas E. L. and Shutilin, S. V. (2001). Possible dynamics and thermal causes for the recent decrease in sea ice in the Arctic Basin. *6th Conference on Polar Meteorology and Oceanography, San Diego, 14–18 May* (pp. 17–20).

Maksimov, I. V. (1959). Long-term lunar–solar tides in the high-latitude seas of the Earth. *Uchenye Zapiski LVIMU [Scientific Notes of the Leningrad High Engineering Marine School]*, **13**, 3–38 [in Russian].

Malyutina, M. V. and Kussakin, O. G. (1996). Additions to Polar Sea bathyal and abyssal Isopoda (Crustacea: Malacostraca), Part 3: Asellota: Munnopsidae. *Zoosystematica Rossica*, **5**, 12–27.

Manual for Hydrometeorological Stations and Posts (1969) (Vol. 3, No. 1, 307 pp.). Gidrometeoizdat, Leningrad [in Russian].

Manual on the Hydrological Studies in Oceans and Seas (1977) (726 pp.). Gidrometeoizdat, Leningrad [in Russian].

Manual on the Procedures of Biological Analysis of Sea Water and Bottom Sediments (1977) (191 pp.). Gidrometeoizdat, Leningrad [in Russian].

Manual on the Procedures of Chemical Analysis of Sea Water (1977) (208 pp.). Gidrometeoizdat, Leningrad [in Russian].

Manual on Solar Radiation Observation for Hydrometeorological Stations (1971). Gidrometeoizdat, Leningrad (220 pp.) [in Russian].

Marchenko, A. S. (1964). Circular and elliptic forms of the two-dimensional normal law in the wind air–climatology research. *Trudy NII Aeroclimatologii [Proceedings of Air–Climatology Research Institute]*, **25**, 131–208 [in Russian].

Markhaseva, E. L. (1980). Calanoida of the genus *Janschnovia*, nom. n. (Derjugina Jaschnov. Nom. N. peaeocc.) (Calanoida: Aetideidae). *Innovations in the Systematic of Marine Invertebrates* (pp. 63–76). Nauka, Leningrad [in Russian].

Marshunova, M. S. (1983). Peculiarities of Arctic Ocean radiation regime formation. *Proceedings of the AARI*, **381**, 20–32 [in Russian].

Marshunova, M. S. and Chernigovsky, N. T. (1965). *Climate of the Soviet Arctic: Radiation Regime* (200 pp.). Gidrometeoizdat, Leningrad [in Russian].

Marshunova M. S. and Chernigovsky, N. T. (1971). *Radiation Regime of the Foreign Arctic* (180 pp.). Gidrometeoizdat, Leningrad [in Russian]. (English translation available in Marshunova M. S. and Chernigovskiy, N. T. (1978), *Radiation Regime of the Foreign Arctic* (189 pp.), published for the Office of Polar Programs and NSF by the Indian National Scientific Documentation Centre, New Delhi – Contract NSF C-466.)

Marshunova, M. S. and Mishin, A. A. (1994). *Handbook of the Radiation Regime of the Arctic Basin: Results from the Drift Stations* (66 pp.). Gidrometeoizdat, St. Petersburg [in Russian]. (English translation available in Technical Report APL, UW TR 9413, December 1994.)

Marshunova, M. S. and Radionov, V. F. (1995). Dynamics of the Arctic radiation climate. *Problemy Arktiki i Antarktiki [Problems of the Arctic and Antarctic]*, **69**, 64–73 [in Russian].

Maykut, T. A., Thorndike A. S., and Untersteiner, N. (1972). AIDJEX scientific plan. *AIDJEX Bulletin*, **15**, 67.

McLaren, A. S., Bourke, R. H., Walsh, J. E., and Weaver, R. L. (1994). Variability in sea ice thickness over the North Pole from 1958 to 1992. In: O. M. Johannessen, R. D Muench, and J. E. Overland (eds.), *Polar Oceans and Their Role in Shaping the Global Environment* (pp. 363–371). American Geophysical Union, Washington, DC.

Melnikov, I. A. (1976). Hydrobiological investigations in the central part of the Arctic Basin (Spring 1976). *Oceanology*, **16**(6), 1034–1035 [in Russian].

Melnikov, I. A. (1978). Hydrobiological research in the central Arctic Basin (Spring 1978). *Oceanology*, **18**(6), 1132 [in Russian].

Melnikov, I. A. (1981). Hydrological investigations in the Arctic Basin (NP-22, December 1979–April 1980). *Oceanology*, **21**(2), 379 [in Russian].

Melnikov, I. A. (1989). *Ecosystem of Arctic Sea Ice* (191 pp.). Institute of Oceanography, USSR Academy of Sciences, Moscow [in Russian].

Melnikov, I. A. and Tsinovskiy, V. D. (1978). Hydrobiological investigations in the Arctic Basin on NP-23 (May–October 1977). *Oceanology*, **18**(2), 378–379 [in Russian].

Melnikov, I. A., Zhitina, L. S., and Kolosova, E. G. (2000). *Response of the Marine Ecosystem to Global Changes in the Arctic: The Upper Ocean* (Vol. 2, pp. 12–21). Arctic Regional Center, Vladivostok [in Russian].

Mikhalevsky, P. N., Muench, R. D., and DiNapoli, F. R. (1991). *Acoustic Measurements of Arctic Ocean Warming*. SAIC Proposal for NSF.

Milyaev, N. A. (1961). Simultaneous geomagnetic variations in the Arctic and in the Antarctica. *Informative Bulletin of the Soviet Antarctic Expedition*, **30**, 27–31 [in Russian].

Mironov, Ye. U. (1985). Large-scale variability of ice thickness in the Arctic Basin and adjoining seas (205 pp.). Doctoral thesis, AARI, Leningrad [in Russian].

Mironov, Ye. U. and Uralov, N. S. (1991). *Year-to-year Variations of Ice Transport from the Arctic Basin through the Straits of the Canadian Arctic Archipelago and Fram Strait* (IAHS No. 208, pp. 128–141). International Association of Hydrological Sciences, Oxford, UK.

Mosby, H. (1932). Sunshine and radiation. *The Norwegian North Polar expedition with 'Maud' 1918–1925: Scientific Research* (Vol. 1a, No. 7). Geofysisk Institutt, Bergen.

Moskalev, L. I. (1980). Benthic fauna of the outer part of the shelf of the Chukchi Sea. *Ecological Investigations of the Shelf* (pp. 73–79). Moscow [in Russian].

Musina, A. A. (1960). Hydrochemical characteristics of the Arctic Basin. *Proceedings of the AARI*, **218**, 5–58 [in Russian].

Nansen, F. (1901). *In the Dark of Night and Ice: The Norwegian Expedition on board 'Fram' to the North Pole* (Vol. 2, 308 pp.). O. N. Popova, St. Petersburg [in Russian].

Nansen, F. (1902). Oceanography of the North Polar basin. *The Norwegian North Polar Expedition 1893–1896: Scientific Results* (Vol. 3, 88 pp.). Longwance, Green and Co., London.

Brestkin, S., Yulin, A., Karklin, V., Ashik, I., Gudkovich, Z., Karelin, I., Klyachkin, S., Makarov, E., Sapershtein, E., Sergeeva, I., *et al.* (1998) Natural conditions along the selected routes (INSROP Working Paper No. 121, 31pp.). Fridtjof Nansen Institute, (INSROP), Oslo.

Nazintsev, Yu. P. (1964). The heat budget of the multiyear ice cover surface in the Central Arctic. *Proceedings of the AARI*, **267**, 110–126 [in Russian].

Nazintsev, Yu. P. (1971a). Snow accumulation on ice cover in the Kara Sea. *Proceedings of the AARI*, **303**, 185–190 [in Russian].

Nazintsev, Yu. P. (1971b). The estimate of lateral melting of drift ice floes. *Proceedings of the AARI*, **303**, 180–184 [in Russian].

Nekrasov, B. Y., Shirochkov, A. V., and Shumilov, I. A. (1982). Investigation of the irregular structure of the polar ionosphere using oblique incidence soundings. *Journal of Atmospheric and Terrestrial Physics*, **44**, 769–772

Nikiforov, Ye. G. (1957). Ice concentration changes connected with its dynamics. *Problemy Arktiki [Problems of the Arctic]*, **2**, 59–72 [in Russian].

Nikiforov, Ye. G. (1995). Oceanographic investigations in the Arctic. *Problemy Arktiki i Antarktiki [Problems of the Arctic and Antarctic]*, Jubilee Edition, **70**, 150–171 [in Russian].

Nikiforov, Ye. G. and Shpaykher, A. O. (1980). *Regularities in the Formation of Large-scale Fluctuations of the Hydrological Regime in the Arctic Ocean* (270 pp.), Gidrometeoizdat, Leningrad [in Russian].

Nikiforov, Ye. G. and Timokhov, L. A. (1974). Some problems with ice cover dynamics. *Proceedings of the AARI*, **316**, 4–17 [in Russian].

Nikiforov, Ye. G. and Timokhov, L. A. (1978). Mathematical models of ice cover deformation. *Proceedings of the AARI*, **354**, 69–79 [in Russian].

Nikiforov, Ye. G., Blinov, N. I., and Lukin, V. V. (1977). New data on the nature of intermediate waters in the Arctic Basin. *Proceedings of the AARI*, **338**, 17–24 [in Russian].

Nikiforov, Ye. G., Blinov, N. I., and Lukin, V. V. (1979). Some results of field research work as part of the 'POLEX-North-76' program. *POLEX-North-76: Scientific Results* (Vol. 1, pp. 129–147). Gidrometeoizdat, Leningrad [in Russian].

Nikitin, M. M. (1969). Nonperiodic currents of the Arctic Basin. *Proceedings of the AARI*, **117**, 101 [in Russian].

Nikolayev, Yu. V. (1963). The theory of airmass transformation over the sea. *Problemy Arktiki i Antarktiki [Problems of the Arctic and Antarctic]*, **13**, 35–43 [in Russian].

Nikolayeva, A. Ya. and Shesterikov, N. P. (1970). A method for ice condition calculation (on the example of the Laptev Sea). *Proceedings of the AARI*, **292**, 143–217 [in Russian].

Nikolayev, Yu. V., Makshtas, A. P., and Ivanov, B. V. (1986). Problems of studying the marginal zones of the Arctic seas. *Proceedings of the AARI*, **406**, 33–37 [in Russian].

Nikolsky, A. P. (1956). The second zone of enhanced intensity of geomagnetic disturbances in the near-pole region. *Proceedings of the AARI*, **83**, 15–22 [in Russian]

Obukhov, A. M. (1941). Energy distribution in the spectrum of turbulent flow. *Reports of the USSR Academy of Science, Geography and Geophysics*, **4–5**, 453–463 [in Russian].

Oceanographic Atlas of the Polar Seas (1958). U.S. Navy Hydrographical Office, Washington, DC (*Part II: Arctic*).

Ohl, A. I. (1969). Indices of geomagnetic activity and their heliogeophysical essence. *Proceedings of the AARI*, **289**, 28–32 [in Russian].

Okolodkov, Yu. B. (1989). Ice flora of the East Siberian Sea (May 1987 collections). *Novitates Systematicae Plantarum non Vascularium*, **26**, 36–41 [in Russian].

Okolodkov, Yu. B. (1992). Cryopelagic flora of the Chukchi, East Siberian and Laptev Seas. *Proceedings of the NIPR Symposium on Polar Biology* (Vol. 5, pp. 28–43).

Orvig, S. (ed.) (1970). *Climates of the Polar Regions*. Elsevier, Amsterdam.

Ovsiyenko, S. N. (1976). The numerical modeling of ice drift. *Reports of the USSR Academy of Science, Physics of Atmosphere and Ocean*, **12**(1), 1201–1206 [in Russian].

Ovsiyenko, S. N. and Efroimson, V. O. (1983). The theory of dynamic and thermal redistribution of ice at sea. *Studies of the Ice Cover of North-western Seas*. Nauka, Moscow [in Russian].

Ozmidov, R. V. (1968). *Horizontal Turbulence and Turbulent Exchange in the Ocean* (199 pp.). Nauka, Moscow [in Russian].

Pasternak, Ph. A. (1980). Pennatularia *Umbellula encrinus* (L.) from the Canadian Hollow of the Arctic Ocean. *Biology of the Central Arctic Basin* (pp. 236–239). Nauka, Moscow [in Russian].

Pavshtiks, E. A. (1977). Seasonal changes of the age content of the Calanoida populations in the Arctic Basin. *Marine Fauna Research*, **19**(27), 56–72 [in Russian].

Pavshtiks, E. A. (1980). Several regularities in the life of plankton of the central Arctic Basin (pp. 142–154). *Biology of the Central Arctic Basin*. Nauka, Moscow [in Russian].

Petrov, V. M. and Frolov, I. Ye. (1984). The mechanism behind the formation of the warm streak in the pycnocline layer of the Arctic seas. *Meteorologiya i Gidrologiya [Meteorology and Hydrology]*, **11**, 110–112 [in Russian].

Petryashev, V. V. (2004). Misids (Crustacea, Mysidacea) of the Eurasian Subbasin of the Arctic Basin and adjacent seas: Barents, Kara and Laptev. In: *Marine Fauna Researc – Fauna and Ecology of the Laptev Sea and Adjacent Abyssal Regions of the Arctic Basin. Part. I.* (**54**(62), pp. 124–145) [in Russian].

Pivovarov, S. V. (2000). *Chemical Oceanography of the Russian Arctic Seas* (88 pp.). Gidrometeoizdat, St. Petersburg [in Russian].

Pivovarov, S. V., Ivanov, V. V., Jones E.P., and Swift, J. T. (1997). The dissolved oxygen minimum in the Atlantic water in the Eurasian subbasins of the Arctic Ocean. *Polar Processes and Global Climate* (Draft Summary Report at Rosario, Orcas Island, WA, 3–6 November, pp. 169–171). U.S. Geological Survey, Reston, VA.

Pogosyan, Kh. P. (1972). *General Circulation of the Atmosphere* (379 pp.). Gidrometeoizdat, Leningrad [in Russian].

Pokrovskiy, O. M. and Timokhov, L. A. (2002). Reconstruction of winter fields of temperature and salinity in the Arctic Ocean. *Oceanology*, **42**(6), 822–830 [in Russian].

Polyakov, I. V., Proshutinsky, A. Y., and Johnson, M. A. (1999). Seasonal cycles in two regimes of Arctic climate. *Journal of Geophysical Research*, **104**(C11), 25761–25788.

Polyakov, I. V., Alekseev, G. V., Timokhov, L. A., Bhatt, U. S., Colony, R. L., Simmons, H. L., Walsh, D., Walsh, J. E., and Zakharov, V. F. (2004). Variability of the intermediate Atlantic water of the Arctic Ocean over the last 100 years. *Journal of Climate*, **17**(23), 4485–4497.

Porubayev, V. S. (2000). The influence of dynamic and thermal factors on seasonal changes in mean ice thickness in the Arctic Basin. *Meteorologiya i Gidrologiya [Meteorology and Hydrology]*, **11**, 73–79 [in Russian].

Pozdnyshev, S. P. (1990). Calculation of wind speed in non-stationary baric structures as applied to the problems of short-term ice forecast. *Proceedings of the AARI*, **418**, 89–99 [in Russian].

Prik, Z. M. (ed.) (1961). *The Climatological Handbook of the Soviet Arctic* (Vol. II, 308 pp.). Morskoy Transport, Leningrad [in Russian].

Prik, Z. M. (1965). *The Climate of the Soviet Arctic: Meteorological Regime* (279 pp.). Gidrometeoizdat, Leningrad [in Russian].

Prik, Z. M. (1974). Climatological processing of the meteorological observations from the drifting stations. *Proceedings of the AARI*, **328**, 4–21 [in Russian].

Proshutinsky, A. Yu. (1993). *Fluctuations in Arctic Ocean Level* (216 pp.). Gidrometeoizdat, St. Petersburg [in Russian].

Proshutinsky, A. Y. and Johnson, M. A. (1997). Two circulation regimes of the wind-driven Arctic Ocean. *Journal of Geophysical Research*, **102**, 12493–12514.

Radionov, V. F. (ed.) (1997). *Technological Handbook of the Climate of Russia (Arctic Region): Solar Radiation* (230 pp.). Gidrometeoizdat, St. Petersburg [in Russian]. (English translation is available in Arctic Climatology Project, Environmental Working Group (2000), *The Arctic Meteorology and Climate Atlas*, edited by F. Fetterer and V. Radionov, National Snow and Ice Data Center, Boulder, CO, CD-ROM.)

Radionov, V. F., Bryazgin, N. N., and Aleksandrov, E. I. (1996). *Snow Cover of the Arctic Basin* (124 pp.). Gidrometeoizdat, St. Petersburg [in Russian]. (English translation available from Polar Science Center, University of Washington, Seattle, WA 98105, Tech. Rep. APL-UW, TR 9701.)

Ragozin, A. I. and Chukanin, K. I. (1959). Mean trajectories and drift speeds of the pressure systems in the Eurasian Arctic and Sub-Arctic. *Proceedings of the AARI*, **217**, 35–64 [in Russian].

Regulation of Presidium of the USSR Academy of Sciences of 16 May 1974: On the assumed names of physical-geographic objects located outside the USSR (1974). *Bulletin of the USSR Academy of Sciences*, **9**, 125–126 [in Russian].

Rigor, I. G. and Wallace, J. M. (2004). Variations in the age of Arctic sea ice and summer sea ice extent. *Geophysical Research Letters*, **31**, L09401, doi:10.1029/2004GL019492.

Rishbeth, H. and Garriott, O. K. (1969) *Introduction to Ionospheric Physics* (304 pp.). Academic Press, New York and London.

Romanov, I. P. (1992). *Ice Cover of the Arctic Basin* (212 pp.). I. P. Romanov, St. Petersburg [in Russian].

Romanov, I. P., Konstantinov, Yu. B. and Kornilov, N. A. (1997). *Drifting Stations 'North Pole' (1937–1991)* (226 pp.). Gidrometeoizdat, St. Petersburg [in Russian].

Rothrock D. A. and Maykut, G. A. (1999). Thinning of Arctic sea ice cover. *Geophysical Research Letters*, **26**(23), 3469–3472.

Rozhkov, V. A. (2001). *The Theory of Estimating the Probability Characteristics of Random Values and Functions with Hydrometeorological Examples* (340 pp.). Gidrometeoizdat, St. Petersburg [in Russian].

Rusanov, V. P., Yakovlev, N. I., and Buynevich, A. G. (1979). Hydrochemical analysis of the Arctic Ocean. *Proceedings of the AARI*, **355**, 144 pp.

Schlitzer, R. (2003). *Ocean Data View*. Available online at ⟨*http://www.awi-bremerhaven.de/GEO/ODV*⟩

Semtner, A. J. (1976). A model for the thermodynamic growth of sea ice in numerical investigations of climate. *J. Phys. Oceanogr.*, **6**, 379–389.

Semtner, A. J. (1987). A numerical study of sea ice and ocean circulation in the Arctic. *J. Phys. Oceanogr.*, **17**, 1077–1099.

Sen'ko, P. K. (1958). On the unusual local features of geomagnetic variations nearby Mirny Station. *Informative Bulletin of the Soviet Antarctic Expedition Proceedings of the AARI*, **1**, 15–19 [in Russian].

Shchuka, T. I. (1969). Ground surface atmospheric pressure variations during precipitation of high-energy particles. *Proceedings of the AARI*, **289**, 54–57 [in Russian]

Shesterikov, N. P. (1963). Some peculiarities of ice growth in the area of Mirny. *Problemy Arktiki i Antarktiki [Problems of the Arctic and Antarctic]*, **13** [in Russian].

Shirochkov, A. V. and Maurits, S. A. (1986). Impact of solar proton flares on variations of the stratospheric temperature above the Antarctica. *Proceedings of the AARI*, **405**, 125–131 [in Russian]

Shirochkov, A. V. and Nagurny, A. P. (1992). On the problem of a connection between ozone dynamics at high latitudes and magnetospheric processes. *Journal of Atmospheric and Terrestrial Physics*, **54**, 835–840

Shirochkov, A. V., Makarova, L. N., Verbin, Yu. P., and Semenov, G. A. (1998). Investigations of the effects of solar proton events by data of VLF phase records at high-latitude radio paths. *Geomagnetism and Aeronomy*, **38**, 101–107 [in Russian]

Shirshov, P. P. (1938). Oceanographic observations. *Proceedings of the USSR Academy of Sciences*, **XIX**(8), 569–580 [in Russian].

Shirshov, P. P. (1944). Scientific results of the 'North Pole' station. *General Meeting of the USSR Academy of Sciences, 14–17 February* (pp. 110–140) [in Russian].

Shpaykher, A. O. (1969). Atlantic and Pacific water advection as a factor of Arctic Basin climate changes. *Reports of the USSR Academy of Science, Geography*, **3**, 29–38 [in Russian].

Shpaykher, A. O., Konstantinov, Yu. B., and Rachkov, V. S. (1968). Temperature and salinity characteristics of the surface waters in fractures in the Arctic Basin. *Problemy Arktiki i Antarktiki [Problems of the Arctic and Antarctic*], **29**, 35–41 [in Russian].

Shuleykin, V. V. (1938). Ice floe drift. *Reports of the USSR Academy of Science*, **19**(8), 582–587 [in Russian].

Shuleykin, V. V. (1953). *Physics of the Sea* (990 pp.). USSR Academy of Science Publishers, Moscow [in Russian].

Shvets, M. Ye. (1946). On the hydromechanical theory of ice floe drift. *Meteorologiya i Gidrologiya [Meteorology and Hydrology]*, **6**, 58–68 [in Russian].

Shy, T. L. and Walsh, J. E. (1996). North Pole ice thickness and association with ice ocean history 1977–1992, 1979–1986. *Geophysical Research Letters*, **23**, 2975–2978.

Sirenko, B. I. (2001). List of species of free-living invertebrates of the Eurasian Arctic seas and adjacent deep waters. *Marine Fauna Research*, **51**(59), 131.

Sirenko, B. I., Denisenko, S. G., Deubel, H., and Rachov, E. (2004) Deep water communities of the Laptev Sea and adjacent parts of the Arctic Ocean. *Marine Fauna Research*, **54**(62), 28–73.

Smagin, V. M. (1995). Hydrochemical investigations of the oolar oceans. *Problemy Arktiki i Antarktiki [Problems of the Arctic and Antarctic]*, **70**, 183–192 [in Russian].

Smagin, V. M., Timokhov, L. A., and Colony, R. (2003). Interannual variability of hydrochemical elements in the Arctic Ocean. *ACSYS Final Conference, St. Petersburg* (abstract, pp. 133).

Smetannikova, A. V. (ed.) (1983). *The Radiation Regime of the Greenland and Norwegian Seas* (64 pp.). Gidrometeoizdat, Leningrad [in Russian].

Smirnov, V. B. (ed.) (1972). *Oblique Sounding of the Ionosphere* (112 pp.). Gidrometeoizdat, Leningrad [in Russian]

Smirnov, V. N. (1972). Ice cover fluctuations caused by the internal waves of the Arctic Ocean. *Reports of the USSR Academy of Science*, **206**(5), 1105–1108 [in Russian].

Smirnov, V. N. (1996). *The Dynamic Processes in Sea Ice* (162 pp.). Gidrometeoizdat, St. Petersburg [in Russian].

Solodovnikov, G. K., Sinelnikov, V. M., and Krochmalnikov, E. B. (1988). *Remote Sounding of the Earth's Ionosphere by Using Signals from Satellite Radio Beacons*. (156 pp.). Nauka, Moscow [in Russian].

Spichkin, V. A. (1963). Role of melting in forming the relief of the multiyear ice upper surface. *Problemy Arktiki i Antarktiki [Problems of the Arctic and Antarctic]*, **14**, 71–73 [in Russian].
Steel, V. and Boyd, T. (1998). Retreat of the cold halocline layer in the Arctic Ocean. *Journal of Geophysical Research*, **103**, 10419–10435.
Sverdrup, H. U. (1928). The wind-drift of the ice on the North Siberian shelf. *The Norwegian North Polar Expedition with 'Maud' 1918–1925: Scientific Research* (Vol. 4, No. 16, pp. 46–62). Geofysisk Institutt, Bergen.
Swift, J. H., Aagaard, K., Timokhov, L., and Nikiforov, Ye. G. (2004). Long-term variability of Arctic Ocean waters: Evidence from a reanalysis of the EWG data set. *Journal of Geophysical Research.*
Sytinsky, A. D. and Tripolnikov, V. P. (1964). Some results of studies of natural fluctuations of ice floes in the Central Arctic. *Reports of the USSR Academy of Science*, **4**, 210–212 [in Russian].
The Climatic Atlas of the Arctic (1963). Morskoy Transport, Leningrad [in Russian].
Thompson, D. W. J. and Wallace, J. M. (1998). The Arctic Oscillation signature in the wintertime geopotential height and temperature fields. *Geophysical Research Letters*, **25**, 1297–1300.
Thorndike, A. and Colony, R. (1982). Sea ice motion in response to geostrophic winds. *Journal of Geophysical Research*, **87**(8), 5852–5854.
Thorndike, A. S. and Colony, R. (1983). *Objective Analysis of Atmospheric Pressure and Sea Ice Motion over the Arctic Ocean* (10 pp.). POAC, Helsinki.
Timerev, A. A. (1976). Reflecting properties of the underlying surface of the polar regions. *Proceedings of the AARI*, **328**, 106–115 [in Russian].
Timerev, A. A. (1981). *Illumination Regime of the Soviet Arctic* (101 pp.). Gidrometeoizdat, Leningrad [in Russian].
Timofeev, V. T. (1960). *Waters of the Arctic Basin* (191 pp.). Gydrometeoizdat, Leningrad [in Russian].
Timokhov, L. A. (1970). On the theory of ice floe rotation in the ice cover at various concentrations. *Problemy Arktiki i Antarktiki [Problems of the Arctic and Antarctic]*, **33**, 46–56 [in Russian].
Timokhov, L. A. and Kheysin, D. Ye. (1987). *Sea Ice Dynamics* (272 pp.). Gidrometeoizdat, Leningrad [in Russian].
Treshnikov, A. F. (1959). The surface waters of the Arctic Basin. *Problemy Arktiki [Problems of the Arctic]*, **7**, 5–14 [in Russian].
Treshnikov, A. F. (1979). Aims and results of the national field research experiment 'POLEX-North-76'. *POLEX-North-76: Scientific Results* (Vol. 1, pp. 7–15). Gydrometeoizdat, Leningrad [in Russian].
Treshnikov, A. F. (ed.) (1985). *Atlas of the Arctic* (204 pp.). GUGiK, Moscow [in Russian].
Treshnikov, A. F. and Baranov, G. I. (1972). *Structure of Arctic Basin Water Circulation* (158 pp.). Gidrometeoizdat, Leningrad [in Russian].
Treshnikov, A. F., Balakshin, L. L., Belov, N. A., Demenitskaya, R. M., Dibner, V. D., Karasik, A. M., Shpaikher, A. O., and Shurgaeva, N. D. (1967). Geographical names of the main parts of the bottom relief of the Arctic Basin. *Problemy Arktiki i Antarktiki [Problems of the Arctic and Antarctic]*, **27**, 5–15 [in Russian].
Troshichev, O. A. (ed.) (1986). *Ionospheric–Magnetospheric Disturbances in High Latitudes* (256 pp). Gidrometeoizdat, Leningrad [in Russian].
Tsinovskiy, V. D. (1980). Ichthyofauna of the abyssal basins of the central Arctic Basin. *Biology of the Central Arctic Basin* (pp. 214–218). Nauka, Moscow [in Russian].

Tsinovskiy, V. D. and Melnikov, I. A. (1980). The presence of *Liparis koefoedi* (Liparidae: Osteichthyes) in waters of the central Arctic Basin. *Biology of the Central Arctic Basin* (pp. 211–214). Nauka, Moscow [in Russian].

Tsurikov, V. L. (1976). *Liquid Phase in Sea Ice* (211 pp.). Nauka, Moscow [in Russian].

Tyshko, K. P., Cherepanov, N. V., and Fedotov, V. U. (2000). *The Crystalline Structure of Sea Ice Cover* (68 pp.). Gidrometeoizdat, St. Petersburg [in Russian].

Ugryumov, A. I. and Korovin, V. P. (2004). *To the North Pole on an Ice Floe: History of Polar Drifting Stations* (125 pp.). Gidrometeoizdat, St. Petersburg [in Russian].

Untersteiner N. and Badgley F. I. (1965). The roughness parameters of sea ice. *Journal of Geophysical Research*, **70**, 4573–4577.

Usachev, P. I. (1949). Polar ice microflora. *Proceedings of the Oceanography Institute of the USSR Academy of Sciences*, **3**, 216–259 [in Russian].

Usachev, P. I. (1961). Phytoplankton at the North Pole (from the collections of P. P. Shirshov on the first drifting station NP-1 in 1937–1938 under the supervision of I. D. Papanin). *Proceedings of the All-Union Hydrobiological Society*, **11**.

Ushakov, P. V. (1982). Polychaeta of the subclass Aphroditiformia of the Arctic Basin and northwestern part of the Pacific Ocean (Aphroditidae and Polynoidae families). *USSR Fauna, New Series – Ed. 126: Polychaeta* (Vol. 2, pp. 271). Nauka, Leningrad [in Russian].

Vagapov, R. Kh., Gavrilo, V. P., Kozlov, A. I., Lebedev, G. A., and Logvin, A. I. (1993). *Remote-sensing Methods of Sea Ice Study* (342 pp.). Gidrometeoizdat, St. Petersburg [in Russian].

Vangengeim, G. Ya. (1952). The basis of the macrocirculation method for long-term meteorological forecasts for the Arctic. *Proceedings of the AARI*, **34**, 314 [in Russian].

Vasilenko, S. V. (1988). A new deep-sea species *Hemilamprops canadensis* sp. n. (Crustacea: Cumacea) from the Canadian Basin of the Arctic Basin. *Zoological Journal*, **67**(6), 945–949 [in Russian].

Vernadskiy, V. I. (1938). Some future tasks for ice research in the Arctic seas. *Proceedings of the USSR Academy of Sciences*, **19**(8), 619–622 [in Russian].

Vinje, T. (1992). Sea ice variability in the nordic seas. *Proceedings of International Conference on the Role of the Polar Regions in Global Change* (Vol. 1, pp. 23–27).

Vinje, T. and Finnekasa, O. (1986). *Ice Transport through the Fram Strait* (Skr. No. 186, 40 pp.). Norsk Polarinstitutt, Oslo.

Vinogradov, N. D., Ivanov, V. V., Pryamikov, S. M., *et al.* (1999). The climate change scenario in the Barents Sea region through 2050. *Proceedings of the AARI*, **441**, 195–201 [in Russian].

Virketis, M. A. (1957). Some data on zooplankton of the central part of the Arctic Basin. *Observational Material from Scientific Research Stations NP-3 and NP-4 in 1954–1955* (Vol. 1). Marine Transport, Leningrad [in Russian].

Virketis, M. A. (1959). Material on zooplankton of the central Arctic Basin. In: *Observational material from Scientific Research Stations NP-3 and NP-4 in 1955–1956* (Volume 2, pp. 133–144). Marine Transport, Leningrad [in Russian].

Vize, V. Yu. (1933). Ice drift in the Polar Basin. *Notes of the SHI*, **10**, 327–336 [in Russian].

Vize, V. Yu. (1939). *The Seas of the Soviet Arctic* (568 pp.). Glavsevmorput, Leningrad [in Russian].

Vize, V. Yu. (1940). *The Climate of the Soviet Arctic Seas* (124 pp). Glavsevmorput, Leningrad [in Russian].

Vize, V. Yu. (1944a). *Basis of the Long-term Ice Forecasts for the Arctic Seas*. Glavsevmorput, Leningrad. (Also in *Trudy Arkticheskogo Nauchno-issledovatel'skogo Institute [Proceedings of the ARI]*, **190** [in Russian].)

Vize, V. Yu. (1944b). Hydrometeorological conditions in the area of the ice edge in Arctic seas. *Trudy Arkticheskogo Nauchno-issledovatel'skogo institute [Proceedings of the ARI]*, **184**, 125–151 [in Russian].

Volkov, N. A., Gudkovich, Z. M., and Uglev, V. D. (1971). The results of ice drift non-uniformity studies in the Arctic Basin. *Proceedings of the AARI*, **303**, 76–88 [in Russian].

Vorobyev, V. N. (1966). Lunar–solar semi-monthly and monthly tide in the seas of the Soviet Arctic. *Reports of the USSR Academy of Science*, **167**(5), 1039–1041 [in Russian].

Vorobyev, V. N. and Gudkovich, Z. M. (1976). Interannual variability in the ice drift and currents of the Arctic Basin. *Proceedings of the AARI*, **319**, 23–32 [in Russian].

Vovk, V. J. and Shirochkov, A. V. (1978). Use of data of oblique sounding of the ionosphere for validation of numerical calculations of ray trajectories at high-latitude radio paths. *Geomagnetism and Aeronomy*, **18**(6), 946–950 [in Russian].

Vovk, V. J. and Shirochkov, A. V. (1988). Impact of auroral absorption on the radio channel characteristics in case of oblique sounding of the ionosphere. *Proceedings of the AARI*, **411**, 58–67 [in Russian].

Vovk, V. J., Smirnov, V. B., and Kosterin, I. N. (1978). Investigation of HF radiowave propagation at polar radio paths by means of oblique sounding. *Proceedings of the AARI*, **351**, 20–24 [in Russian].

Voynov, G. N. (1999). *Tidal Phenomena in the Kara Sea* (110 pp.). Russian Geographical Society, St. Petersburg [in Russian].

Wadhams P. (1990). Evidence for thinning of the Arctic ice cover north of Greenland. *Nature*, **345**, 795–797.

Wadhams P. (1994) Sea ice thickness changes and their relation to climate. In: *The Polar Oceans and Their Role in Shaping the Global Environment* (pp. 337–361). American Geophysical Union, Washington, DC.

Warren, S. G., Rigor, I. G., Untersteiner, N., Radionov, V. F., Bryazgin, N. N., Aleksandrov, Ye. I., and Colony, R. (1999). Snow depth on Arctic sea ice. *Journal of Climate*, **12**, 1814–1829.

Washington, W. M. and VerPlank, L. (1986). *A Description of Coupled General Circulation Models of the Atmosphere and Oceans Used for Carbon Dioxide Studies* (Tech. Note NCAR TN-27126 pp.).

Watanabe, K. (1967). Drift velocities of ice measured from air and separately computed values of their wind-induced and current-induced components: Studies on sea ice in the Okhotsk Sea. *Oceanography Magazine*, **14**(1).

Westman, J. (1903). Mesures de l'intensité de la radiation solaire faites en 1899 et en 1900 à la station d'hivernage Suédoise à la baie de Ttreurenberg, Spitzberg. *Miss. Scient. pour la Mesure d'un Arc de Méridien en Spitzberg Entreprises en 1899–1902*, **II**, sec VIII-e B. Radiation solaire. Stockholm.

Winchester, D. (1962). Studies of Arctic sea ice motion in the Canadian Arctic on the basis of hydrometeorological factors. *Sea Ice Studies Abroad* (pp. 70–83). Gidrometeoizdat, Leningrad [in Russian].

Worthington, L. V. (1953). Oceanographic results of Project Skijump I and Skijump II in the Polar Sea, 1951–1952. *EOS, Transactions of the American Geophysical Union*, **34**(4), 543–551.

Yakovlev, G. N. (1957). The radiation regime of the Central Arctic. *Proceedings of the AARI*, **207**, 115 pp. [in Russian].

Yakovlev, G. N. (1960). Snow cover on drifting ice in the Central Arctic. *Problemy Arktiki i Antarktiki [Problems of the Arctic and Antarctic]*, **3**, 65–76 [in Russian].
Yanishevsky, Yu. D. (1957). *Actinometric Instruments and Methods of Measurement* (416 pp.). Gidrometeoizdat, Leningrad [in Russian].
Zakharov, V. F. (1966). Role of the flow polynyas in the hydrological and ice regime of the Laptev Sea. *Okeanologiya [Oceanology]*, **6**(6), 1014–1022 [in Russian].
Zakharov, V. F. (1977) Surface Arctic waters as the factor of ice cover stability. *Proceedings of the AARI*, **346**, 122–134 [in Russian].
Zakharov, V. F. (1981). *Arctic Ice and Modern Natural Processes* (136 pp.). Gidrometeoizdat, Leningrad [in Russian].
Zakharov, V. F. (1996). *Sea Ice in the Climatic System* (213 pp.). Gidrometeoizdat, St. Petersburg [in Russian].
Zavyalova, I. N. (2000). The climatology of Arctic clouds. *Proceedings of the AARI*, **440**, 45–65 [in Russian]. (English translation is available in Arctic Climatology Project, Environmental Working Group (2000), *Arctic Meteorology and Climate Atlas*, edited by F. Fetterer and V. Radionov, National Snow and Ice Data Center, Boulder, CO – CD-ROM.)
Zenkevich, L. A. (1966). Peculiarities of the biological regime of the Arctic Basin and native Arctic seas. *Proceedings of the VNIRO*, **60**, 19–26 [in Russian].
Zezina, O. N. (1980). About deep sea existence of brachiopods in the Arctic Basin. *Biology of the Central Arctic Basin* (pp. 240–241). Nauka, Moscow. [in Russian].
Zhigalov, L. N. (1960). Some peculiarities of variations in the vertical component of the geomagnetic field in the Arctic Ocean. *Geomagnetic Explorations* (Vol. 5, pp. 73–77). Nauka, Moscow [in Russian]
Zhirkov, I. A. (1980). Polychaetus fauna of the Canadian Basin abyss. *Biology of the Central Arctic Basin* (pp. 229–236). Nauka, Moscow [in Russian].
Zhulina, E. M., Kishcha, P. V., Lukashkin, V. M., and Shirochkov, A. V. (1983). *Additional Energetic Losses at High-latitude Radio Paths* (208 pp.). Nauka, Moscow [in Russian].
Zubakin, G. K. (1987). *Large-scale Variability of the Ice Cover of the North-European Basin Seas* (160 pp.). Gidrometeoizdat, Leningrad [in Russian].
Zubov, N. N. (1945). *Arctic Ice* (360 pp.). Moscow [in Russian].
Zuyev, A. N. (1983). Numerical model for calculation of ice drift speed and redistribution at sea. *Proceedings of the AARI*, **385**, 19–26 [in Russian].

Index

Color plates

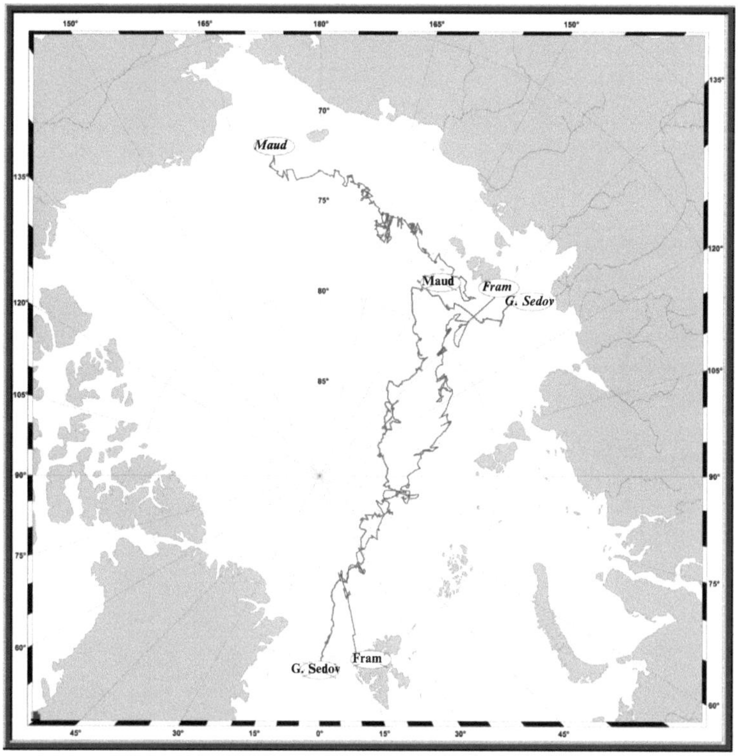

Figure 1.11. Layout of the drift of the *Fram*, *Maud* and *G. Sedov*. (Initial positions in italic.)

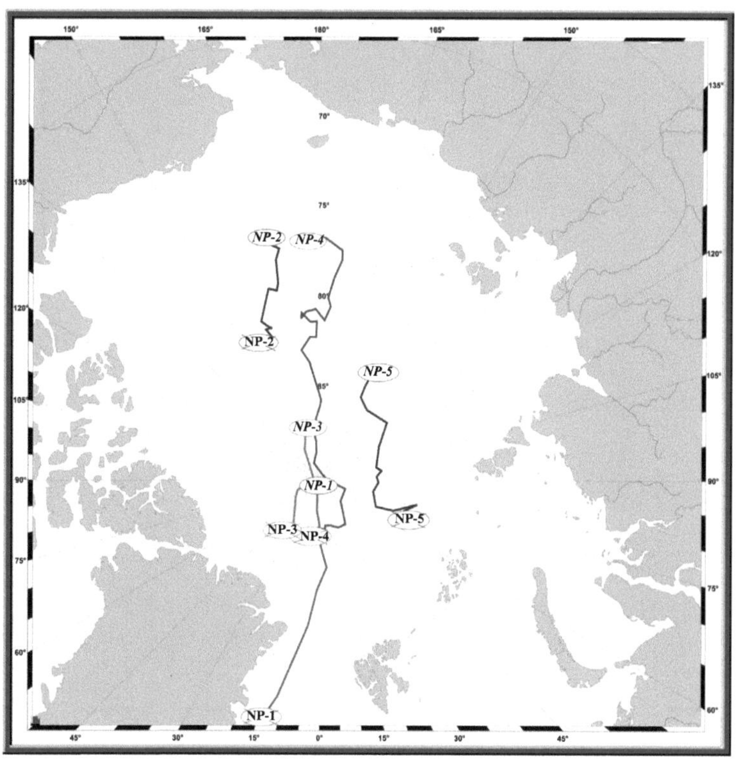

Figure 1.12. Layout of the drift of the 'NP-1'–'NP-5' stations. (Initial positions in italic.)

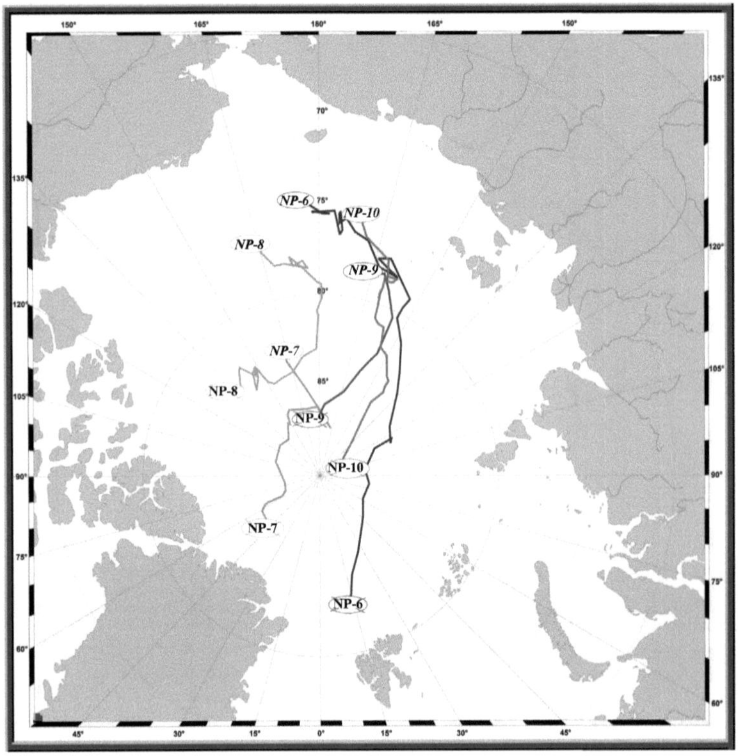

Figure 1.13. Layout of the drift of the 'NP-6'–'NP-10' stations. (Initial positions in italic.)

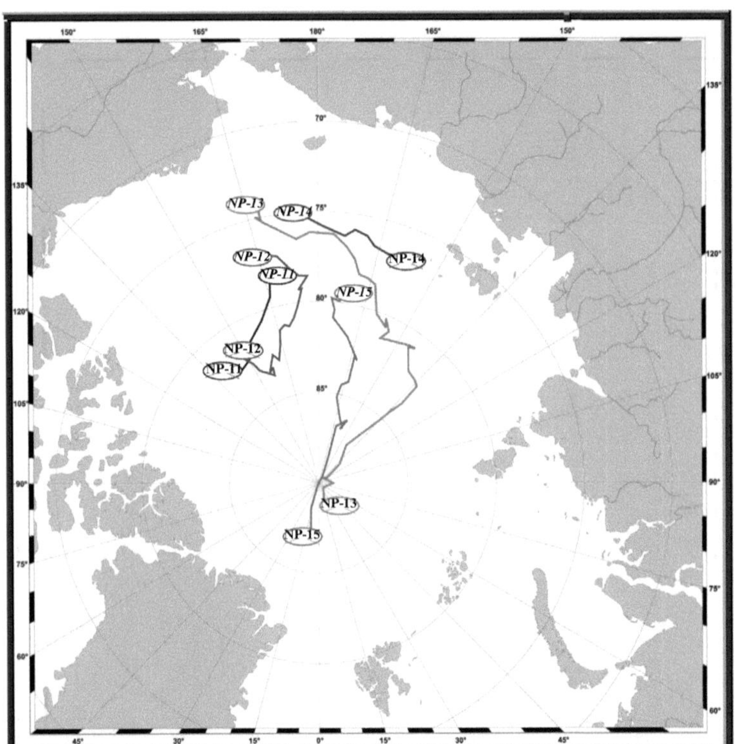

Figure 1.14. Layout of the drift of the 'NP-11'–'NP-15' stations. (Initial positions in italic.)

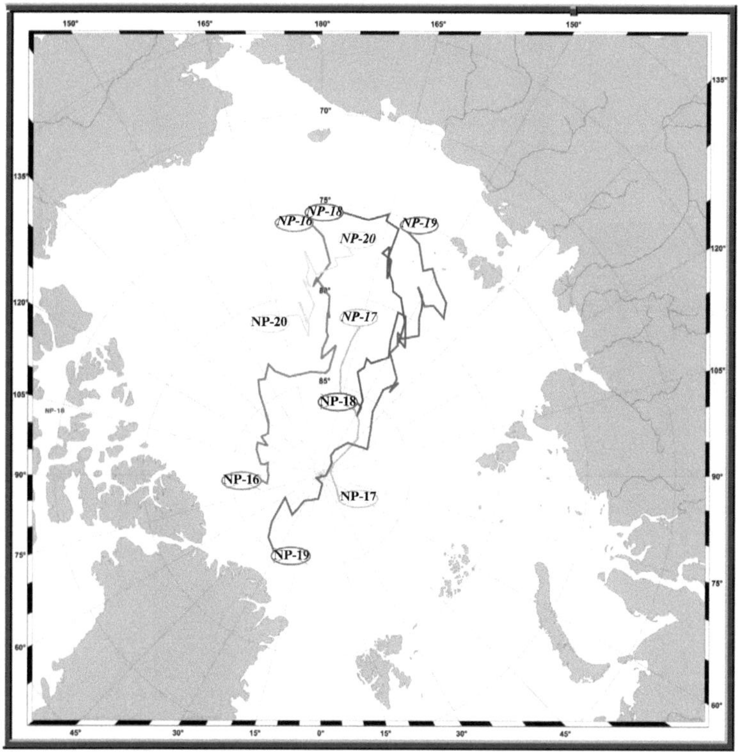

Figure 1.15. Layout of the drift of the 'NP-16'–'NP-20' stations. (Initial positions in italic.)

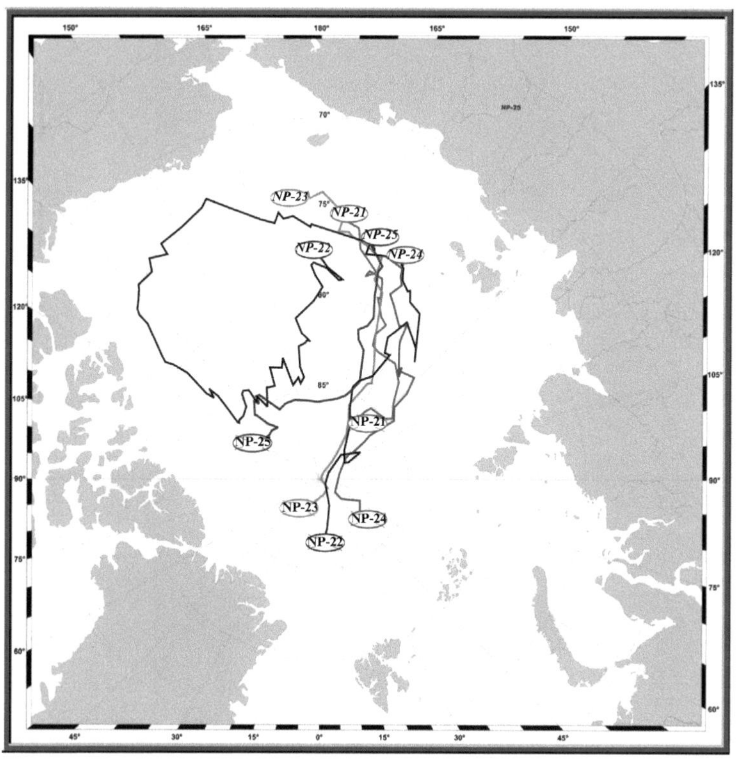

Figure 1.16. Layout of the drift of the 'NP-21'–'NP-25' stations. (Initial positions in italic.)

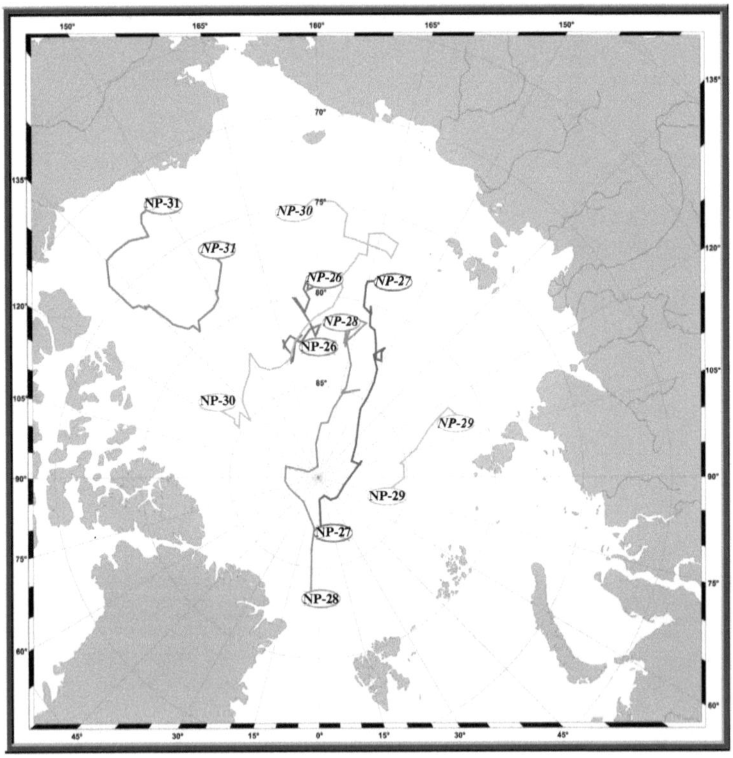

Figure 1.17. Layout of the drift of the 'NP-26'–'NP-31' stations. (Initial positions in italic.)

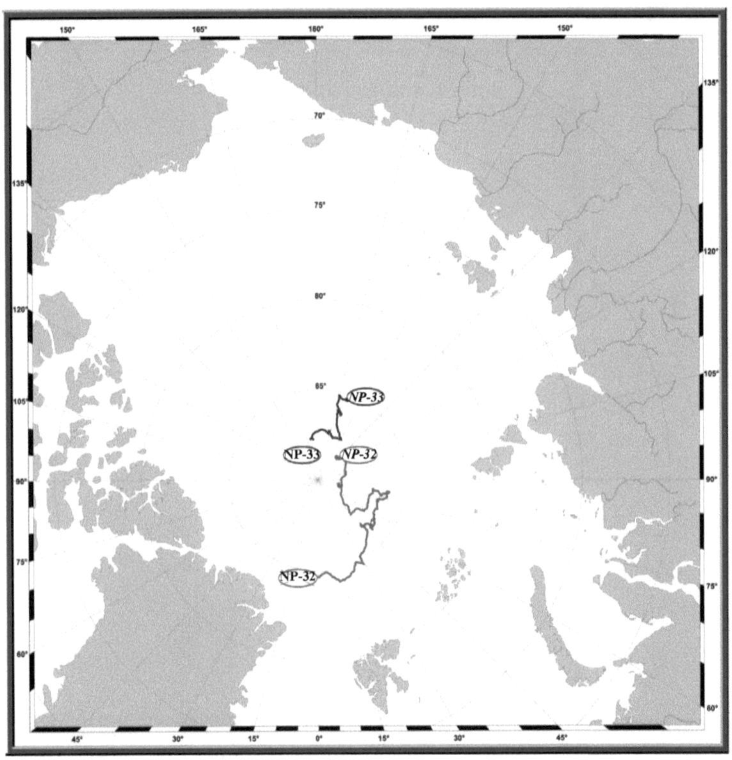

Figure 1.18. Layout of the drift of the 'NP-32'–'NP-33' stations. (Initial positions in italic.)

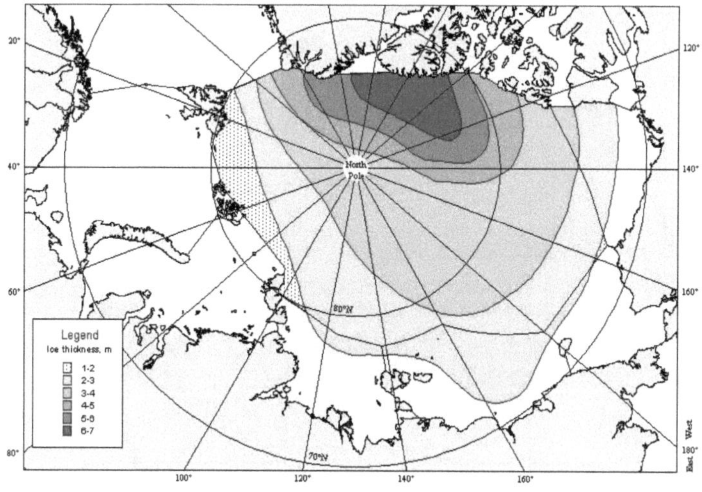

Figure 3.2. Average ice thickness in the Arctic basin.

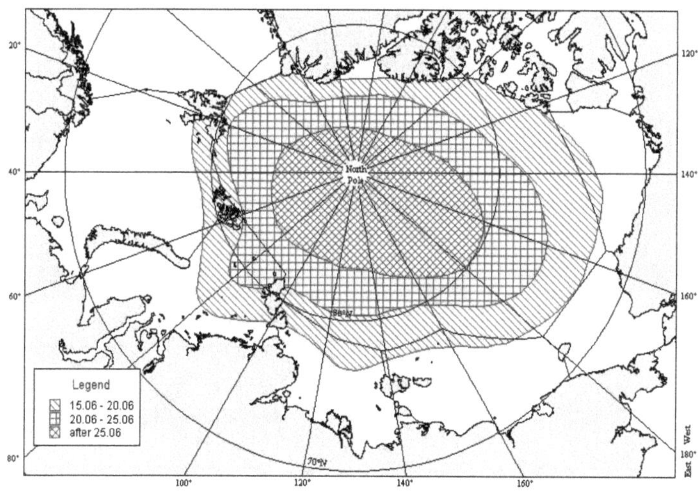

Figure 3.3. Average dates of the onset of melting in the Arctic basin.

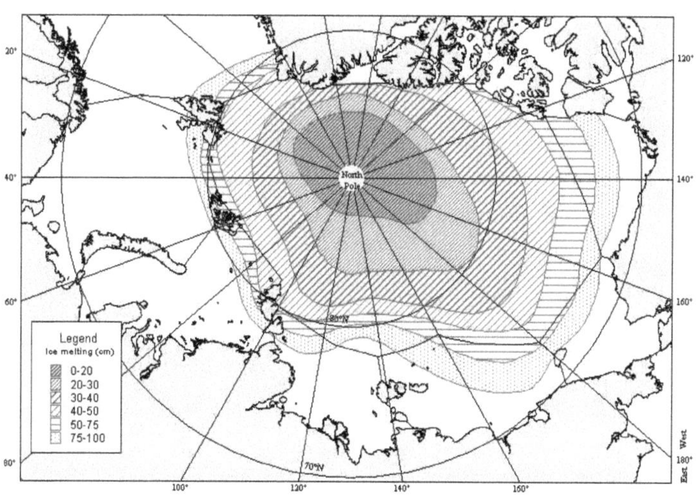

Figure 3.4. Average value of icemelt over summer (cm) in the Arctic basin.

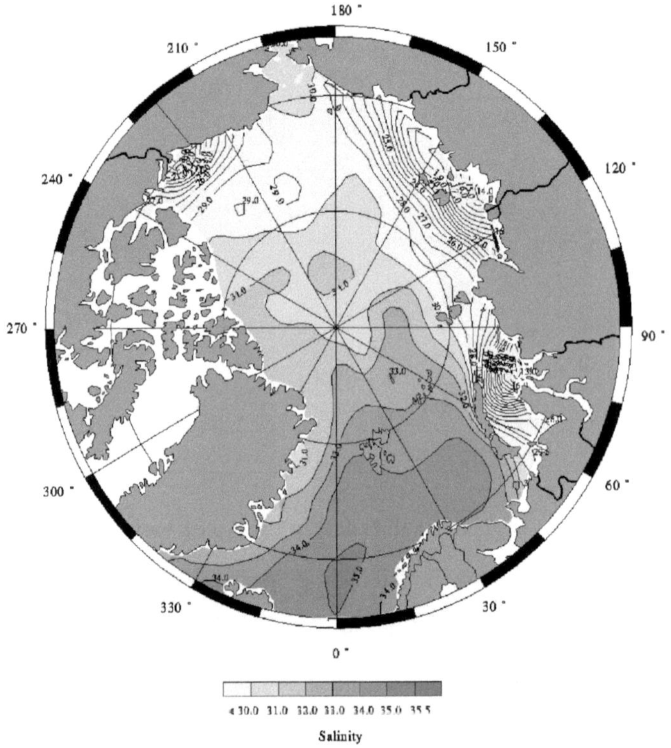

(a)

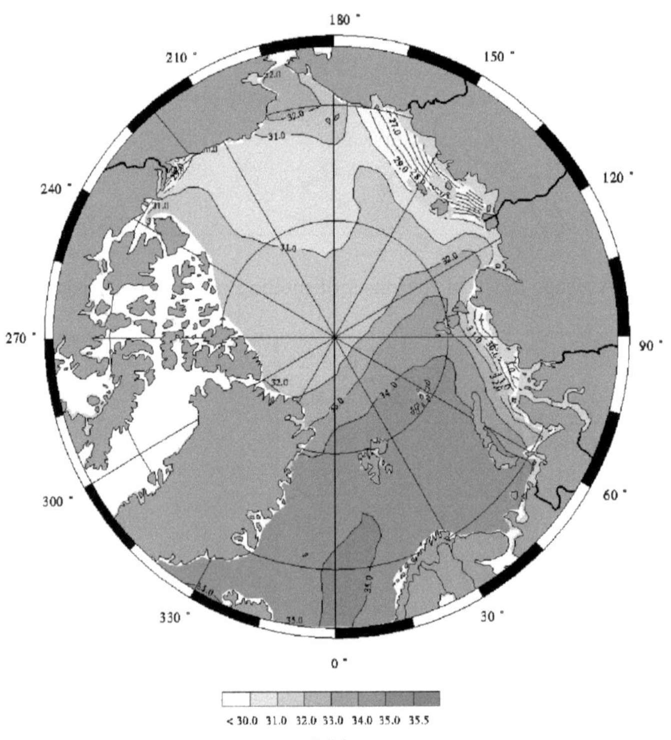

(b)

Figure 4.8. (a, b) Mean water salinity (psu) and (c, d) temperature (°C) for the years 1950–1989 at 5 m for the winter and summer periods.
From *Joint US-Russian Atlas of the Arctic Ocean* (1997, 1998).

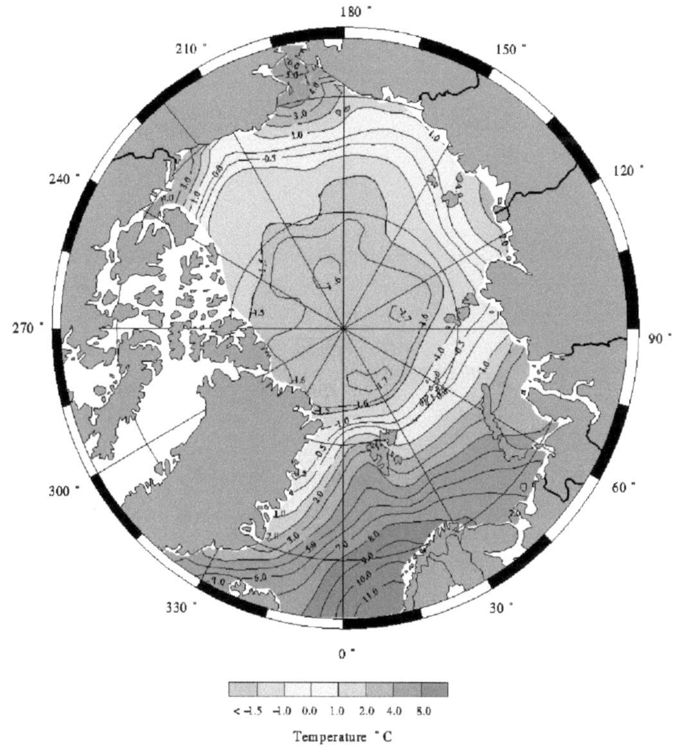
180 °
210 °
150 °
240 °
120 °
270 °
90 °
300 °
60 °
330 °
30 °
0 °
< -1.5 -1.0 0.0 1.0 2.0 4.0 8.0
Temperature °C

(c)

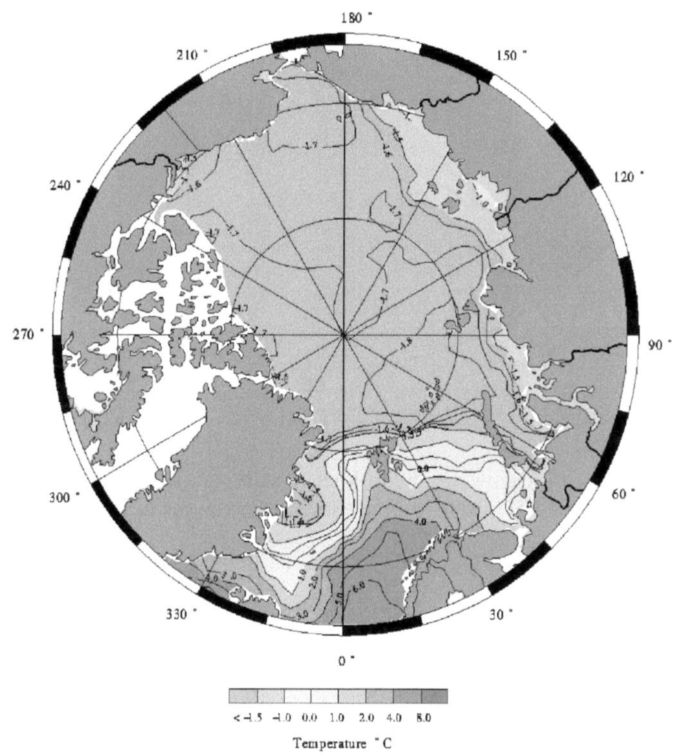
180 °
210 °
150 °
240 °
120 °
270 °
90 °
300 °
60 °
330 °
30 °
0 °
< -1.5 -1.0 0.0 1.0 2.0 4.0 8.0
Temperature °C

(d)

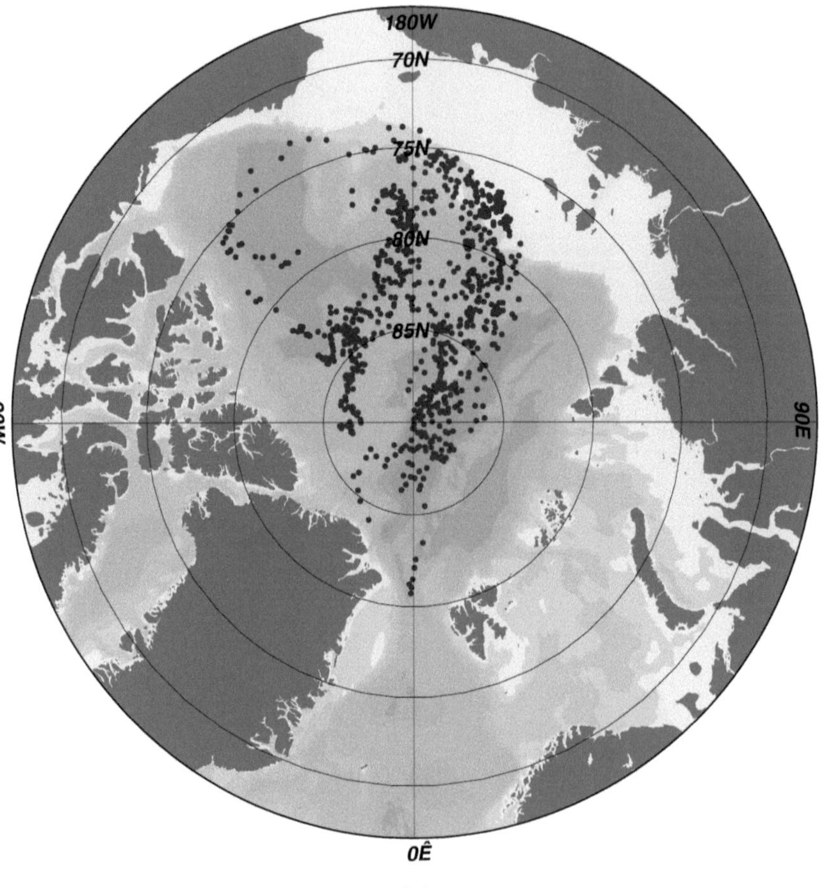

(a)

Figure 4.15. Location of deepwater hydrological stations conducted at the 'NP' drifting stations during the years 1937–1991 (a) and 'Sever' expeditions in the years 1941–1993 (b).

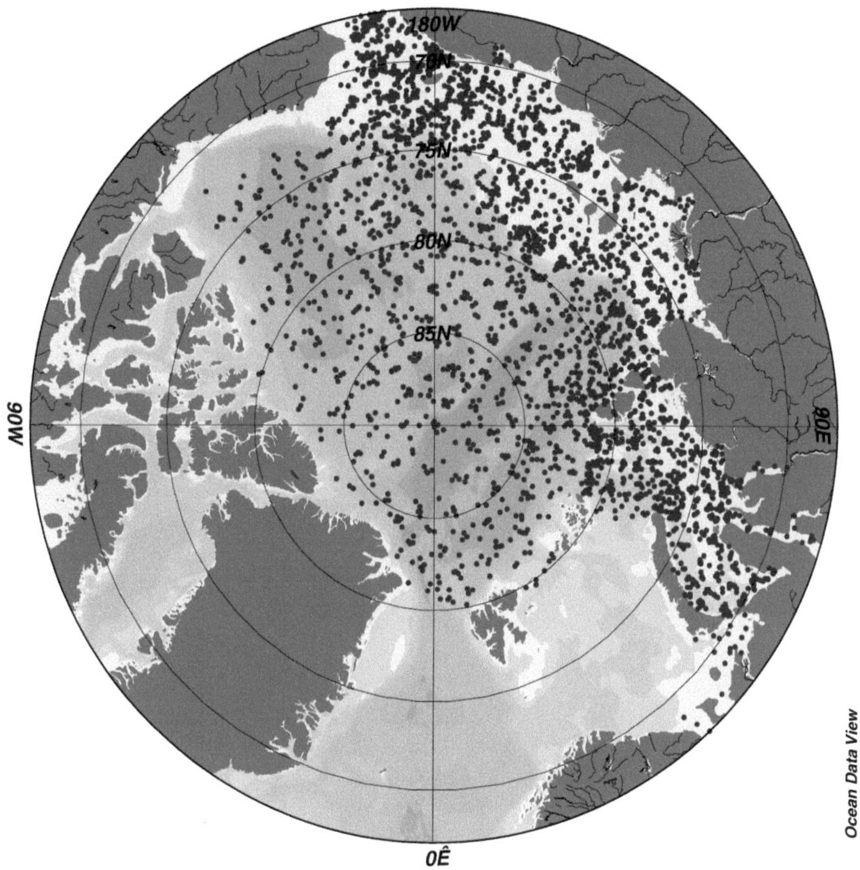

(b)

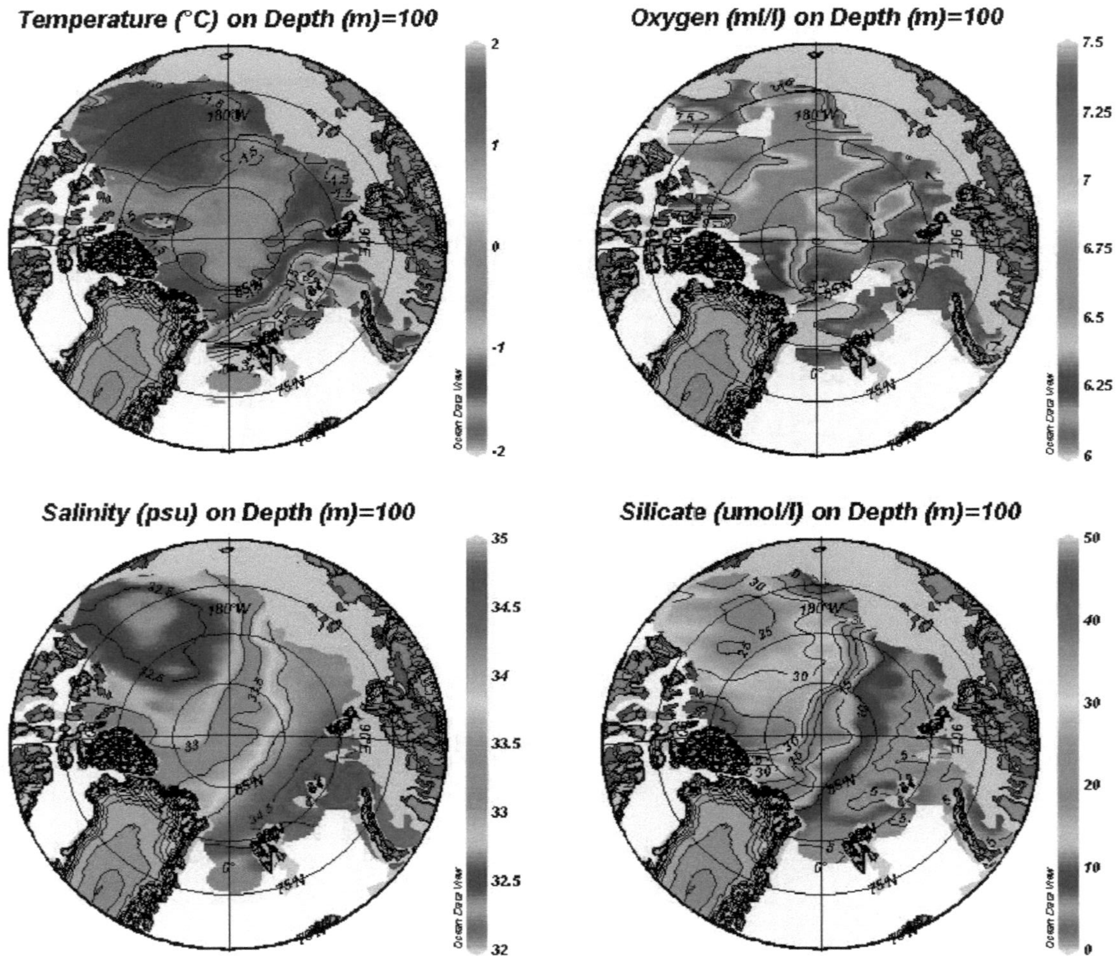

Figure 4.21. Mean temperature, salinity, oxygen and silicon values at 100 m in the Arctic Ocean (1965–1989).

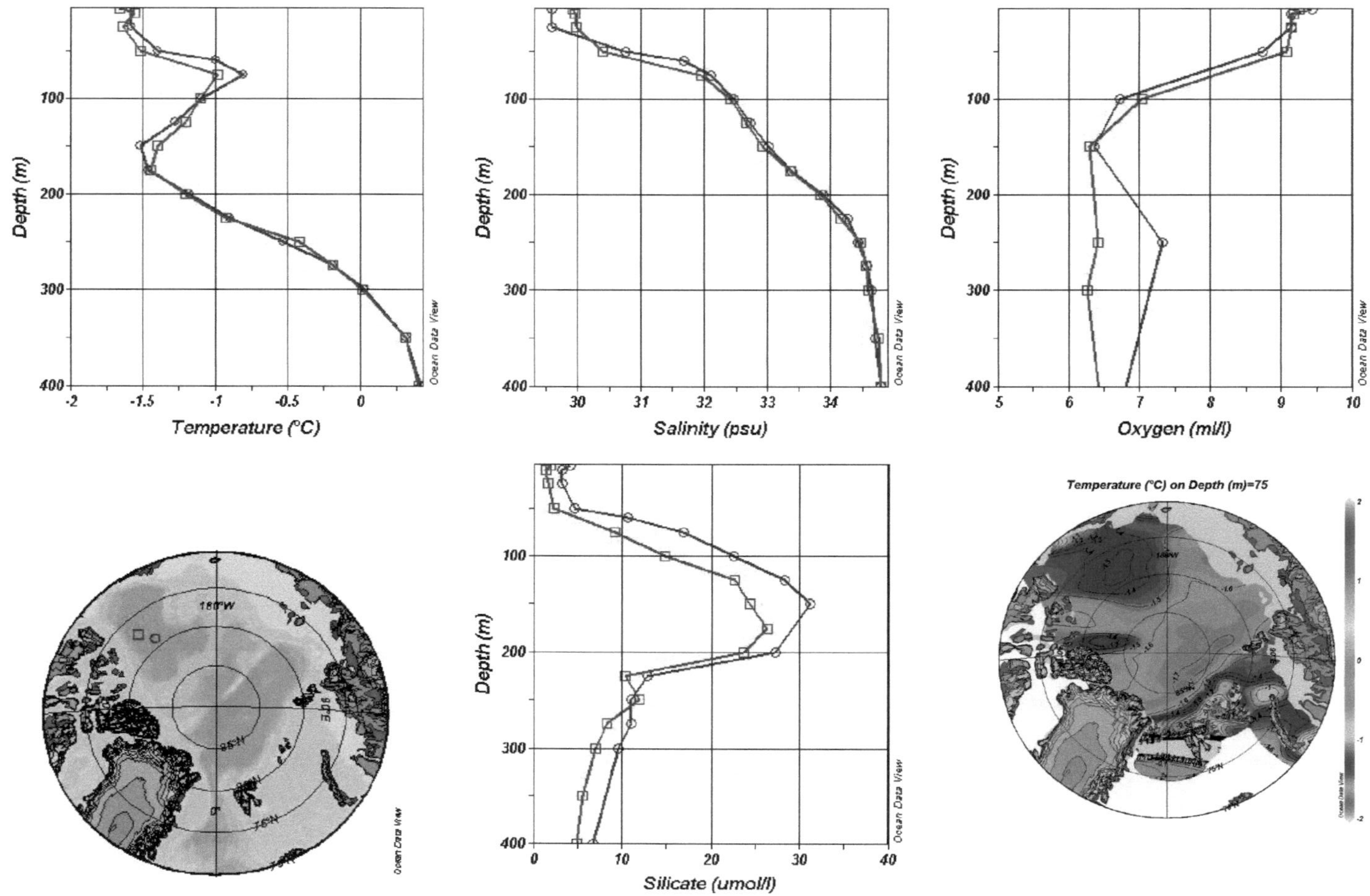

Figure 4.22. Vertical distribution of temperature, salinity, oxygen and silicon (blue line – August, red line – December, 1989), and distribution of temperature at 75 m in the Arctic Ocean (1965–1989)

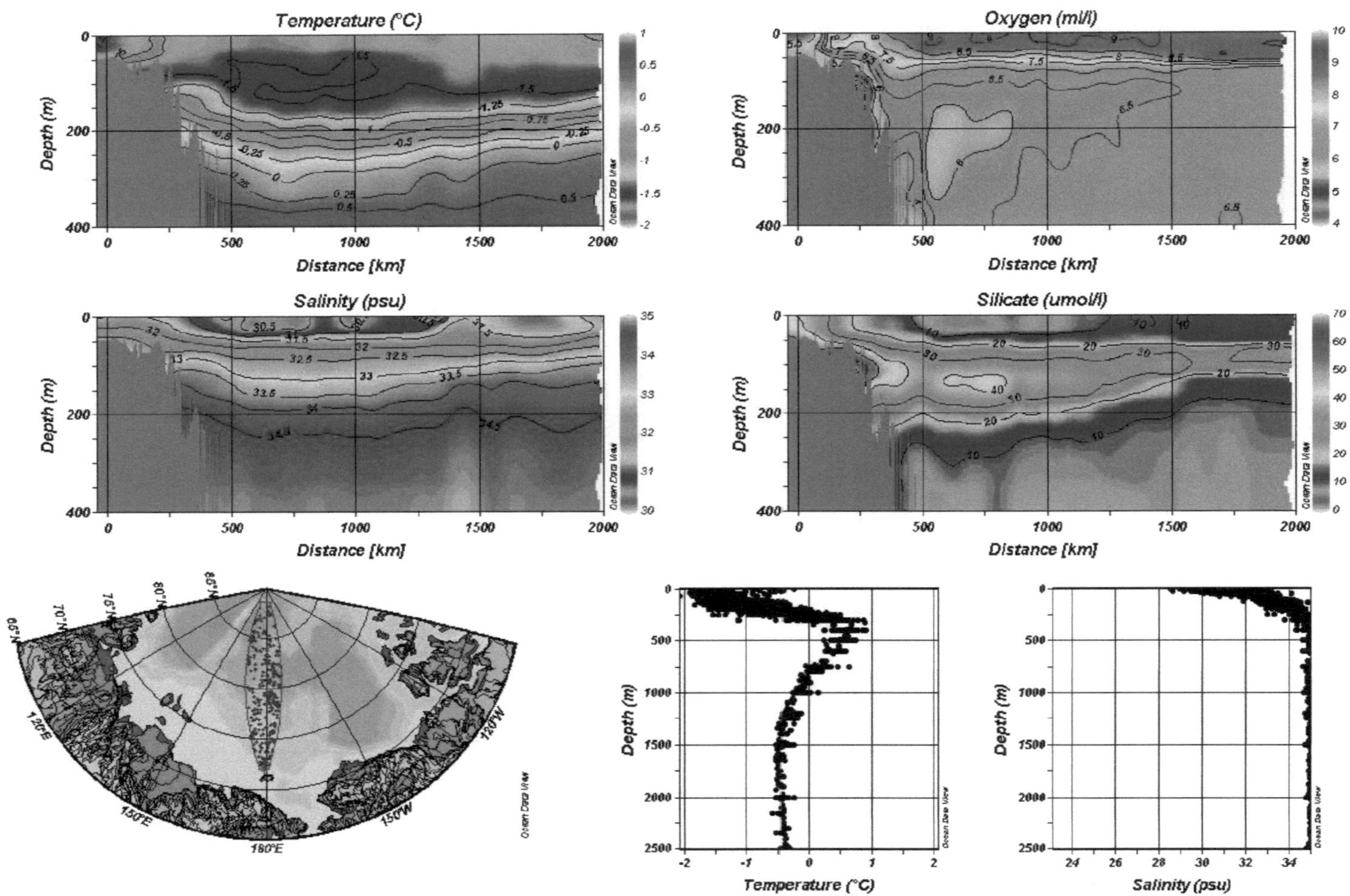

Figure 4.23. Section along the meridian 180°E..

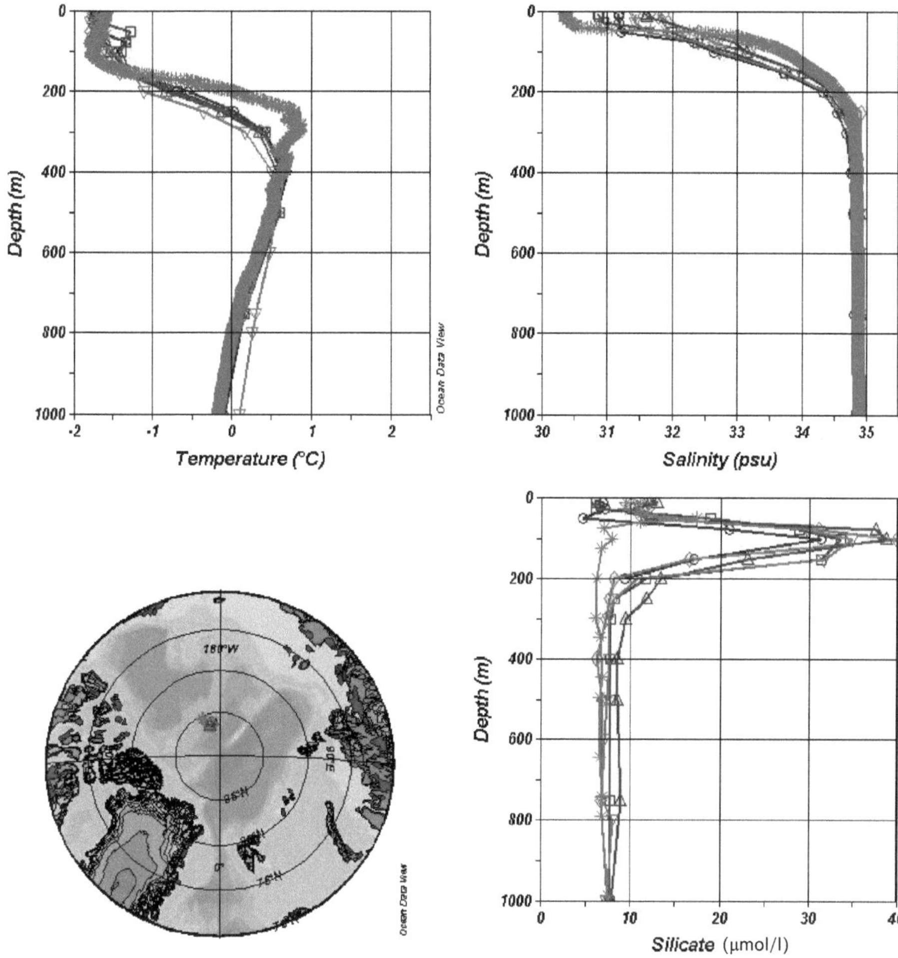

Figure 4.24. Vertical distribution of temperature, salinity and silicate in 1973–1977 and 1998.

MIX
Papier aus verantwortungsvollen Quellen
Paper from responsible sources
FSC® C105338

If you have any concerns about our products,
you can contact us on
ProductSafety@springernature.com

In case Publisher is established outside the EU,
the EU authorized representative is:
Springer Nature Customer Service Center GmbH
Europaplatz 3, 69115 Heidelberg, Germany

Printed by Libri Plureos GmbH
in Hamburg, Germany